BIOLOGY OF CARDIOVASCULAR AND METABOLIC DISEASES

BIOLOGY OF CARDIOVASCULAR AND METABOLIC DISEASES

CHAYA GOPALAN, PH.D., FAPS

Department of Applied Health, School of Education, Health & Human Behavior; Department of Nurse Anesthesiology, School of Nursing, Southern Illinois University Edwardsville, Edwardsville, IL, United States

ERIK KIRK, PH.D.

Department of Applied Health, School of Education, Health & Human Behavior, Southern Illinois University Edwardsville, Edwardsville, IL, United States

ELSEVIER

ACADEMIC PRESS

An imprint of Elsevier

Academic Press is an imprint of Elsevier
125 London Wall, London EC2Y 5AS, United Kingdom
525 B Street, Suite 1650, San Diego, CA 92101, United States
50 Hampshire Street, 5th Floor, Cambridge, MA 02139, United States
The Boulevard, Langford Lane, Kidlington, Oxford OX5 1GB, United Kingdom

Library of Congress Cataloging-in-Publication Data
A catalog record for this book is available from the Library of Congress

British Library Cataloguing-in-Publication Data
A catalogue record for this book is available from the British Library

ISBN 978-0-12-823421-1

For information on all Academic Press publications
visit our website at https://www.elsevier.com/books-and-journals

Publisher: Stacy Masucci
Acquisitions Editor: Katie Chan
Editorial Project Manager: Billie Jean Fernandez
Production Project Manager: Omer Mukthar
Cover Designer: Christian J. Bilbow

Typeset by STRAIVE, India

Dedication

To my family and students who have inspired me in my teaching and writing.
– Dr. Chaya Gopalan

To my family, mentors, and the many students I have had the privilege to teach.
– Dr. Erik Kirk

Contents

5. Atherosclerosis

6. Coronary artery disease

7. Renal physiology

8. Blood pressure, hypertension, and exercise

9. Cellular metabolism

Companian Website:
https://www.elsevier.com/books-and-journals/book-companion/9780128234211

About the authors

Dr. Chaya Gopalan received her bachelor's and master's degrees from Bangalore University, India, and PhD from the University of Glasgow, Scotland, for her research on opioid peptides in the regulation of the release of luteinizing hormone. She continued her work on galanin and other neuropeptides as a postdoctoral research fellow at Michigan State University. Dr. Gopalan wanted to follow her passion for teaching. It started as an adjunct position at Maryville University in St. Louis, which led to tenure-track positions at St. Louis Community College and St. Louis College of Pharmacy, and now at Southern Illinois University Edwardsville (SIUE). She has been teaching in the areas of anatomy, physiology, and pathophysiology at both graduate and undergraduate levels for health professional programs. Dr. Gopalan has been practicing evidence-based teaching using team-based learning, case-based learning, and most recently, the flipped classroom methods. Besides her passion for teaching, Dr. Gopalan has kept up with lab research in neuroendocrine physiology. Currently, she is working on two research projects: one on the role of gonadal steroid hormones in the sexual dimorphism of the brain and the other on obesity, intermittent fasting, and physical and mental exhaustion. Dr. Gopalan has received many teaching awards, including the Arthur C. Guyton Educator of the Year Award from the American Physiological Society (APS), Outstanding Two-Year College Teaching Award by the National Association of Biology Teachers, and Excellence in Undergraduate Education Award by SIUE. She has also received several grants, including an NSF-IUSE, an NSF-STEM Talent Expansion Program, and the APS Teaching Career Enhancement awards.

Besides teaching and research, Dr. Gopalan enjoys mentoring not only her students but also her peers. She regularly conducts workshops and participates in panel discussions related to higher education. Dr. Gopalan is very active in the teaching section of the APS and has served on many committees. She has published numerous manuscripts and case studies and contributed several textbook chapters and question banks for textbooks and board exams.

Dr. Erik Kirk received his bachelors from Drury University and his PhD from the University of Kansas. His research and clinical experience are in the areas of exercise, metabolism, and obesity. Dr. Kirk is Certified Clinical Exercise Physiologist through the American College of Sports Medicine and a Certified Strength and Conditioning Specialists and Certified Personal Trainer through the National Strength and Conditioning Association. Dr. Kirk teaches courses in advanced exercise physiology, cardiovascular and respiratory physiology, biology of cardiovascular and metabolic diseases, and pathophysiology and treatment of obesity. Dr. Kirk's research interest involves evaluating how obesity contributes to cardiovascular and metabolic diseases. His work focuses on understanding the normal adipose tissue physiology, the alterations in fat metabolism associated with obesity and diabetes, and how weight loss improves the metabolic problems caused by obesity. He is also involved in conducting clinical trials to evaluate the safety and efficacy of obesity therapies such as physical activity and nutrition. Dr. Kirk also studies the effects of physical activity and diet on learning in preschool children. He studies how incorporating physical activity into academic lessons and altering dietary intake can reduce obesity and improve early learning skills in preschool children. Dr. Kirk has many peer-reviewed publications in obesity-related journals and has presented more than the past 20 years at national conferences related to cardiovascular and metabolic diseases. His greatest joy, however, is teaching and mentoring students, both in and out of the classroom.

Preface

Teaching health science majors about the biology of cardiovascular and metabolic disease in a straightforward yet concise way has always been our passion. It is essential for students who would work in health care to learn about the significant cardiovascular and metabolic diseases they would encounter in their careers. In addition, students must have a good understanding of the many metabolic pathways/mechanisms associated with these conditions to provide quality support. For example, only recently have we learned about the role of inflammation and the interaction with specific genes in contributing to various diseases. It is also vital for students to understand how diet and exercise could prevent or treat these conditions. These are just some of the critical areas for health science students to learn as they will someday work in various areas such as exercise science, nutrition, dietetics, medicine, nursing, physical therapy, occupational therapy, and athletic training, among others. While the amount and type of treatment vary by profession, all health professionals have patients with one or more common conditions. We strongly feel that health science students would benefit by having a basic understanding of the more common cardiovascular and metabolic diseases. Even if they are not directly treating the condition(s), practitioners should still understand how and why they developed since many of their patients will have these conditions. However, upon researching the best textbook for a basic understanding of the more common cardiovascular and metabolic diseases, we found that the choices were minimal. Some excellent pathophysiology books go into great detail over various diseases, but these textbooks are too advanced for an introductory course. We also found that they contain many details over conditions that, although necessary, were not very common. Therefore we decided to develop an introductory textbook focused on the more common cardiovascular and metabolic diseases in the health sciences. Many conditions are not included, but the goal in this introductory book is to focus on the more common diseases and be as clear yet concise in the approach. We feel we have accomplished that goal with this textbook. We hope you and your students do too!

Organization

The book is organized into two units: Cardiovascular diseases and metabolic diseases. Each contains chapters relevant to these two units. To help with the comprehension of the material, the authors have included various features and aids to reinforce concepts and enhance learning. The following is an overview of each feature and aid:

Chapter objectives: Objectives are listed at the beginning of each chapter to help identify key concepts.

Illustrations and figures: Numerous illustrations, figures, charts, and tables clarify and enhance the essential concepts and information.

CHAPTER

1

The heart

OUTLINE

Biology of Cardiovascular and Metabolic Diseases
https://doi.org/10.1016/B978-0-12-823421-1.00011-1

OBJECTIVES

- Review the location, size, and external a anatomy of the heart.
- Locate the internal details of the heart including the chambers, valves, and layers within the wall.
- Follow the flow of blood that reaches the heart in a step-by-step manner until it ejects blood.
- Summarize the characteristics of the cardiac muscle fiber.
- Compare the electrical activity within a cardiac conductive (pacemaker) cell to that of a cardiac contractile cell.
- Locate the parts of the cardiac conduction system. Name the normal pacemaker of the heart.
- Define the P wave, QRS complex, T wave, PR and ST segments, and PR and QT intervals in a normal electrocardiogram.
- Highlight the events that occur in a normal cardiac cycle beginning with the depolarization of the pacemaker cells and ending with the next round of pacemaker depolarization. Correlate the electrical events to the mechanical, pressure, and blood volume changes that occur in one complete cardiac cycle.
- Explain how cardiac output is measured. Describe how each component of cardiac output is regulated.
- State the Frank-Starling law of the heart.
- Define preload and afterload.
- Compare the mechanisms by which the parasympathetic and sympathetic nervous systems affect cardiac output.

1.1 Introduction

A human heart pumps approximately 108,000 times per day, more than 39 million times in 1 year, and nearly 3 billion times during a 75-year life span at the normal heart rate (HR) of 75 beats per minute (bpm). Each chamber of the heart ejects approximately 70 mL of blood per contraction in a resting adult. This would equal to 5.25 L of blood per minute and approximately 14,000 L per day. Over 1 year, this would equal to 10,000,000 L or 2.6 million gallons of blood sent through roughly 60,000 miles of vessels.

1.2 Location of the heart

The heart is situated within the thoracic cavity, medially between the lungs, in a dedicated space known as the **mediastinum** (Fig. 1.1). The heart is separated from the other structures within the mediastinum by a tough membrane known as the **pericardium**, or **pericardial sac**, which sits in its own space known as the **pericardial cavity**.

1.3 The pericardium

The **pericardium**, which translates as "around the heart," is a double-layered connective tissue membrane that surrounds the heart like a sac and, therefore, is also called as a

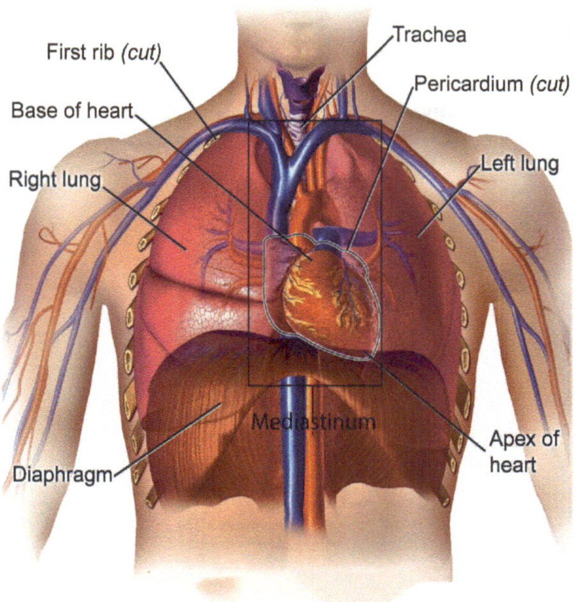

FIG. 1.1 Location of the heart in the thoracic cavity [1]. The heart is situated within the thoracic cavity, medially between the lungs in the mediastinum.

pericardial sac (Fig. 1.2). The outer sturdy **parietal pericardium** consists of dense connective tissue that protects the heart and maintains its position within the thoracic cavity. The inner **visceral pericardium**, or **epicardium**, is attached to the heart and is part of the heart wall. There is a space between the visceral pericardium and the parietal pericardium and is referred to as the pericardial cavity. The pericardium secretes a small amount of serous fluid that fills the pericardial cavity and coats the two pericardial layers which serves as a lubricant to reduce friction as the heart expands and contracts.

1.4 Layers of the heart wall

The wall of the heart is composed of three layers of unequal thickness. From superficial to deep, these layers are the **epicardium**, **myocardium**, and **endocardium** (see Fig. 1.2). The outermost layer, the epicardium, is also the innermost layer of the pericardium, the visceral pericardium. The thickest layer of the heart is the myocardium. As the name suggests, the myocardium consists of cardiac muscle fibers along with its supply of blood vessels and nerve fibers to help the heart pump blood. It is the contraction of the myocardium that pumps blood through the heart and into the major arteries to be distributed throughout the body. As shown in the figure (Fig. 1.3), the pattern is unique to cardiac muscle where the muscle fibers swirl and spiral around the chambers of the heart and form a figure 8 pattern between the atria and the ventricles. This swirling pattern allows the heart to withstand stress and pump blood effectively.

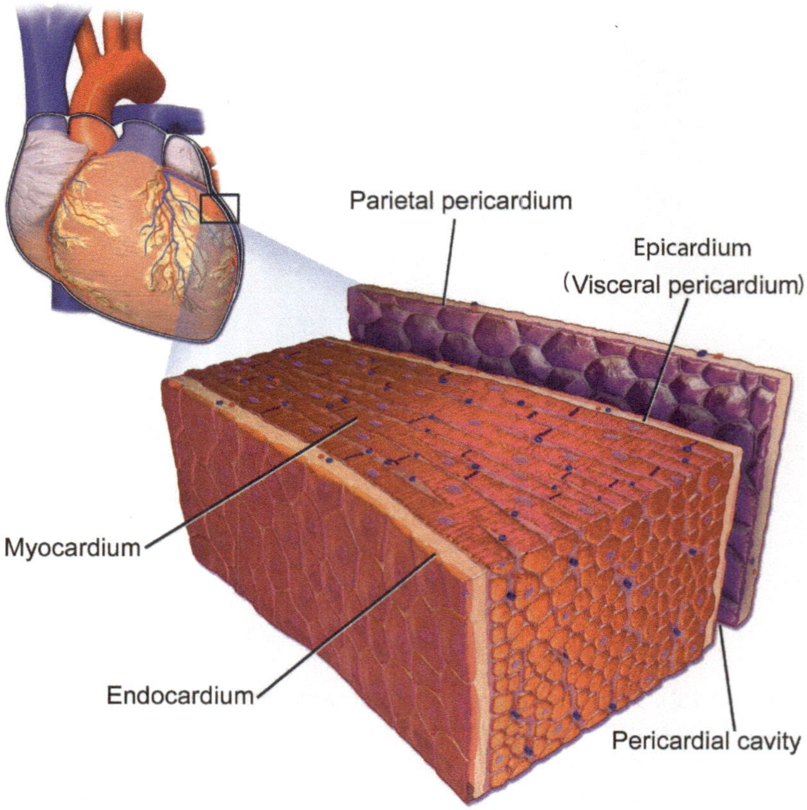

FIG. 1.2 Pericardial membranes and layers of the heart wall [2].

Although the ventricles on the right and left sides pump the same amount of blood per contraction, the muscle of the left ventricle is much thicker compared to the right because the left ventricle must generate a very high pressure to overcome resistance required to pump blood into the long **systemic circuit**. On the other hand, the right ventricle does not need to generate as much pressure since the **pulmonary circuit** is shorter and provides less resistance. Fig. 1.4 illustrates the differences in muscular thickness between the two ventricles. The innermost layer of the heart wall, the **endocardium**, is a thin layer of connective tissue. The endocardium lines the chambers and covers the heart valves.

1.5 External anatomy of the heart

The adult heart weighs approximately 250–350 g (9–12 oz) and measures 12 cm (5 in.) in length, 8 cm (3.5 in.) wide, and 6 cm (2.5 in.) in thickness. A well-trained athlete may have a considerably larger heart since exercise results in the addition of muscle proteins to pump more blood. It is important to note, however, that the thickened heart is not always due to

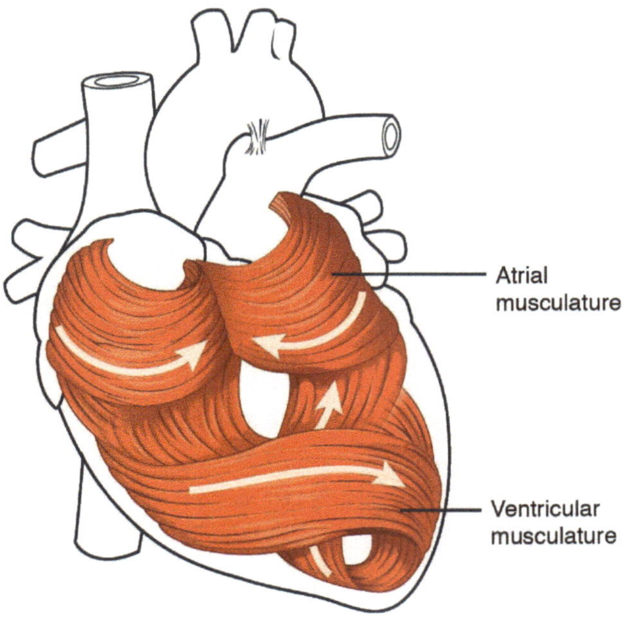

FIG. 1.3 Musculature of the heart [3]. The swirling pattern of cardiac muscle tissue contributes significantly to the heart's ability to pump blood effectively.

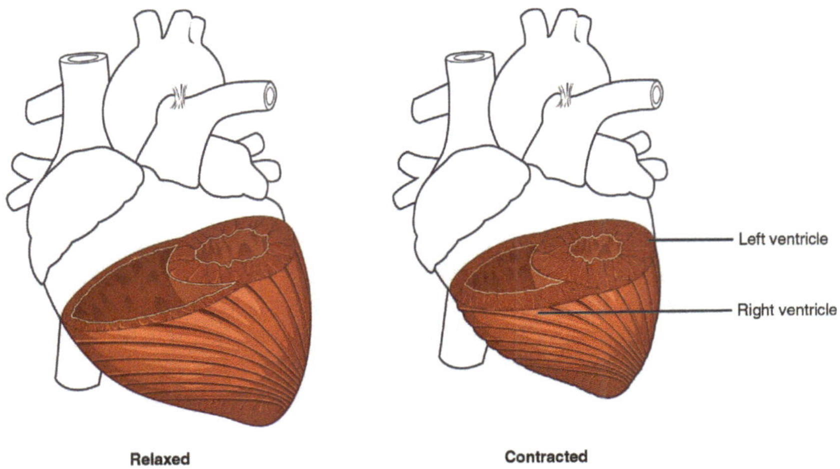

FIG. 1.4 Differences in the right and left ventricular muscle thickness [4].

exercise. It can result from abnormal conditions such as **hypertrophic cardiomyopathy**, which may cause sudden death in apparently otherwise healthy young people.

The heart is broader at the superior surface and is referred to as the **base** which tapers off at the **apex** (see Fig. 1.5). The base is at the level of the third costal cartilage whereas the apex is between the fourth and fifth ribs. Cardiac muscle (myocardium) is nourished from the right and left coronary arteries. These arteries branch off into smaller and smaller arteries, delivering oxygenated blood and nutrients to the myocardium.

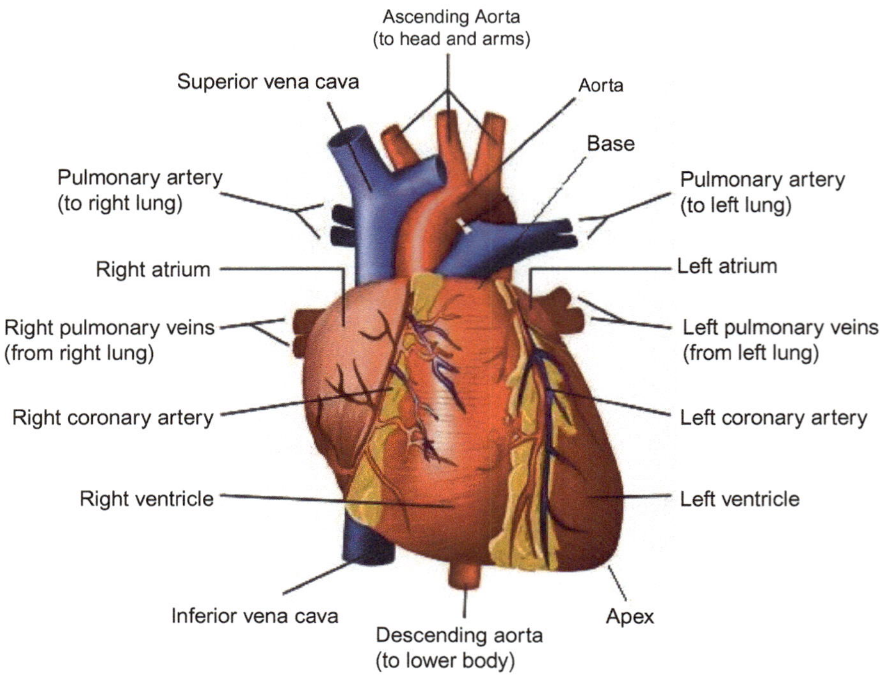

FIG. 1.5 External anatomy of the anterior part of the heart [5].

1.6 Internal anatomy of the heart

The heart is divided into four chambers by the septa, or walls. Located between the two atria is the **interatrial septum**. The septum situated between the two ventricles is called the **interventricular septum**. The interventricular septum is substantially thicker compared to the interatrial septum to allow the ventricles to generate high pressure when they contract. The **atrioventricular septum**, as the name suggests, is found between the atria and the ventricles. It houses four openings that allow blood to move from the atria into the ventricles and from the ventricles into the pulmonary trunk and aorta. Located in each of these openings between the atria and ventricles is a **valve**, a specialized structure that ensures one-way flow of blood. The valves between the atria and ventricles are known as the **atrioventricular** (AV) **valves** (Fig. 1.6). The right AV valve is known as the **tricuspid valve** and the left AV valve is known as the **mitral valve**.

1.6.1 Chambers of the heart

1.6.1.1 Right atrium

The right atrium receives deoxygenated blood from the systemic circulation. The two major systemic veins, the **superior** and **inferior venae cavae**, and the large coronary vein called the **coronary sinus** that drains the myocardium, empty into the right atrium (Figs. 1.5 and

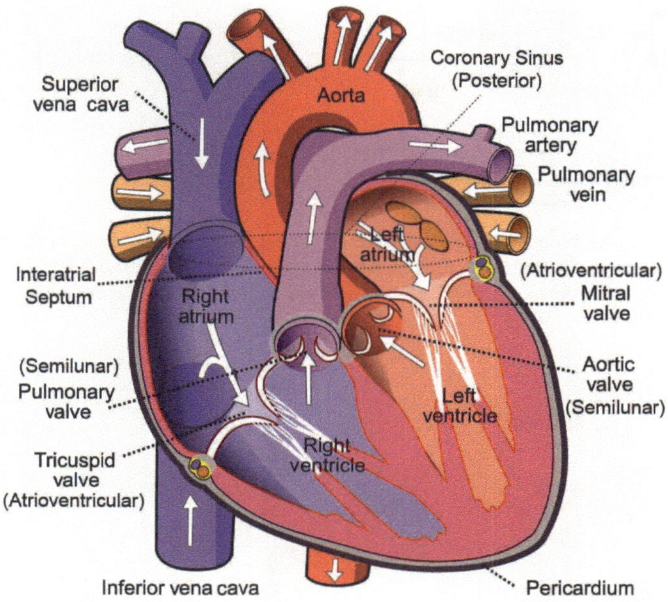

FIG. 1.6 Internal anatomy of the heart [6]. This anterior view of the heart shows the four chambers, the major vessels, and their early branches, as well as the valves. The presence of the pulmonary trunk and aorta covers the interatrial septum, and the atrioventricular septum is cut away to show the atrioventricular valves. *Arrows* represent flow of blood through the heart.

1.6). The superior vena cava drains blood from regions above the diaphragm and the inferior vena cava drains blood from areas below the diaphragm. The coronary sinus drains most of the coronary veins that return systemic blood from the heart.

1.6.1.2 Right ventricle

The right ventricle receives blood from the right atrium through the tricuspid valve, a valve found in the opening of the atrioventricular septum. When the right ventricle contracts, the pressure within the ventricular chamber increases causing the blood to flow toward the pulmonary trunk and the right atrium (Figs. 1.7 and 1.8). The tricuspid valve closes to prevent any potential backflow into the right atrium.

1.6.1.3 Left atrium

After the exchange of gases between the alveoli and the pulmonary capillaries, oxygenated blood returns to the left atrium via four pulmonary veins (Figs. 1.7 and 1.8). Most of the filling occurs while the atria are relaxed.

1.6.1.4 Left ventricle

The blood from the left atrium flows into the left ventricle through the bicuspid (mitral) valve which is situated in the opening of the atrioventricular septum on the left side of the heart. The left ventricle ejects blood into the aorta through the aortic semilunar valve (Figs. 1.7 and 1.8).

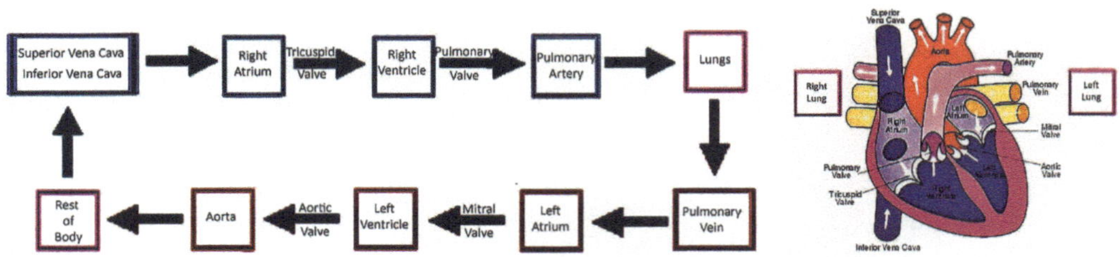

FIG. 1.7 A flow chart showing the circulation of blood through the heart as well as between the heart and the lungs. *White arrows* within the heart and *black arrows between boxes* represent flow of blood through the heart. *Blue boxes* indicate deoxygenated blood, *red boxes* indicate oxygenated blood, and *purple boxes* indicate sites of gas exchange.

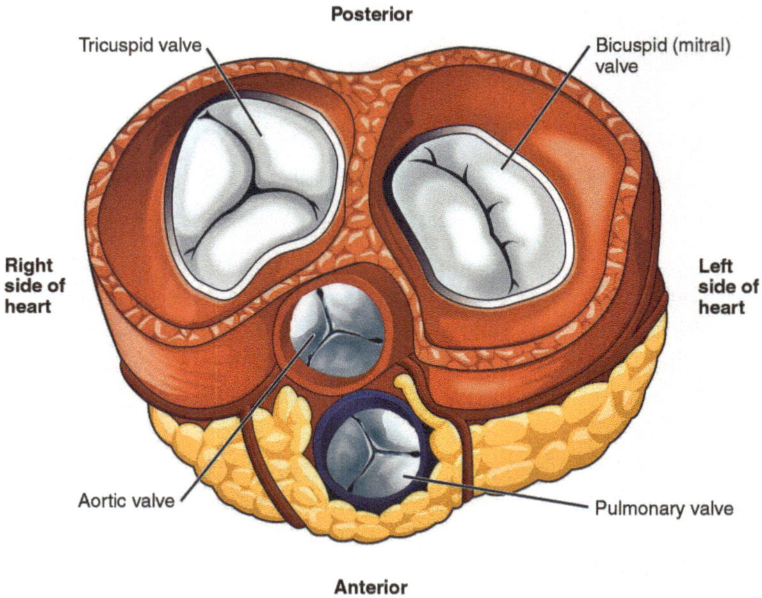

FIG. 1.8 Heart valves [7]. The atrioventricular valves include the tricuspid valve and mitral valve. The semilunar valves are the pulmonary and mitral valves, https://www.heartandstroke.ca/heart-disease/conditions/valvular-heart-disease

1.6.1.5 Valves of the heart

Valves are specialized structures that ensure one-way flow of blood. There are four valves in total within the heart. The opening between the right atrium and right ventricle is guarded by a tricuspid valve or the right AV valve. It consists of three flaps, or leaflets, made of endocardium along with additional connective tissue. Each flap of the valve is attached to strong strands of connective tissue, the **chordae tendineae** (Figs. 1.9B and 1.10B). There are several

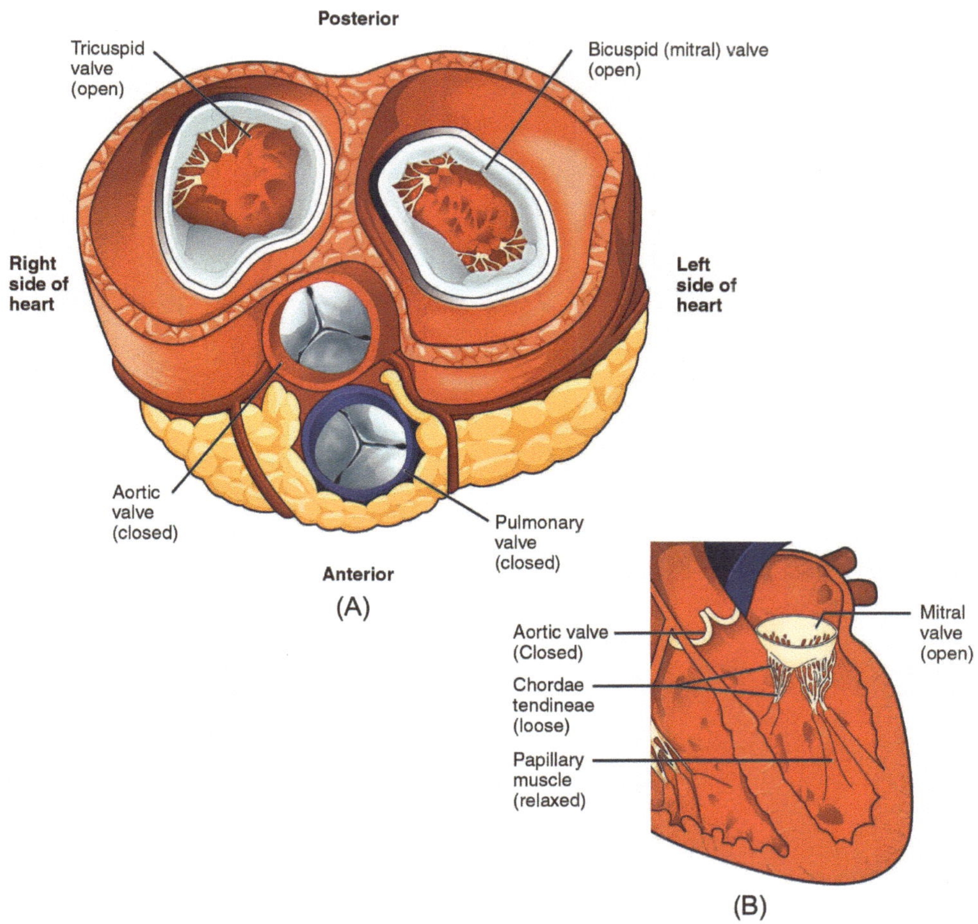

FIG. 1.9 Blood flow from the left atrium to the left ventricle [7]. (A) A transverse section through the heart illustrates the four heart valves. The two atrioventricular valves are open (Tricuspid and Bicuspid or Mitral); the two semilunar valves (Aortic and Pulmonary) are closed. The atria and vessels have been removed. (B) A frontal section through the heart illustrates blood flow through the mitral valve. When the mitral valve is open, it allows blood to move from the left atrium to the left ventricle. The aortic semilunar valve is closed to prevent the backflow of blood from the aorta to the left ventricle.

chordae tendineae associated with each of the flaps which in turn are attached to a **papillary muscle** (Figs. 1.9B and 1.10B) that extends from the inferior ventricular surface. There are three papillary muscles in the right ventricle which correspond to the three sections of the valves.

Located at the opening between the left atrium and left ventricle is the mitral valve, also called the left AV valve. Structurally, this valve consists of two cusps which are attached by chordae tendineae to two papillary muscles that project from the inferior wall of the left ventricle.

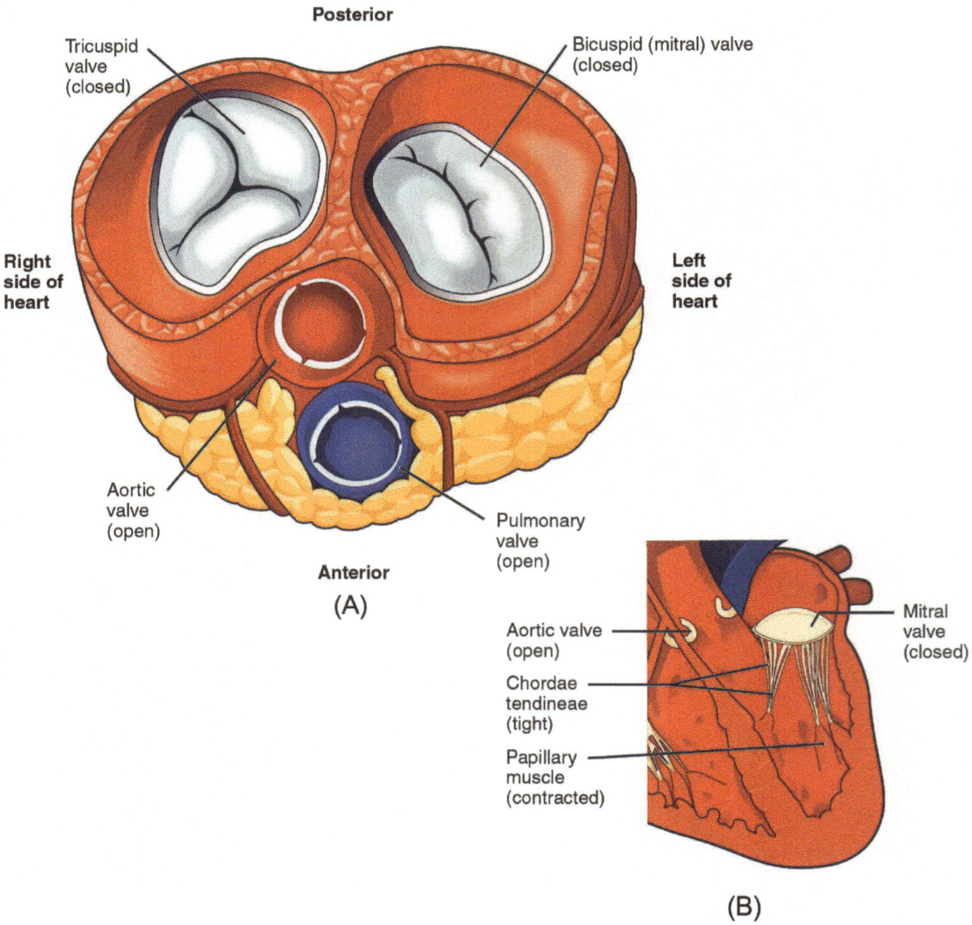

FIG. 1.10 Blood flow from the left ventricle into the aorta and pulmonary artery [8]. (A) A transverse section through the heart illustrates the four heart valves during ventricular contraction. The two atrioventricular valves are open (Tricuspid and Mitral); the two semilunar valves (Aortic and Pulmonary) are closed. The atria and vessels have been removed. (B) A frontal view shows the closed mitral (bi-leaflet) valve that prevents the backflow of blood into the left atrium. The aortic semilunar valve is open to allow blood to be ejected into the aorta.

Emerging from the right ventricle at the base of the pulmonary trunk is the **pulmonary semilunar valve** or the **pulmonary valve**; it is also known as the **pulmonic valve** or the **right semilunar valve** which prevents backflow of blood from the pulmonary trunk into the right ventricle. The pulmonary valve is made of three small flaps of endocardium reinforced with connective tissue. When the ventricle relaxes, the pressure difference causes blood to flow back into the right ventricle from the pulmonary trunk. This flow of blood fills the pocket-like flaps of the pulmonary valve, causing the pulmonary valve to close producing an audible sound.

At the base of the aorta is the **aortic semilunar valve**, or the **aortic valve**, which prevents backflow from the aorta into the left ventricle. It is also composed of three flaps. When the ventricle relaxes and blood attempts to flow back into the left ventricle from the aorta, blood will fill the cusps of the valve, causing it to close and producing an audible sound.

The AV valves are closed while the two semilunar valves are open (Fig. 1.10). This occurs when the ventricles contract to eject blood into the pulmonary trunk and aorta. Closure of the two AV valves prevents blood from being forced back into the atria. In Fig. 1.9A, the two AV valves are open and the two semilunar valves are closed. This occurs when both atria and ventricles are relaxed and filling of the ventricles from the atria occurs.

When the ventricles begin to contract, the pressure within the ventricles rises, and blood flows toward the area of lower pressure, which is initially in the atria. This backflow causes the cusps of the tricuspid and mitral valves to close (Fig. 1.9A). During the relaxation phase of the cardiac cycle, the papillary muscles are also relaxed and the tension on the chordae tendineae is slight (see Fig. 1.9B). However, as the ventricles contract, so do the papillary muscles. This creates tension on the chordae tendineae (see Fig. 1.9B), helping to hold the cusps of the AV valves in place and preventing them from being blown back into the atria.

Unlike the AV valves, there are no papillary muscles or chordae tendineae associated with the semilunar valves. When the ventricles relax causing a change of pressure, it forces the blood toward the ventricles, the blood presses against these cusps and seals the openings. Visit http://openstaxcollege.org/l/heartvalve to observe an echocardiogram of actual heart valves opening and closing.

1.7 Cardiac muscle and electrical activity

1.7.1 Structure of cardiac muscle

Cardiac muscle fibers, or cardiomyocytes, like skeletal muscle fibers, are striated and have a single, central nucleus. However, two or more nuclei may be found occasionally. **Transverse (**or **T) tubules**, the folds of the **sarcolemma** (plasma membrane), penetrate deep into the interior of the cell, allowing electrical impulses to reach the inner parts of the cell. **Sarcoplasmic reticulum** (endoplasmic reticulum) stores Ca^{2+} but is not sufficient for muscle contraction. Additional Ca^{2+} must come from the extracellular fluid. Mitochondria are present in large numbers in the cardiac muscle fibers to provide energy for muscle contraction (Fig. 1.11A).

Unlike skeletal muscle fibers, cardiac muscle fibers branch freely. A junction between two adjoining cells is marked by a specialized structure called an **intercalated disc** (Fig. 1.11A), which helps spread depolarization between adjacent cells (Fig. 1.11B). The sarcolemmas from adjacent cells bind together at the intercalated discs and possess gap junctions to allow passage of ions between the cells (Fig. 1.11C). **Desmosomes** are special cell-to-cell junctions found at the intercalated discs to provide additional support and stability for the cardiac muscle fibers.

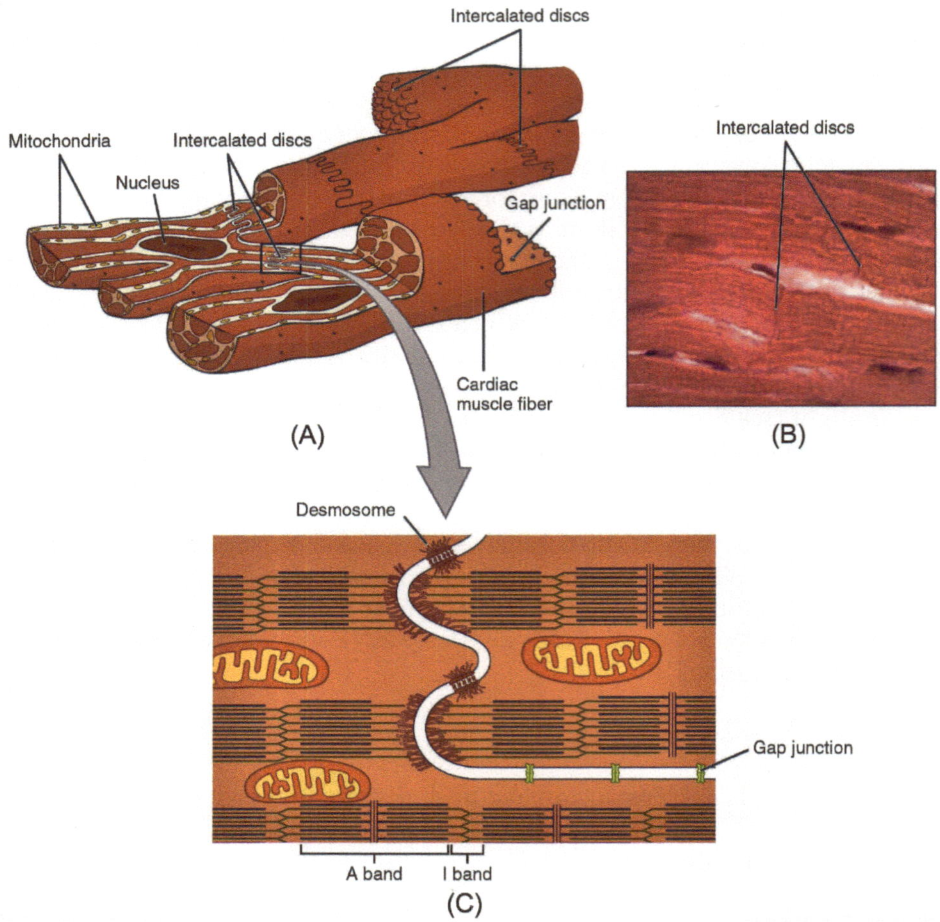

FIG. 1.11 Cardiac muscle [9]. (A) Cardiac muscle cells have myofibrils composed of myofilaments arranged in sarcomeres, T tubules to transmit the impulse from the sarcolemma to the interior of the cell, numerous mitochondria for energy, and intercalated discs that are found at the junction of different cardiac muscle cells. (B) A photomicrograph of cardiac muscle cells shows the nuclei and intercalated discs. (C) An intercalated disc connects cardiac muscle cells and consists of desmosomes and gap junctions.

There are two major types of cardiac muscle fibers: **myocardial contractile cells** and **myocardial conducting cells (pacemaker cells)**. The myocardial contractile cells constitute the bulk (99%) of the cells in the atria and ventricles and are responsible for contractions that pump blood through the body. The myocardial conducting cells make up the remaining 1% to form the **conduction system** of the heart. The myocardial conducting cells are generally much smaller than the contractile cells and have few of the myofibrils or filaments needed for contraction.

1.8 Cardiac muscle metabolism

Fatty acids and glucose from the blood supply are broken down within the mitochondria to release energy in the form of ATP. Both fatty acid droplets and glycogen, the storage form of glucose, are stored within the sarcoplasm to provide additional nutrient supply.

1.9 Conduction system of the heart

Cardiac muscle has the ability to initiate its electrical potential that spreads rapidly from cell to cell to trigger the contractile mechanism. This property is known as **autorhythmicity**. Every cardiac muscle fiber is capable of generating its own electrical impulse. The cell with the higher inherent rate of depolarization sets the pace, and the impulse spreads from the faster to the slower cell to trigger a contraction. The five key components of the cardiac conduction system include the **sinoatrial (SA) node**, the **AV node**, the **AV bundle** or **bundle of His**, the **AV bundle branches**, and the **Purkinje fibers** (Fig. 1.12).

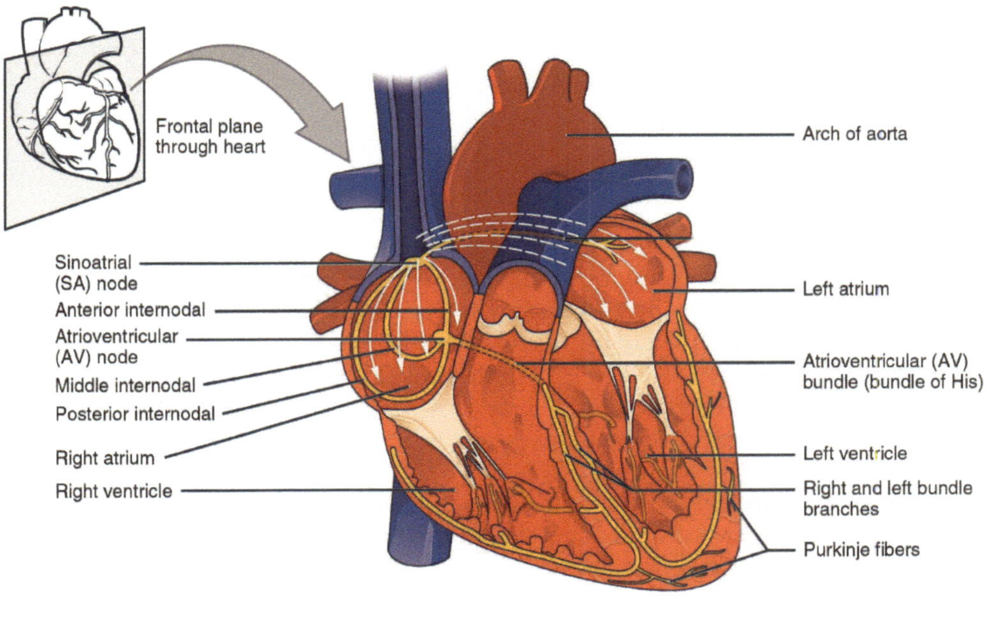

FIG. 1.12 Conduction system of the heart [10]. The initiation of an action potential at the sinoatrial (SA) node spreads throughout the atria reaching the atrioventricular (AV) node, AV bundle (bundle of His), bundle branches, and ultimately Purkinje fibers. The contractile cells then begin contraction from the superior to the inferior portions of the atria, efficiently pumping blood into the ventricles.

1.9.1 Sinoatrial node

Sinoatrial node is a specialized group of myocardial conducting cells located in the superior and posterior walls of the right atrium close to the opening of the superior vena cava. The SA node has the highest inherent rate of depolarization and therefore referred to as the **pacemaker** of the heart. It initiates the sinus rhythm or normal electrical pattern. This impulse spreads from the SA node to the atrial myocardial contractile cells and the AV node. The internodal pathways that connect contractile cells consist of three bands (anterior, middle, and posterior; Fig. 1.12). The impulse takes approximately 50 milliseconds (ms) to travel between the SA node and the AV node.

1.9.2 Atrioventricular node

The AV node is a second set of myocardial conductive cells, located in the inferior portion of the right atrium as part of the atrioventricular septum. As the impulse reaches the atrioventricular septum, the connective tissue surrounding the heart, referred to as **fibrous skeleton**, prevents the impulse from spreading into the myocardial cells in the ventricles except the AV node.

There is a delay in the spreading of depolarization from the AV node to the AV bundle (see Fig. 1.13, step 3). This delay in transmission is partially due to the small diameter of the cells of the node, which slows the impulse. Thus, it takes the impulse approximately 100 ms to pass through the node. This pause helps atrial myocardium to complete their contraction that pumps blood into the ventricles before the impulse is transmitted to the cells of the ventricle.

1.9.3 Atrioventricular bundle (bundle of His), bundle branches, and Purkinje fibers

AV bundle, or **bundle of His**, originates at the AV node and proceeds through the interventricular septum before dividing into two **AV bundle branches**, commonly called the **left** and **right bundle branches**. According to their locations, the left bundle branch supplies the left ventricle, and the right bundle branch supplies the right ventricle. Since the left ventricle is much larger compared to the right, the left bundle branch is also considerably larger compared to the right. Both bundle branches descend and reach the apex of the heart where they connect with the Purkinje fibers (see Figs. 1.12 and 1.13, step 4). This passage takes approximately 25 ms.

The Purkinje fibers are myocardial conductive cells that spread the impulse to the myocardial contractile cells in the ventricles. They extend throughout the myocardium from the apex of the heart toward the atrioventricular septum and the base of the heart. The Purkinje fibers have a fast conduction rate, and the electrical impulse reaches all of the ventricular muscle cells in about 75 ms (see Fig. 1.13, step 5 and 6). Since the electrical impulse arrives at the apex through the bundle branches, the contraction also begins at the apex and travels toward the base of the heart, similar to squeezing a tube of toothpaste from the bottom. This allows the blood to be pumped out of the ventricles into the aorta and pulmonary trunk. The total time elapsed from the initiation of the impulse in the SA node until depolarization of the ventricles is approximately 225 ms.

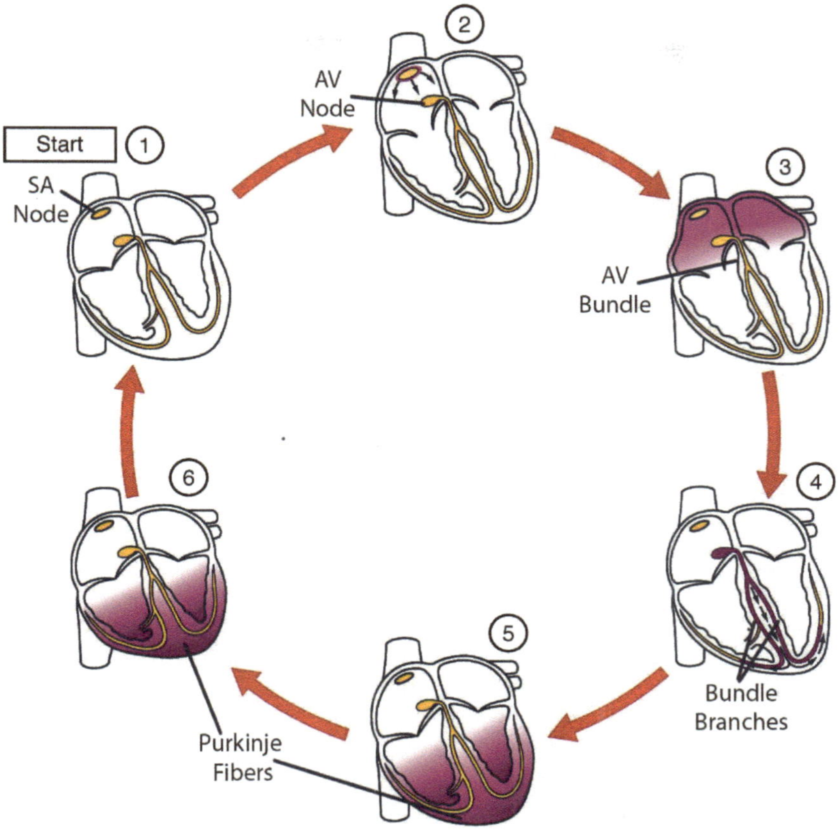

FIG. 1.13 Cardiac conduction [11]. (1) The sinoatrial (SA) node and the remainder of the conduction system are at rest. (2) The SA node initiates the action potential, which sweeps across the atria. (3) After reaching the atrioventricular node (AV), there is a delay of approximately 100ms that allows the atria to complete pumping blood before the impulse is transmitted to the atrioventricular bundle. (4) Following the delay, the impulse travels through the atrioventricular bundle and bundle branches to the Purkinje fibers (5 and 6).

1.10 Membrane potentials in cardiac conductive cells

The resting membrane potential of conductive fibers in the heart ranges from −50 to −60mV. Action potentials are considerably different between cardiac conductive cells (pacemaker cells) such as the SA node and AV node and cardiac contractile cells (nonpacemaker) such as the Purkinje cells. Cardiac conductive cells do not have a stable resting membrane potential because they consist of a series of Na^+ leak channels (referred to as funny channels) which allow a slow influx of Na^+ and therefore raise membrane potential slowly from an initial value of −60mV to about −40mV (orange line). Slow Ca^{2+} channels are also open during this period resulting in a **spontaneous depolarization** (or **prepotential depolarization**). The shift in membrane potential to its threshold level of −40mV causes voltage-gated fast Ca^{2+} channels to open and Ca^{2+} enter the cell in large numbers, further depolarizing it at a more

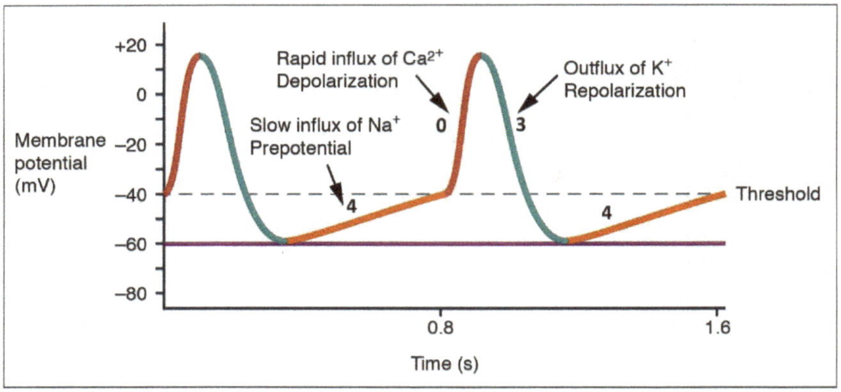

FIG. 1.14 Action potential in the sinoatrial node [12]. The prepotential (phase 4) is due to a slow influx of Na^+ until the threshold is reached followed by a rapid depolarization (phase 0) and repolarization (phase 3). The prepotential (phase 4) accounts for the membrane reaching threshold and initiates the spontaneous depolarization and contraction of the cell. Note the lack of a resting membrane potential.

rapid rate until the membrane potential reaches a value of approximately +5 mV (red line). This phase of rapid depolarization is referred to as **phase 0**. Such sudden change in membrane potential allows Ca^{2+} channels to close and voltage-gated K^+ channels to open, allowing an outflux of K^+ bringing a repolarization state. This particular phase is referred to as **phase 3**. When the membrane potential reaches approximately −60 mV (resting membrane potential), K^+ channels close and Na^+ channels open once again, and the prepotential phase begins as in the previous cycle and is referred to as **phase 4**. This phenomenon explains the **autorhythmicity** properties of cardiac muscle (Fig. 1.14).

1.11 Membrane potentials in cardiac contractile cells

The action potential pattern is different in cardiac contractile muscle fibers which demonstrate a much more stable resting phase than conductive cells. The resting membrane potential is approximately −80 mV for atrial myofibers and −90 mV for the ventricular muscle fibers. The nature of the action potential is very different in the cardiac contractile cells; there is a rapid depolarization, followed by a **plateau phase** and then repolarization. This phenomenon accounts for the long refractory periods required for the cardiac muscle cells to pump blood effectively before they are capable of firing for a second time. These cardiac myocytes do not normally initiate their own electrical potential, although they are capable of doing so, but are triggered by an impulse from a conductive cell (SA node) to initiate an action potential.

When stimulated by an action potential in the pacemaker cell, voltage-gated Na^+ channels open in a positive-feedback manner allowing a rapid influx of Na^+ to raise the membrane potential to approximately +30 mV, at which point the voltage-gated Na^+ channels close. The rapid depolarization period typically lasts 3–5 ms (orange line; **phase 0**). Depolarization is followed by the plateau phase. During the initial period of the plateau phase, only the K^+

channels are open and membrane potential declines relatively slowly (**phase 1**). The opening of the slow Ca^{2+} channels, allowing Ca^{2+} to enter the cell maintains a membrane potential of approximately 0 mV. During this period, K^+ channels are also open, allowing K^+ to exit the cell while Ca^{2+} are moving inward. Thus, this period consists of the movements of Ca^{2+} inward whereas K^+ outward and it lasts approximately 175 ms (purple line; **phase 2**). Next, the Ca^{2+} channels close and K^+ channels remain open, allowing more K^+ to exit the cell initiating a repolarization phase (**phase 3**). The repolarization phase lasts approximately 75 ms. At this point, membrane potential drops until it reaches resting levels and K+ channels close (**phase 4**) once more and the cycle repeats. The entire event lasts between 250 and 300 ms (Fig. 1.15).

After an action potential is initiated, the cardiac muscle cell is unable to initiate another action potential for some time, and this period of time is referred to as the **refractory period**, which lasts 250 ms in duration and helps protect the heart. The cardiac refractory period is separated into an **absolute refractory period** and a **relative refractory period**. During the absolute refractory period, a new action potential cannot be elicited. On the other hand, during the relative refractory period, a new action potential can be generated given a situation such as a strong stimulus.

The action potential in the contractile cells of the cardiac muscle is compared with that of the skeletal muscle fiber in Fig. 1.15B. Both the duration of an action potential and refractory period are much shorter in the skeletal muscle fiber compared to the contractile cells of the cardiac muscle. Fig. 1.15B also compares the nature of contraction that follows each action potential. Since the refractory period is much longer in the cardiac muscle fiber, the contraction cycle is such that the entire cycle must be completed prior to a new contraction cycle unlike in the skeletal muscle where a new contraction may be initiated before the completion of the previous one due to the shorter refractory period. This phenomenon is referred to as tetanus which is commonly seen in the skeletal muscle but not in the cardiac muscle. Cardiac muscle cells undergo twitch type of contractions with long refractory periods followed by brief relaxation periods. The relaxation is essential so the heart can fill with blood for the next cycle. The refractory period is very long to prevent the possibility of **tetany**, a condition in which muscle remains involuntarily contracted. In the heart, tetany is not compatible with life, since it would prevent the heart from pumping blood.

1.11.1 Comparative rates of conduction system firing

The pattern of slow depolarization, followed by rapid depolarization and repolarization, is seen in the SA node and a few other conductive cells in the heart. The SA node serves as the pacemaker since it reaches the threshold faster than any other component of the conduction system. It initiates the impulses spreading to the other conducting cells. The SA node, without nervous or endocrine control, initiates a heart impulse approximately 80–100 times per minute. Although each component of the conduction system is capable of generating its own impulse, the rate progressively slows from the SA node to the Purkinje fibers.

Without the SA node, the AV node would generate a heart rate of 40–60 bpm. If the AV node also was incapable of serving as a pacemaker, the AV bundle would fire at a rate of

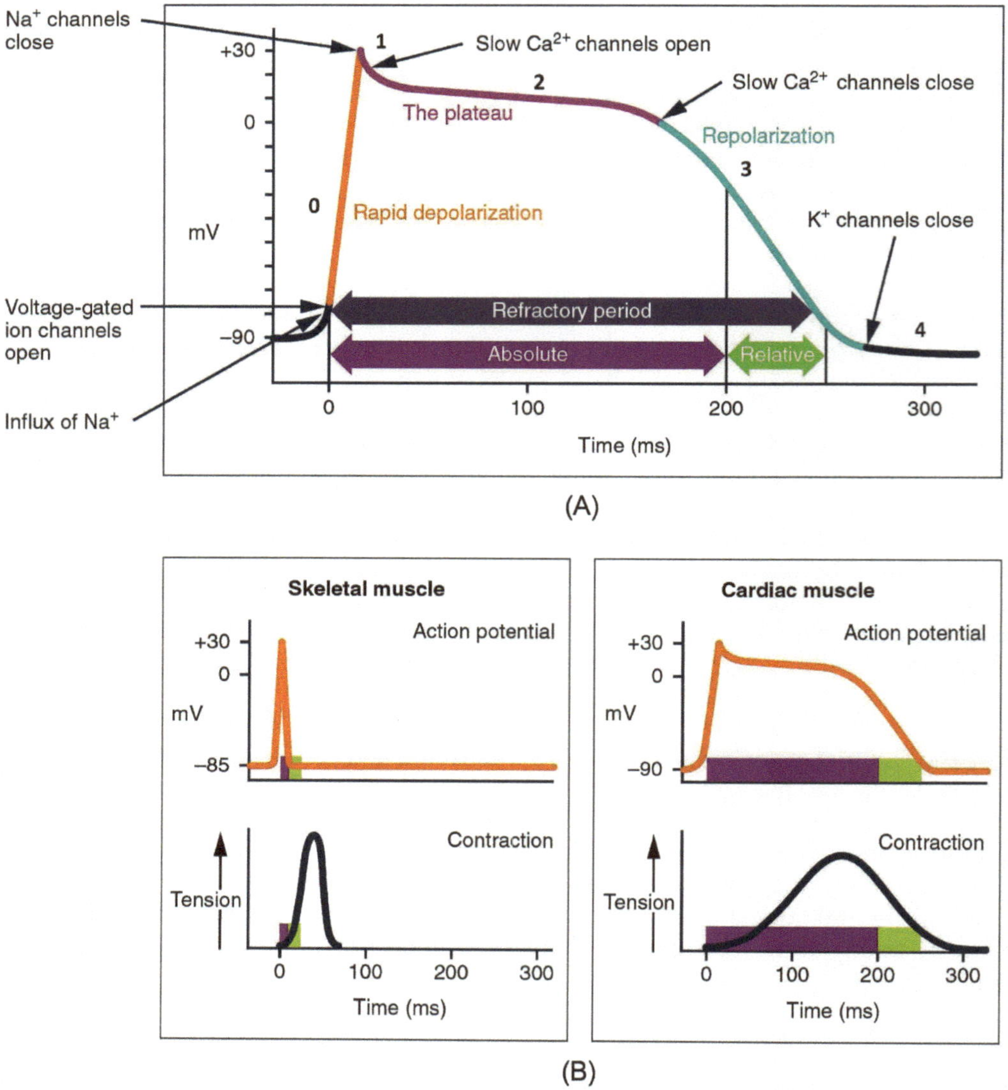

FIG. 1.15 Action potential in the cardiac contractile cells [13]. (A) Note the long plateau phase due to the influx of Ca^{2+}. The extended refractory period allows the cell to fully contract before another electrical event can occur. The numbers 0, 1, 2, 3, and 4 indicate the phases of the action potential. (B) The action potential for heart muscle is compared to that of skeletal muscle.

approximately 30–40 impulses per minute. The bundle branches would have an inherent rate of 20–30 impulses per minute, and the Purkinje fibers would fire at about 15–30 impulses per minute. While a few exceptionally trained aerobic athletes demonstrate resting HR in the range of 30–40 bpm (the lowest recorded figure is 28 bpmfor Miguel Indurain, a cyclist), for most individuals, rates lower than 50 bpmwould indicate **bradycardia**, a slow HR. As rates fall much below this level, the heart would be unable to maintain adequate perfusion of blood to vital tissues, resulting in organ failure or multisystem failure and ultimately death.

1.12 Electrocardiogram

One can record the electrical signals of the heart by the placement of electrodes in certain parts of the body. The tracing of the electrical signal from the heart is called the **electrocardiogram (ECG)**, also abbreviated as **EKG** (K stands for *kardiology*, the German term for cardiology). A typical ECG is not a recording from a single muscle fiber but the overall change in the flow of electrical current and reveals a detailed picture of the heart function thus serving as an important clinical diagnostic tool. The term "lead" typically describes the voltage difference between two of the electrodes. The standard electrocardiograph uses 3, 5, or 12 leads. The greater the number of leads an electrocardiograph uses, the more detailed information the ECG provides. The 12-lead electrocardiograph uses 10 electrodes placed in the various locations on the patient's body as shown in Fig. 1.16.

The major points on the ECG are the **P wave**, the **QRS complex**, and the **T wave** (Fig. 1.17). The small P wave represents the depolarization of the atria whereas the large QRS complex represents the depolarization of the ventricles. QRS complex is a much stronger electrical signal because of the larger size of the ventricular cardiac muscle. The T wave represents

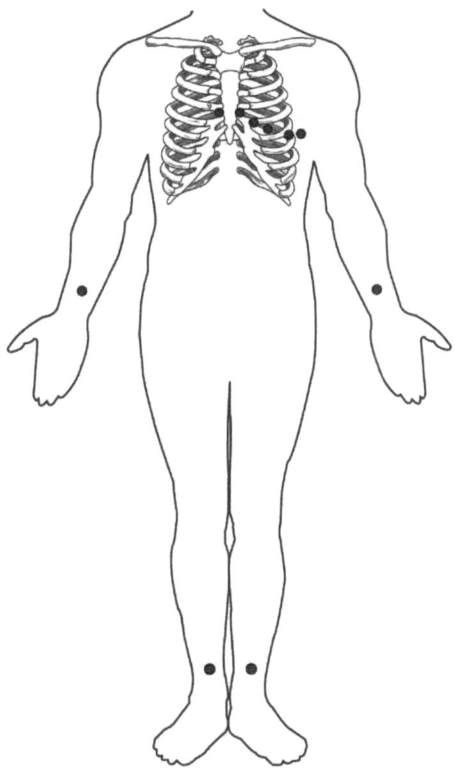

FIG. 1.16 Standard placement of ECG leads in a 12-lead ECG, 6 electrodes are placed on the chest, and 4 electrodes are placed on the limbs [14].

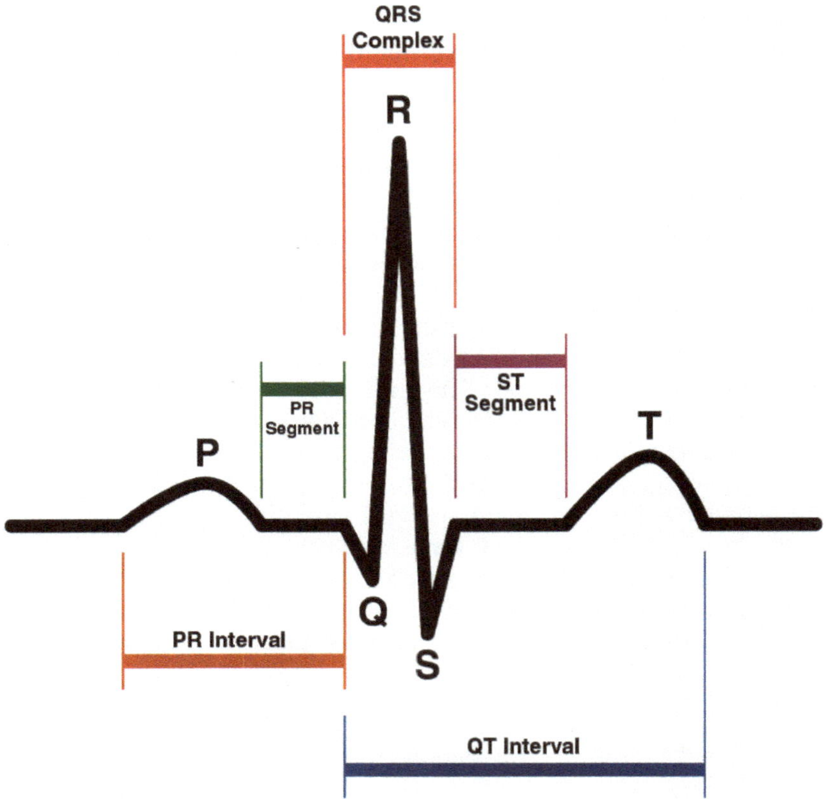

FIG. 1.17 Electrocardiogram [15]. A normal tracing shows the P wave, QRS complex, and T wave. Also indicated are the PR, QT, QRS, and ST intervals, and the P-R and S-T segments.

repolarization of the ventricles. The atrial repolarization occurs during the QRS complex but is masked by the QRS complex of the ECG since it is larger than the repolarization wave of the atria.

The major segments and intervals of an ECG tracing are also depicted in Fig. 1.17. Segments are defined as the regions between two waves. Intervals include one segment and one or more waves. For example, the PR segment begins at the end of the P wave and ends at the beginning of the QRS complex whereas the PR interval includes the P wave and the PR segment. The atria begin contracting approximately 25 ms after the start of the P wave. Similarly, the ventricles begin to contract as the QRS reaches the peak of the R wave. The PR interval is clinically more relevant, as it measures the duration from the beginning of atrial depolarization (the P wave) to the initiation of the QRS complex.

1.13 Cardiac cycle

The pumping action of the heart is an ongoing process where the events occur in a cyclical fashion. Each set of these repeating events is referred to as one **cardiac cycle** or one **heartbeat**.

Each cycle is divided into two major events. The period of contraction the heart undergoes to pump blood is called **systole**. The period of relaxation that occurs as the chambers fill with blood is called **diastole**. Although both atria and ventricles undergo systole and diastole, ventricular systole and ventricular diastole are the major events compared to atrial systole and atrial diastole based on the roles they play in the actual pumping of blood.

The heart could be described as having two pumps, the right and the left. Both pumps fill at the same time, pump at the same time, and relax at the same time. The volume of blood handled by the two pumps is also the same. However, the pressure between the two pumps varies. The left pump generates a pressure that is much greater compared to the right pump due to its purpose of perfusing blood throughout the body. The right pump having one-third of the pressure of the left pump pushes blood into the lungs which are situated close to the heart. While the left pump deals with oxygenated blood, the right pump contains deoxygenated blood.

1.13.1 Phases of the cardiac cycle

1.13.1.1 Atrial filling

In this particular phase, both the atria and the ventricles are relaxed and all four valves are closed. Ventricles would have the least amount of blood (50–60 mL) referred to as **end-systolic volume**. Deoxygenated blood flows into the right atrium from the superior and inferior venae cavae as well as from the **coronary sinus**, a major vein draining blood from the heart. Similarly, the left atrium receives oxygenated blood from the four pulmonary veins. Blood flow into the atria from the veins depends on pressure differences between the two. As blood flows into the atria, the pressure begins to rise within these chambers due to accumulation of blood, which, in turn, pushes open the AV valves. Thus, there is passive filling of blood into the ventricles for which gravity also contributes. Approximately 70%–80% of the blood fills the ventricles in this manner.

1.13.1.2 Atrial systole and diastole

The spreading of depolarization from the SA node through the atria and subsequent contraction that follows atrial muscle fibers raises pressure within the atria further, causing the atria to pump remaining blood into the ventricles and this event is referred to as **atrial systole**. At the start of atrial systole, the ventricles are already filled with 70%–80% of their capacity due to the passive inflow of blood. Atrial contraction also referred to as the "atrial kick" contributes to the remaining 20%–30% of filling (Figs. 1.18 and 1.19). Atrial systole lasts approximately 100 ms. Soon after atrial systole, the atrial muscle fibers return to atrial diastole.

1.13.1.3 Ventricular systole

Ventricular systole (Figs. 1.18 and 1.19) follows depolarization of the ventricles and it lasts a total of 270 ms. Initially, as the ventricular muscles contract, the pressures within the chambers increase, but they are not sufficient to trigger the opening of semilunar valves. The opening of semilunar valves requires a pressure greater than in the arteries they are connected with. The pressures in the ventricles, however, rise above that of the pressure within the atria that are now relaxed and in diastole. This increase in pressure causes blood to flow back

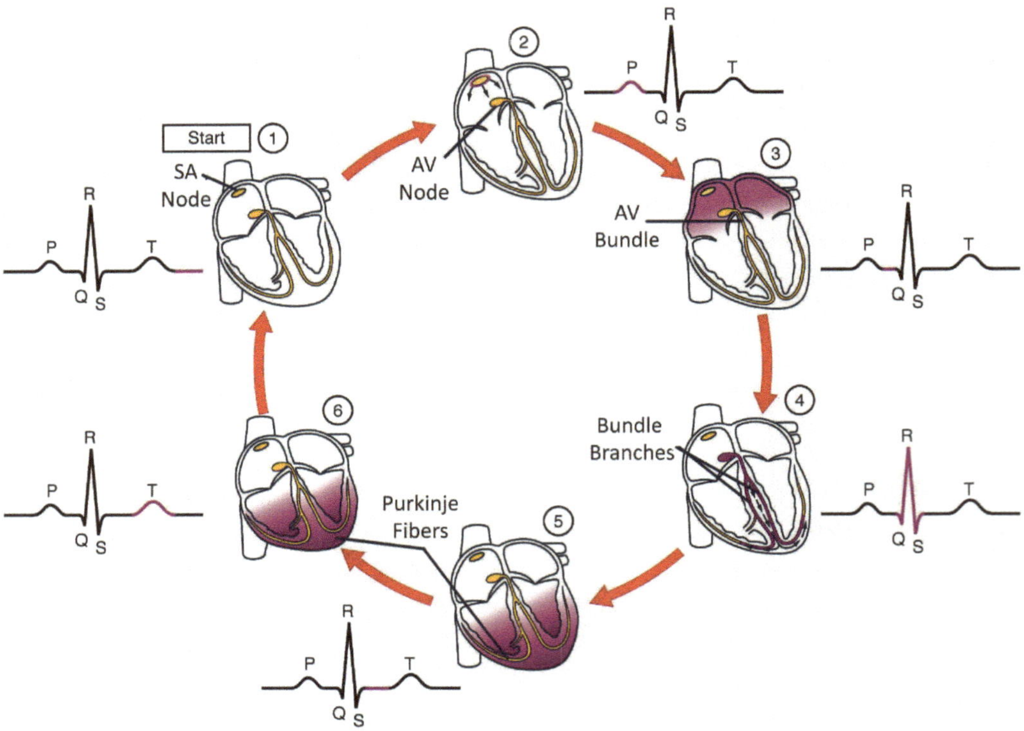

FIG. 1.18 ECG tracing correlated to the cardiac cycle [16]. This diagram depicts the relationship between an ECG tracing and the electrical and mechanical events during one cardiac cycle. (1) Depolarization begins in the SA node. (2) Atrial depolarization spreads from the SA node toward the AV node. (3) PR interval reflects the time the electrical impulse takes to travel from the SA node through the AV node. (4) The QRS complex represents the rapid depolarization of the right and left ventricles. (5) The ST segment connects the QRS complex and the T wave; it represents the period when the ventricles are depolarized. (6) The T wave represents the repolarization of the ventricles.

toward the atria, which causes the closing of the tricuspid and mitral valves. A distinct sound is heard during sudden closing of the AV valves and is referred to as the **first heart sound**, "**lubb**." Since blood is not being ejected from the ventricles at this early stage, the volume of blood within the chamber remains constant when all four valves are closed. This volume is known as the **end-diastolic volume** (EDV) or **preload** which is the largest amount of blood in each ventricle within a cardiac cycle and amounts to approximately 130 mL. The first phase of ventricular systole that occurs while all four valves are closed and the maximum amount of blood is situated in each ventricle is known as **isovolumic or isovolumetric contraction** (see Fig. 1.19).

In the second phase of ventricular systole, the **ventricular ejection phase**, the continued contraction of the ventricular muscle fibers raises the pressure within the ventricles to the point that is greater than the pressures in the pulmonary trunk and the aorta, pushing open the pulmonary and aortic semilunar valves. The pressure generated by the left ventricle would be appreciably greater compared to the pressure generated by the right ventricle since the existing pressure in the aorta is higher. The left ventricle generates a pressure of approximately 120 mmHg whereas the right ventricle exerts a pressure of approximately 35 mmHg.

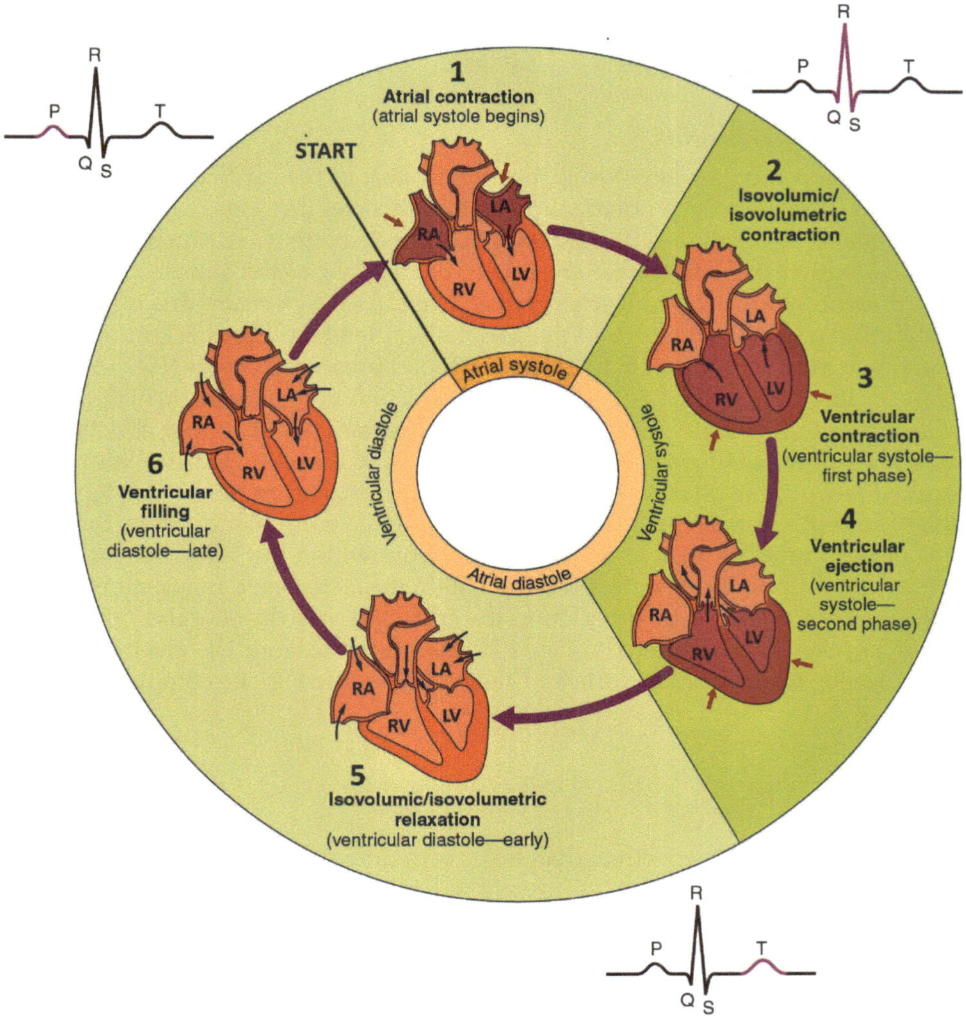

FIG. 1.19 Phases of the cardiac cycle [17]. (1) Atrial contraction forces a small amount of additional blood into relaxed ventricles. (2) Atrial systole ends and atrial diastole begins. (3) Ventricular contraction in phase 1 increases pressure steadily while all four valves are closed (isovolumetric contraction). (4) As ventricular pressure rises and exceeds pressure in the associated arteries (phase 2 of ventricular systole), the semilunar valves open and blood is ejected. (5) Isovolumetric relaxation: As ventricles relax, pressure in the ventricles drops; blood flows back against cusps of semilunar valves and forces them closed. Blood fillsthe relaxed atria. (6) Ventricular filling. Correlations to the ECG are highlighted. *LA*, left atrium; *LV*; left ventricle; *RA*, right atrium; *RV*, right ventricle.

Nevertheless, both ventricles pump the same amount of blood. This quantity of blood released from each ventricle is referred to as **stroke volume** (SV) which is normally in the range of 70–80 mL. There would be 50–60 mL of blood remaining in each ventricle following ejection and this volume of blood is known as **end-systolic volume** (ESV).

Ejection fraction is the portion of the blood that is released from the heart with each contraction. To calculate ejection fraction, SV is divided by EDV. Ejection fraction ranges from approximately 55% to 70%, with a mean of 58%.

1.13.1.4 Ventricular diastole

Ventricular relaxation, or diastole, is the longest part of the cardiac cycle lasting approximately 430 ms. It follows the repolarization of the ventricles (see Figs. 1.18 and 1.19) and also occurs immediately after the ejection of blood. It is divided into two distinct phases, early and late diastole. During the early phase of ventricular diastole, as the ventricular muscle fibers relax, the pressure within the ventricles begins to fall. As the pressure drops below the pressure within the pulmonary trunk and the aorta, blood tends to flow back into the ventricles which is shown as a dicrotic **notch** (small dip) in blood pressure tracings (Fig. 1.21). The semilunar valves close to prevent this backflow of blood into the ventricles which results in the **second heart sound**, "**dub**." Since the AV valves remain closed during this time, there is no additional volume of blood reaching the ventricles, so the early phase of ventricular diastole is called **isovolumic or isovolumetric ventricular relaxation phase** (see Fig. 1.19).

In the second phase of ventricular diastole, called late ventricular diastole, the ventricular myocardium continues to relax and therefore pressure within the ventricles drops even further, eventually dropping below the pressure within the atria. Blood now begins to flow from the atria into the ventricles, pushing open the tricuspid and mitral valves. The semilunar valves remain closed (see Fig. 1.19). Fig. 1.20 takes into account the electrical events and the mechanical events within the atria and the ventricles during one complete cardiac cycle.

1.14 Heart sounds

One of the simplest, yet effective, diagnostic techniques used to assess the state of a patient's heart is **auscultation** which is listening to the heart sounds using a stethoscope. In a normal, healthy heart, there are two audible heart sounds: $\mathbf{S_1}$ and $\mathbf{S_2}$. S_1 is the sound created by the closing of the AV valves during ventricular contraction and is normally described as a "lubb" or the first heart sound. The second heart sound, S_2, is the sound of the closing of the semilunar valves during ventricular diastole and is described as a "dub" (Fig. 1.21). In both cases, as the valves close, the openings guarded by the valves will become smaller, and blood flow through the opening will become more turbulent until the valves are fully closed. There is a third heart sound, $\mathbf{S_3}$, also called 'ventricular gallop' that is heard in early diastole but is rarely heard in healthy individuals [19]. S_3 is often due to the sloshing of blood back and forth in the left ventricle that is very compliant. S_3 may be heard in youth, some athletes, and pregnant women as a normal variant. If the sound is heard later in life, it may indicate congestive heart failure, warranting further tests. The fourth heart sound, $\mathbf{S_4}$, also known as 'atrial gallop' results from the contraction of the atria pushing blood into a stiff or hypertrophic ventricle, indicating failure of the left ventricle. S_4 occurs prior to S_1.

Fig. 1.21 captures the details of events during one cardiac cycle in the left atrium, left ventricle, and the aorta as well as the activities of the mitral valve and the aortic valve. The most dramatic change is seen in the left ventricular pressure (green line) where the pressure ranges

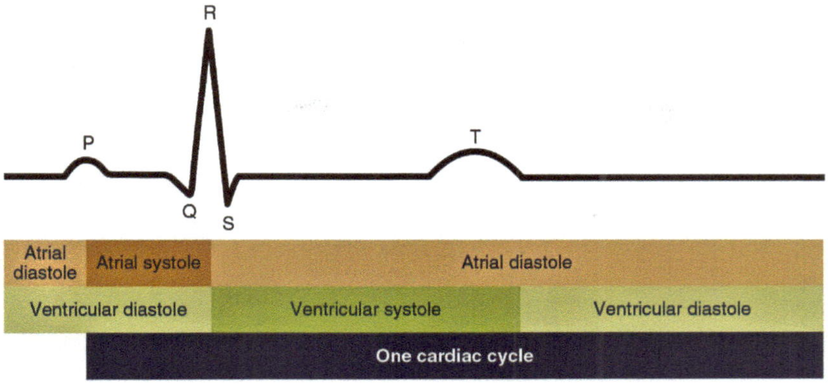

FIG. 1.20 The relationship between the cardiac cycle and the ECG. Initially, both the atria and ventricles are relaxed (diastole) [18]. The P wave represents depolarization of the atria and is followed by atrial contraction (systole). Atrial systole extends until the QRS complex, at which point, the atria relax. The QRS complex represents depolarization of the ventricles and is followed by ventricular contraction. The T wave represents the repolarization of the ventricles and marks the beginning of ventricular relaxation.

from 0 mmHg during early ventricular diastole which increases slightly during its passive filling. The closing of the AV valve and isovolumetric contraction cause a steady increase in pressure to 120 mmHg. The opening of the aortic valve and the release of blood into the aorta causes a decrease in left ventricular pressure back to 0 mmHg. The blood pressure range in the aorta (red line) is from 80 to 120 mmHg. When the blood moves away from the heart, the pressure is lower whereas at the time of ventricular ejection, the pressure escalates to 120 mmHg. The pressure changes in the left atrium are much smaller compared to the pressure changes in the aorta and the left ventricle. Atrial systole results in a small increase in its pressure. The passive filling of the left ventricle while the atrium is filling with blood results in a flat line (Fig. 1.21).

1.15 Heart rate, stroke volume, and cardiac output

1.15.1 Heart rate

HR varies considerably, not only with exercise and fitness levels but also with age. The HR in a newborn is approximately 120 bpmwhich gradually decreases until young adulthood and then increases again with age. An average resting HR is approximately 75 bpmbut could range from 60 to 100 bpm. Maximum HR is normally in the range of 200–220 bpm. As one ages, the ability to generate maximum rate decreases, however. This may be estimated by taking the maximal value of 220 bpmand subtracting the individual's age. For example, a 40-year-old individual would be expected to achieve a maximum rate of approximately 180, whereas a 60-year-old would achieve a HR of 160 bpm.

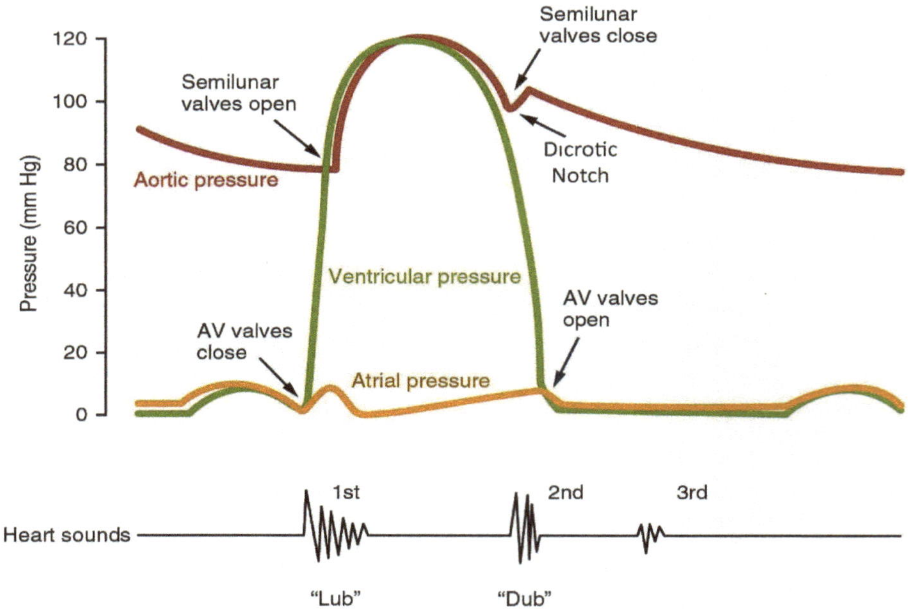

FIG. 1.21 The relationship between the heart sounds and the pressure changes within the left atrium, left ventricle, and the aorta in one cardiac cycle. The x-axis reflects time with a recording of the heart sounds. The y-axis represents pressure [20].

1.15.2 Stroke volume (SV)

SV is the volume of blood ejected by the heart in one heartbeat. While a number of variables are involved, SV is ultimately dependent upon the difference between EDV and ESV. The three primary factors that affect SV are as follows: (1) **preload** or the stretch on the ventricles prior to contraction; (2) **contractility** or the force or strength of the contraction itself; and (3) **afterload**, the pressure the ventricles must generate to pump blood against the resistance in the associated vessels.

1.15.3 Preload

Preload is another way of expressing EDV. The greater the EDV, the greater the preload is. One of the primary factors to consider is filling time which is the duration of ventricular diastole during which filling occurs. The more rapidly the heart contracts, the shorter the filling time becomes, and the lower the EDV and preload are. This effect can be partially overcome by increasing the second variable, contractility, and raising SV, but over time, the heart is unable to compensate for decreased filling time, and preload decreases.

With an increasing ventricular filling, EDV (or preload) increases and the cardiac muscle itself is stretched to a greater degree. At rest, there is a little stretch of the ventricular muscle, and the sarcomeres remain short. With an increased ventricular filling, the ventricular muscle is increasingly stretched and therefore sarcomere length increases. As the sarcomeres reach their optimal lengths, they will contract more powerfully, because more of the myosin heads can bind to actin on the thin filaments, forming more cross bridges and increasing the strength of contraction and SV. If this process continues and the sarcomeres stretch beyond their optimal lengths, the force of contraction would decrease. However, due to the physical constraints of the location of the heart, the excessive stretch is not common in a normal situation.

1.15.4 Contractility

Contractility refers to the force of contraction of the heart muscle, which contributes to SV, and the ESV. The more forceful the contraction is, the greater the SV and the smaller the ESV are. In contrast, less forceful contractions result in smaller SVs and larger ESVs.

The relationship between ventricular stretch and contraction has been stated in the **Frank-Starling mechanism** or simply **Starling's Law of the Heart**. This principle states that, within physiological limits, the force of contraction is directly proportional to the initial length of the muscle fiber. This means that the greater the stretch of the ventricular muscle, the more powerful the contraction is, which, in turn, increases SV. Therefore, by increasing preload, contractility is increased.

Factors that increase contractility are referred to as **positive inotropic factors**, and those that decrease contractility are described as **negative inotropic factors** (ino- = "fiber"; -tropic = "turning toward"). For example, sympathetic stimulation is a positive inotrope, whereas parasympathetic stimulation is a negative inotrope. Higher concentrations of intracellular Ca^{2+} increase the strength of contraction. Excess Ca^{2+} (**hypercalcemia**) also acts as a positive inotropic agent. The drug *digitalis* lowers heart rate and increases the strength of the contraction. It acts as a positive inotropic agent by blocking the sequestering of Ca^{2+} into the sarcoplasmic reticulum which leads to higher intracellular Ca^{2+} levels and greater strength of contraction. Hormones NE and E, thyroid hormones, and glucagon demonstrate positive inotropic effects. Negative inotropic agents include hypoxia (decreased oxygen supply), acidosis, hyperkalemia (excess K^{+}), beta blockers, and Ca^{2+} channel blockers.

1.15.5 Afterload

Afterload refers to the pressure that the ventricles must generate to pump blood effectively against the resistance in the vascular system. Any condition that increases resistance requires a greater afterload to force open the semilunar valves and pump the blood. Damage to the valves, such as **stenosis**, which makes them harder to open, will also increase afterload. Any decrease in resistance decreases the afterload.

SV can be determined using an ECG to record EDV and ESV and calculating the difference [21].

$$SV = EDV - ESV$$

The normal range for SV is 55–100 mL and a mean SV for a resting 70-kg (150-lb) individual would be approximately 70 mL. Several important variables influence this value. They include the size of the heart, physical conditions of the individual, sex, contractility, duration of contraction, preload, and **afterload** or resistance.

1.15.6 Cardiac output

Cardiac output is a measurement of the amount of blood pumped by each ventricle in 1 min. To calculate cardiac output, SV is multiplied by the HR, the number of cardiac cycles per minute. It can be represented mathematically by the following equation: $CO = HR \times SV$ (see Fig. 1.22). The mean CO is 5.25 L/min, with a range of 4.0–8.0 L/min. It must be noted that these numbers refer to cardiac output from each ventricle separately, not the total for the heart.

Physiological conditions that cause HR to increase also trigger an increase in SV in the initial phase. During exercise, the rate of blood returning to the heart increases. However, as the HR rises, there is less time spent in diastole and consequently less time for the ventricles to fill with blood. Even though there is less filling time, SV will initially remain high. However, as HRcontinues to increase, SV gradually decreases due to decreased filling time. Cardiac output will initially stabilize as the increasing HR compensates for the decreasing SV, but at very high rates, cardiac output will eventually decrease as increasing rates are no longer able to compensate for the decreasing SV. Consider this phenomenon in a healthy young individual. Initially, as HR increases from resting to approximately 120 bpm, cardiac output will rise. As HRincreases from 120 to 160 bpm, cardiac output remains stable, since the increase in rate is offset by decreasing ventricular filling time and, consequently, SV. As HRcontinues to rise above 160 bpm, cardiac output actually decreases because SV falls faster than the increase in HR. Although aerobic exercise is critical to maintain the health of the heart, individuals are cautioned to monitor their HRto ensure they stay within the target HRrange of 120–160 bpmso that cardiac output is maintained.

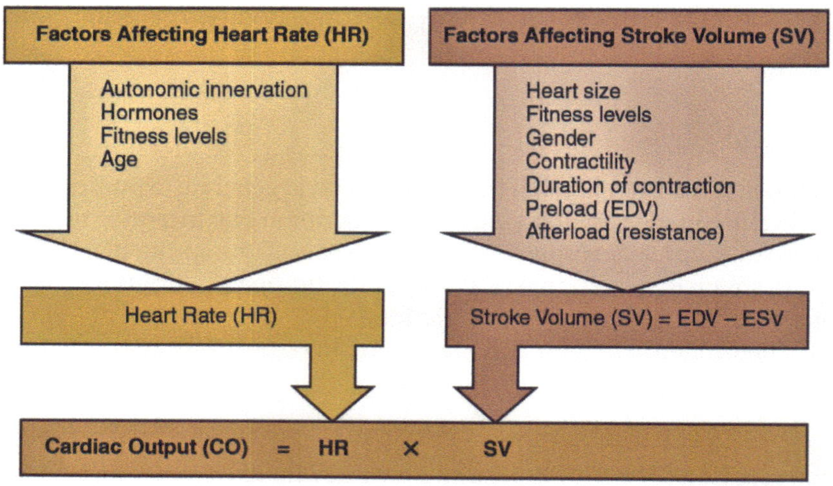

FIG. 1.22 Major factors influencing cardiac output [22]. Cardiac output is influenced by heart rate and stroke volume, both of which are also variable. End-systolic volume (ESV); End-diastolic volume (EDV) or preload.

1.15.7 Cardiac reserve

In healthy young individuals, HRmay increase to 150 bpm during exercise. SV can also increase from 70 mL to approximately 130 mL due to the increased strength of contraction, thus increasing cardiac output to approximately 19.5 L/min, four to five times higher compared to the resting rate. Athletes can achieve even higher levels depending on their levels of training. At their peak performance, they may increase resting cardiac output by seven to eight times. Since the bulk of the heart is made up of muscle, exercising increases its efficiency. The difference between the maximum and resting cardiac output is known as the **cardiac reserve**. It measures the residual capacity of the heart to pump blood.

1.16 Cardiovascular centers

Although the HRis initiated by the pacemaker of the heart, neural input modulates HR. There are two paired cardiovascular centers in the medulla oblongata (Fig. 1.23). Cardiac activity is stimulated via **sympathetic** activation, and decreased via **parasympathetic** stimulation which reaches the heart via **vagus nerve** (cranial nerve X). During rest, both centers provide a basal level of stimulation to the heart, contributing to **autonomic tone**. Normally, vagal stimulation predominates keeping the HR between 70 and 80 bpm. Without the parasympathetic influence, the SA node would initiate a sinus rhythm of approximately 100 bpm.

The cardiac nerves innervate both the SA and AV nodes and send additional fibers to the atria and ventricles. The ventricles are richly innervated by sympathetic fibers compared to parasympathetic fibers. Sympathetic stimulation stimulates the adrenal medulla to secrete E and NE. NE decreases the repolarization period and, therefore, allows a new depolarization and subsequent contraction to occur sooner than normal, resulting in an increase in HR. Its release opens both Na^+ and Ca^{2+} channels, allowing an influx of positivelycharged ions as part of the depolarization phase of the action potential in the pacemaker cells. E and NE brings these effects within the heart by binding to **beta-1 adrenergic receptors**. In addition to their stimulatory effects on HR, E and NE also act on the cardiac muscle cell membrane to increase metabolic rate and the force of contraction. This combination of actions has the net effect of increasing SV and leaving a smaller residual ESV. Some cardiac medications (for example, beta blockers) bring about their effects by blocking these receptors, thereby slowing HR and decreasing the force of contraction. Additionally, any sympathetic stimulation to the venous system will increase venous return to the heart, which contributes to ventricular filling, and EDV.

Parasympathetic impulses travel via the **vagus nerve**. The vagus nerve sends branches to both the SA and AV nodes, and portions of both the atria and ventricles. Parasympathetic stimulation releases the neurotransmitter **acetylcholine (ACh)** at the neuromuscular junction. ACh slows HRby opening K^+ channels which extends repolarization and increases the time before the next spontaneous depolarization, thus delaying the rate of spontaneous depolarization. Since resting HRis considerably less than 100 bpm, it becomes evident that parasympathetic stimulation normally slows HR. This phenomenon is referred to as a negative **chronotropic effect** (*chrono-*, meaning time, and *tropos*, "a turn" are those that change the HR). This is similar to an individual driving a car with one foot on the brake pedal.

FIG. 1.23 Autonomic innervation of the heart [23]. Cardiac center is located in the medulla oblongata of the brain and innervates the heart via the sympathetic innervation that increases cardiac activity and parasympathetic innervation that slows cardiac activity.

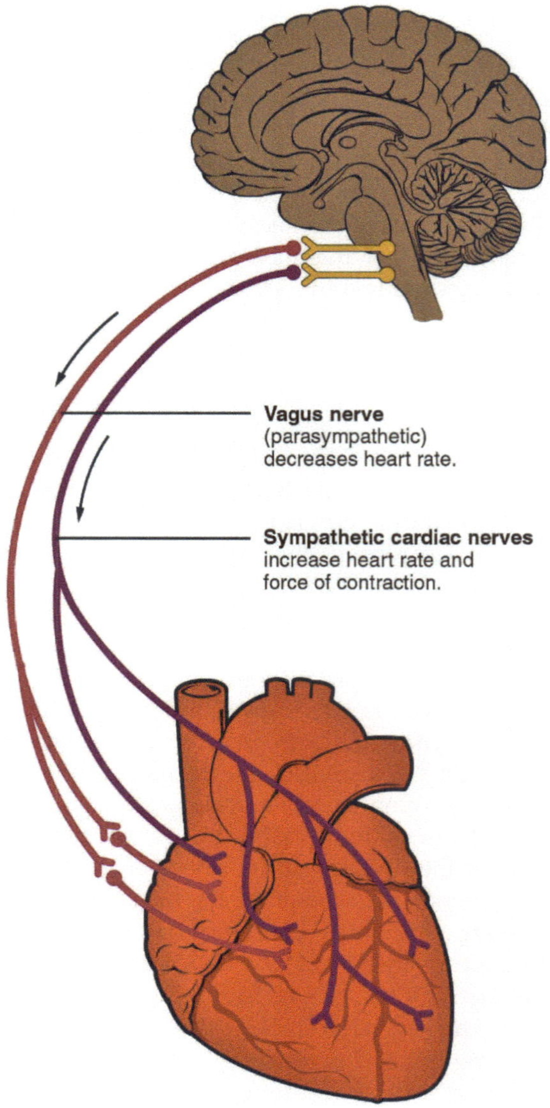

To speed up, one would need to merely remove the foot from the brake and let the engine increase speed. In the case of the heart, decreasing parasympathetic stimulation decreases the release of ACh, which leads to an increase in the HRup to approximately 100 bpm. Any increases beyond this rate would require sympathetic stimulation. Fig. 1.24 illustrates the effects of parasympathetic and sympathetic stimulation on the normal sinus rhythm.

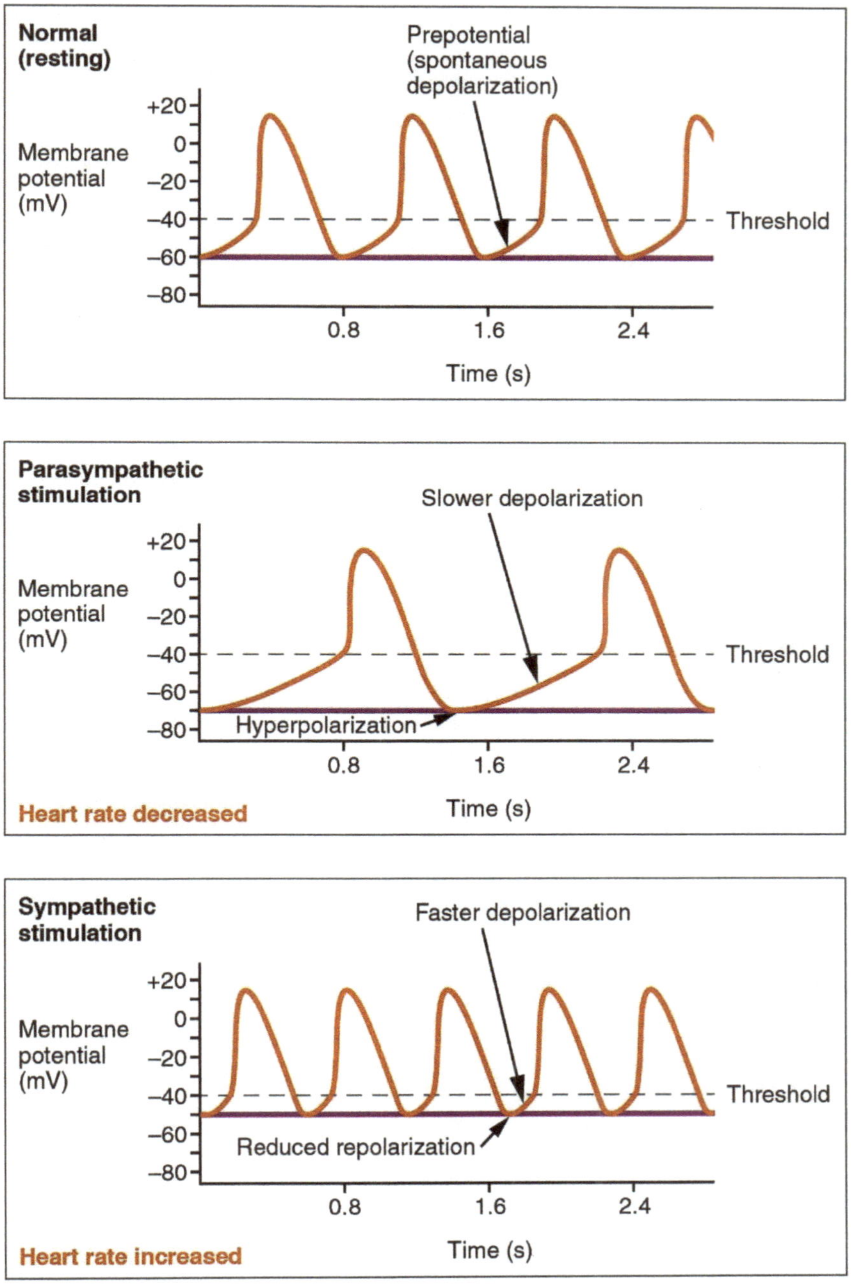

FIG. 1.24 Effects of parasympathetic and sympathetic stimulation on normal sinus rhythm [24]. The wave of depolarization in a normal sinus rhythm shows a stable resting HR. Following parasympathetic stimulation, HR slows whereas following sympathetic stimulation, HR increases.

Since parasympathetic fibers are more widespread in the atria than in the ventricles, the primary site of action is in the atria. Parasympathetic stimulation decreases the atrial kick and reduces EDV, which in turn decreases ventricular stretch and preload, thereby further limiting the force of ventricular contraction.

References

[1] Blausen 0458 Heart ThoracicCavity.png. Wikimedia Commons, the free media repository. Updated October 25, 2020. Accessed February 2, 2021. https://commons.wikimedia.org/w/index.php?title=File:Blausen_0458_Heart_ThoracicCavity.png&oldid=501286362.

[2] Blausen 0470 HeartWall.png. Wikimedia Commons, the free media repository. Updated October 25, 2020. Accessed February 2, 2021. https://commons.wikimedia.org/w/index.php?title=File:Blausen_0470_HeartWall.png&oldid=501284908.

[3] 2006 Heart Musculature.jpg. Wikimedia Commons, the free media repository. Updated October 16, 2020. Accessed February 2, 2021. https://commons.wikimedia.org/w/index.php?title=File:2006_Heart_Musculature.jpg&oldid=491437876.

[4] 2007 Ventricular Muscle Thickness.jpg. Wikimedia Commons, the free media repository. Updated October 5, 2020. Accessed February 2, 2021. https://commons.wikimedia.org/w/index.php?title=File:2007_Ventricular_Muscle_Thickness.jpg&oldid=481555691.

[5] Blausen 0260 CoronaryVessels Anterior.png. Wikimedia Commons, the free media repository. Updated October 10, 2020. Accessed February 2, 2021. https://commons.wikimedia.org/w/index.php?title=File:Blausen_0260_CoronaryVessels_Anterior.png&oldid=485566341.

[6] Blood Circulation.gif. Wikimedia Commons, the free media repository. Updated September 11, 2020. Accessed February 3, 2021. https://commons.wikimedia.org/w/index.php?title=File:Blood_Circulation.gif&oldid=455029158.

[7] 2012 Blood Flow Relaxed Ventricles.jpg. Wikimedia Commons, the free media repository. Updated October 17, 2020. Accessed February 3, 2021. https://commons.wikimedia.org/w/index.php?title=File:2012_Blood_Flow_Relaxed_Ventricles.jpg&oldid=492072908.

[8] 2013 Blood Flow Contracted Ventricles.jpg. Wikimedia Commons, the free media repository. Updated October 20, 2020. Accessed February 3, 2021. https://commons.wikimedia.org/w/index.php?title=File:2013_Blood_Flow_Contracted_Ventricles.jpg&oldid=495431114.

[9] 2017abc Cardiac Muscle.jpg. Wikimedia Commons, the free media repository. Updated September 26, 2020. Accessed February 3, 2021. https://commons.wikimedia.org/w/index.php?title=File:2017abc_Cardiac_Muscle.jpg&oldid=471666846.

[10] 2018 Conduction System of Heart.jpg. Wikimedia Commons, the free media repository. Updated September 18, 2020. Accessed February 3, 2021. https://commons.wikimedia.org/w/index.php?title=File:2018_Conduction_System_of_Heart.jpg&oldid=463222495.

[11] 2019 Cardiac ConductionN.jpg. Wikimedia Commons, the free media repository. Updated September 12, 2020. Accessed February 3, 2021. https://commons.wikimedia.org/w/index.php?title=File:2019_Cardiac_ConductionN.jpg&oldid=456741599.

[12] 2020 SA Node Tracing.jpg. Wikimedia Commons, the free media repository. Updated September 26, 2020. Accessed February 3, 2021. https://commons.wikimedia.org/w/index.php?title=File:2020_SA_Node_Tracing.jpg&oldid=471754249.

[13] 2026 Action Potential Heart Contraction.jpg. Wikimedia Commons, the free media repository. Updated October 9, 2020. Accessed February 3, 2021. https://commons.wikimedia.org/w/index.php?title=File:2026_Action_Potential_Heart_Contraction.jpg&oldid=485318072.

[14] 2021 ECG Placement of Electrodes.jpg. Wikimedia Commons, the free media repository. Updated October 15, 2020. Accessed February 3, 2021. https://commons.wikimedia.org/w/index.php?title=File:2021_ECG_Placement_of_Electrodes.jpg&oldid=490711347.

[15] SinusRhythmLabels.svg. Wikimedia Commons, the free media repository. Updated Septembe 19, 2020. Accessed February 3, 2021. https://commons.wikimedia.org/w/index.php?title=File:SinusRhythmLabels.svg&oldid=464693489.

[16] 2023 ECG Tracing with Heart ContractionN.jpg. Wikimedia Commons, the free media repository. Updated Septembe 5, 2020. Accessed February 3, 2021. https://commons.wikimedia.org/w/index.php?title=File:2023_ECG_Tracing_with_Heart_ContractionN.jpg&oldid=481556462.
[17] 2027 Phases of the Cardiac Cycle.jpg. Wikimedia Commons, the free media repository. Updated October 17, 2020. Accessed February 3, 2021. https://commons.wikimedia.org/w/index.php?title=File:2027_Phases_of_the_Cardiac_Cycle.jpg&oldid=492845651.
[18] 2028 Cardiac Cycle vs Electrocardiogram.jpg. Wikimedia Commons, the free media repository. Updated October 14, 2020. Accessed February 3, 2021. https://commons.wikimedia.org/w/index.php?title=File:2028_Cardiac_Cycle_vs_Electrocardiogram.jpg&oldid=489631495.
[19] H.K. Walker, W.D. Hall, J.W. Hurst, Clinical Methods: The History, Physical, and Laboratory Examinations, Butterworth-Heinemann, 1990.
[20] 2029 Cardiac Cycle vs Heart Sounds.jpg. Wikimedia Commons, the free media repository. Updated October 14, 2020. Accessed February 3, 2021. https://ko.wikipedia.org/wiki/%ED%8C%8C%EC%9D%BC:2029_Cardiac_Cycle_vs_Heart_Sounds.jpg.
[21] C. Kondo, G. Caputo, R. Semelka, E. Foster, A. Shimakawa, C. Higgins, Right and left ventricular stroke volume measurements with velocity-encoded cine MR imaging: in vitro and in vivo validation, AJR Am. J. Roentgenol. 157 (1) (1991) 9–16.
[22] 2031 Factors in Cardiac Output.jpg. Wikimedia Commons, the free media repository. Updated October 12, 2020. Accessed February 3, 2021. https://commons.wikimedia.org/w/index.php?title=File:2031_Factors_in_Cardiac_Output.jpg&oldid=488143995.
[23] 2032 Automatic Innervation.jpg. Wikimedia Commons, the free media repository. Updated October 12, 2020. Accessed February 3, 2021. https://lv.wikipedia.org/wiki/Att%C4%93ls:2032_Automatic_Innervation.jpg.
[24] 2033 Depolarization in Sinus Rhythm.jpg. Wikimedia Commons, the free media repository. Updated October 17, 2020. Accessed February 3, 2021. https://hy.wikipedia.org/wiki/%D5%8A%D5%A1%D5%BF%D5%AF%D5%A5%D6%80:2033_Depolarization_in_Sinus_Rhythm.jpg.

CHAPTER

2

The blood vessels

OUTLINE

OBJECTIVES

- Compare the blood vessel wall of an artery with that of a vein and a capillary.
- List the different types of capillaries and give examples.
- Describe the characteristics of each type of blood vessel: artery, arteriole, capillary, venule, and vein.
- Summarize the mechanisms involved in allowing the blood to flow. Explain the role of

Biology of Cardiovascular and Metabolic Diseases
https://doi.org/10.1016/B978-0-12-823421-1.00004-4

peripheral resistance and blood pressure in the regulation of blood flow.
- Explain various factors associated with blood flow.
- Describe the capillary exchange system. List the forces involved in this process.
- Summarize the role of the lymphatic system in capillary exchange.
- Summarize the role played by the skeletal muscle pump and respiratory pump in relation to venous return.
- Relate exercise in maintaining vascular homeostasis.

2.1 Introduction

Blood vessels serve as the transport medium within the body. They move blood from the heart to the entire body and back to the heart. They also provide a physical site for gases, nutrients, and other substances to be exchanged between the blood and the tissues. Any disturbance associated with blood flow would result in tissue injury and subsequent loss of bodily functions.

2.2 Blood vessel arrangement

Blood vessels are divided into different groups based on their sizes and functions. An **artery** always carries blood away from the heart. Two large arteries, the aorta and the pulmonary artery, begin at the heart. They branch off to become mid-sized and ultimately small-sized arteries. The smaller arteries are called **arterioles**, which continue to branch further and become extensively branched microscopic vessels called **capillaries**, where the exchange of gases, nutrients, and wastes occurs between them and the nearby tissues. Capillaries fuse with other capillaries to form **venules** or small veins. Similarly, venules join to become **veins** or larger blood vessels that return the blood to the heart.

The entire vascular system forms two closed circuits called **systemic** and **pulmonary circuits** (Fig. 2.1). The arteries in the systemic circuit supply blood rich in oxygen to the tissues throughout the body, whereas systemic veins return deoxygenated blood to the heart. In contrast, the pulmonary arteries carry deoxygenated blood exclusively to the lungs for gas exchange. Pulmonary veins carry oxygenated blood from the lungs back to the heart.

2.3 Shared structures of blood vessels

Although there are differences in the structure and functions between the arteries and veins, they share certain features in common. Arteries and arterioles are made up of thicker walls since they are closer to the heart and receive high pressure than the venules and veins (Fig. 2.2). Each type of vessel has a **lumen**, a hollow space in the center to allow blood flow. The lumen is smaller in the arteries than in the veins which helps maintain blood pressure moving through the vascular system. The thicker walls and smaller diameters help arterial

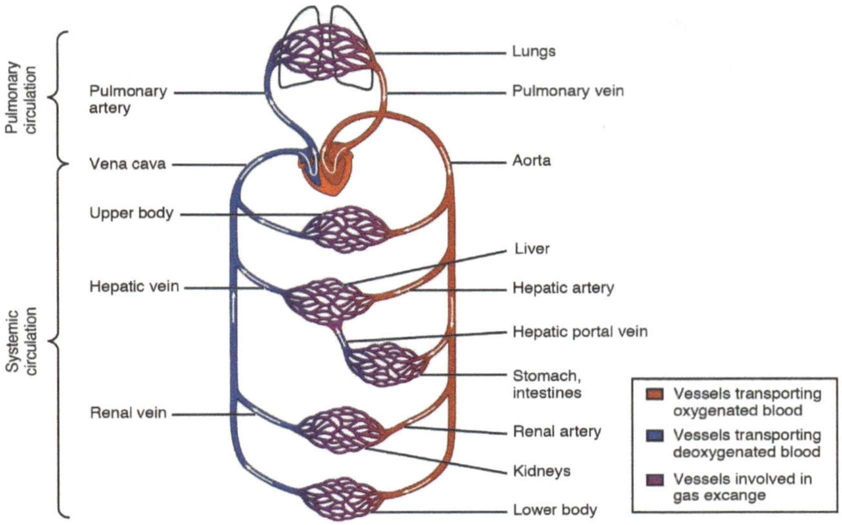

FIG. 2.1 Pulmonary and systemic circuits [1]. The pulmonary circuit transports blood from the right side of the heart to the lungs and back to the heart on the left side. The systemic circuit moves blood from the left side of the heart to the entire body and returns it back to the right side of the heart to repeat the cycle. The *arrows* indicate the direction of blood flow, and the *colors* represent the relative levels of oxygen concentration.

lumens maintain a more rounded appearance in cross-section, unlike that of the lumens of veins, which are larger, and do not maintain a round shape in cross-section.

Tunica intima or **tunica interna** is composed of epithelial and connective tissue layers. It consists of one layer of squamous epithelium called the **endothelium**. This layer has received a great deal of attention in recent years because of its physiological role in releasing several local chemicals in regulating blood flow and inflammation. Right above the endothelium is the **basement membrane** or **basal lamina**, which connects the endothelium to the connective tissue above. The basement membrane provides support to the endothelial lining and is permeable, allowing materials to pass through it. The outer layer of tunica intima contains a small amount of loose connective tissue, primarily made of elastic fibers to provide the vessel with flexibility and collagen fibers for additional strength. There is a thick layer of elastic fibers in larger arteries, referred to as the **internal elastic membrane** or internal elastic lamina at the boundary with the tunica media. The internal elastic membrane is not present in the veins. On the other hand, many veins, particularly in the lower limbs, contain **valves** made from the endothelial tissue extension into the lumen. Valves help prevent the backward flow of blood.

Tunica media is the middle layer of the vessel wall (Fig. 2.2). It is generally the thickest layer in the arteries, and it is much thicker in the arteries than the tunica media in the veins. Tunica media is made of layers of smooth muscle and some connective tissue, which includes primarily elastic fibers. The smooth muscle is arranged in two distinct forms, the inner circular arrangement and the outer longitudinal arrangement. Contraction and relaxation of the circular smooth muscle fibers cause the vessel lumen to decrease and increase its diameter. **Vasoconstriction** refers to the decrease in the lumen size when the smooth muscle in the tunica media contracts, increasing blood pressure, which, in turn, decreases blood flow. On the other hand, **vasodilation** occurs when the smooth muscle relaxes, allowing the lumen to widen, causing a decrease in blood pressure and an increase in blood flow. Sympathetic fibers

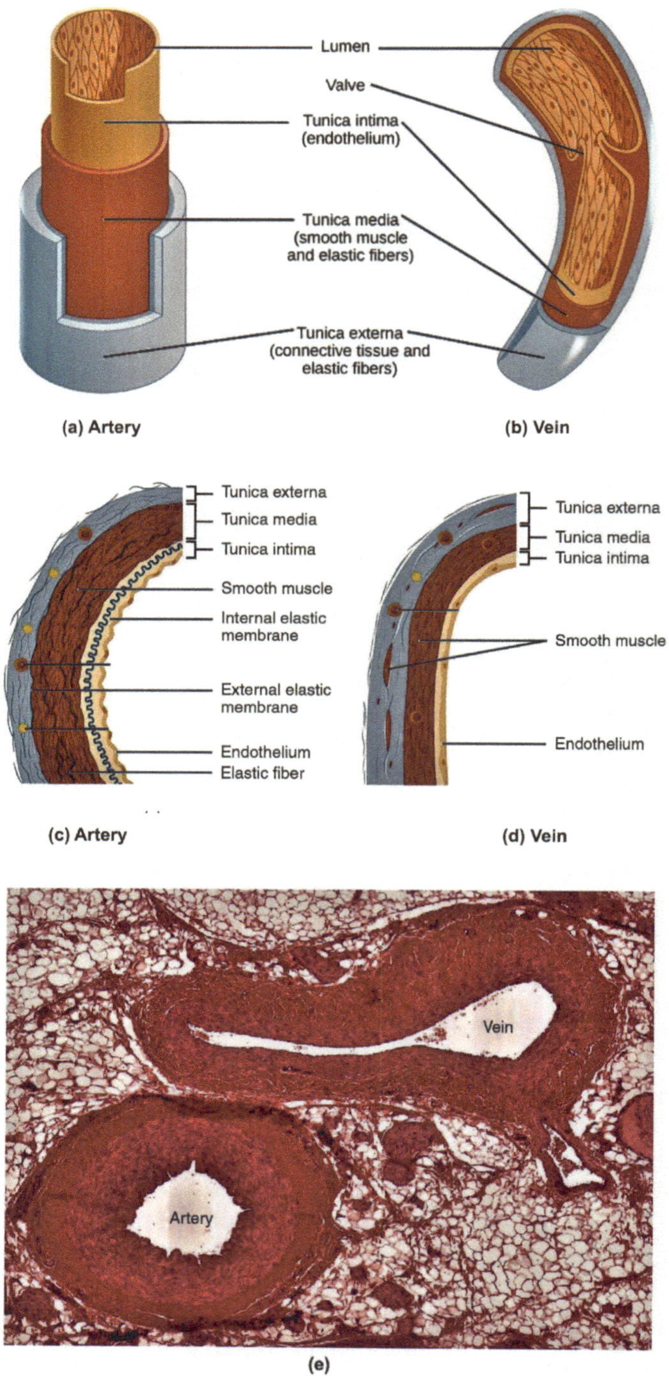

FIG. 2.2 Structure of an artery and a vein [2]. (A) Artery and (B) Vein provide an overview. (C) A detailed view of the wall of an artery and (D) a detailed view of the wall of a vein. (E) A micrograph of a cross-section of an artery and a vein showing the relative differences in thickness. LM × 160. *Micrograph provided by the Regents of the University of Michigan Medical School © 2012.*

that innervate tunica media induce vasoconstriction. Several hormones and other local chemical substances affect vasoconstriction or vasodilation, thus regulating blood flow. The **external elastic membrane** or **external elastic lamina** separates the tunica media from the outer tunica externa in larger arteries but is absent in smaller arteries and veins.

The outer tunic, **tunica externa** (also called the **tunica adventitia**), is a sheath of connective tissue composed primarily of collagen fibers along with bands of elastic fibers. The tunica externa in veins also contains smooth muscle and is often the thickest tunic in veins. The outermost layers of the tunica externa are not as well marked since they blend with the connective tissue outside the vessel, helping to hold the vessel in a relative position.

2.4 Arteries

An **artery** is defined as a blood vessel that conducts blood away from the heart. Arteries have relatively thick walls to withstand high blood pressure. The arteries close to the heart have the thickest walls and contain a greater percentage of elastic fibers in all three of their tunics and therefore are referred to as **elastic arteries** (Fig. 2.3). Elastic fibers expand as blood pumped from the ventricles passes through them and then recoil after the surge has passed. This property of elastic recoil maintains the pressure gradient that drives the blood through the arterial system.

As the arteries branch off and are farther away from the heart, the composition of elastic fibers and smooth muscle changes where the percentage of elastic fibers in the tunica interna decreases and the smooth muscle in its tunica media increases, and these arteries are referred to as **muscular arteries**. The diameter of muscular arteries is about 0.1–10 mm. Having thicker tunica media allows muscular arteries to be essential players in vasoconstriction. Muscular arteries have a limited ability to expand due to the decreased quantity of elastic fibers; they continue to branch further to distribute blood to the arterioles' vast network.

2.5 Arterioles

Arterioles are the smallest arteries with a prominent muscular coat and a diameter of 20–200 μm. They consist of all three tunics, but the thickness of each is greatly diminished (see Fig. 2.3). The muscle fibers in arterioles usually are in a slightly contracted state, causing

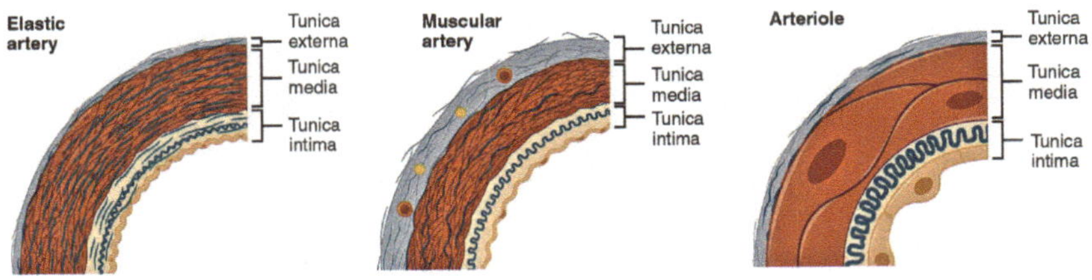

FIG. 2.3 Types of arteries and an arteriole [3]. A comparison of the walls of an elastic artery, a muscular artery, and an arteriole is shown.

arterioles to maintain a consistent muscle tone, also called **vascular tone**. The lumen size decreases in the arterioles averaging 30 μm, which helps generate most systemic vascular resistance. Arterioles are also called **resistance vessels** due to their ability to increase vascular resistance, which is a significant determinant of blood pressure. The diameter of an arteriole's lumen is maintained by vasoconstriction and vasodilation, which, in turn, is regulated by neuronal, hormonal, and local chemical agents that serve as the primary mechanisms for blood flow.

2.6 Capillaries

A capillary is the smallest blood vessel that allows the exchange of substances between the blood and the tissues. This process is referred to as **perfusion**. A capillary lumen diameter varies from 5 to 10 μm; the smallest are barely wide enough for a red blood cell to squeeze through. The capillary wall consists of the endothelial layer surrounded by a basement membrane with occasional smooth muscle fibers (Fig. 2.4).

The arrangement of the endothelial cells varies from one capillary type to another. The **continuous capillary** is where cells are arranged without any space compared to a **fenestrated capillary** where the cells are apart from one another by a specific distance. Continuous capillaries are present in the neural tissue to limit the escape of unknown substances. Fenestrated capillaries are found in the glomerulus, endocrine structures, and small intestine, allowing large substances to move between blood vessels and tissues. **Sinusoids** are specialized capillaries where large spaces exist within the capillary wall to allow blood cells and proteins to be exchanged between the tissues and the blood. The liver, spleen, and bone marrow serve as examples of these capillaries (Fig. 2.4).

A **metarteriole** is a branch of the smallest arteriole that connects it to the capillary bed. It branches further to supply blood to a **capillary bed** that may consist of 10–100 capillaries. A metarteriole is slightly larger than a typical capillary and shares structural characteristics of both an arteriole and a capillary. The smooth muscle lining within the tunica media is not continuous but arranged as rings around the vessel. The circular smooth muscle arrangement surrounding the capillary at the junction between the metarteriole and the capillary is called a

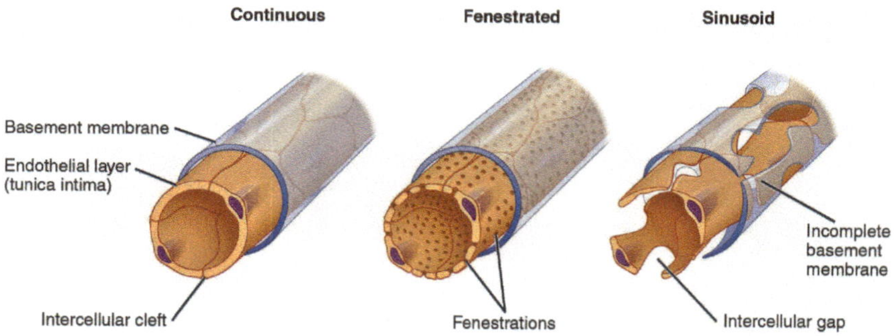

FIG. 2.4 A cross-section of a capillary. Different types of capillaries include continuous capillaries, fenestrated capillaries, and sinusoids [4].

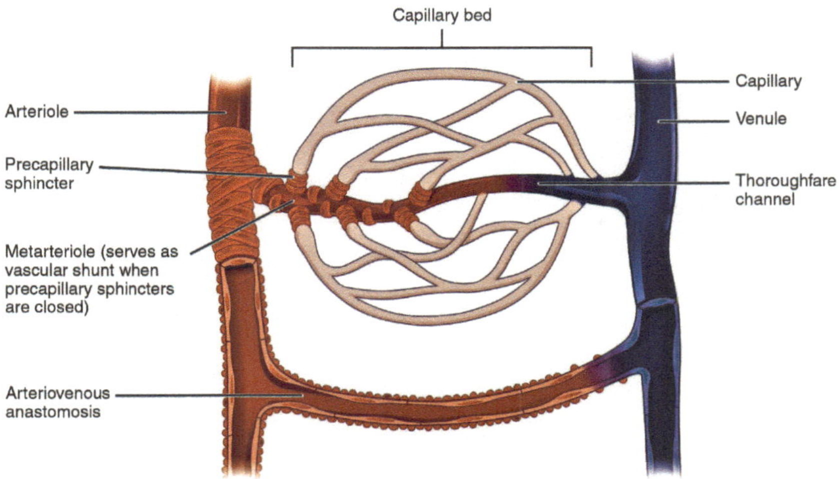

FIG. 2.5 A capillary bed along with a metarteriole and a venule. The precapillary sphincter regulates the flow of blood into the capillary bed. Arteriovenous anastomosis represents alternate blood flow bypassing capillary beds [5].

precapillary sphincter. It tightly regulates blood flow from a metarteriole to the capillaries it supplies. The sympathetic nervous system regulates the smooth muscle serving as the precapillary sphincter. The precapillary sphincters are typically closed and open when the surrounding tissues require O_2 or have excess waste products removed, allowing blood to flow through and exchange to occur before closing once more (Fig. 2.5). The blood will bypass the capillary bed and flow from the metarteriole directly into the venous circulation via a **thoroughfare channel** if all of the precapillary sphincters in a capillary bed are closed. This arrangement is known as a **vascular shunt**, the redistribution of blood. An **arteriovenous anastomosis** is a similar arrangement where there is a direct connection between an artery and a vein, completely bypassing the capillary bed.

A spontaneous oscillation in the blood vessel wall referred to as **vasomotion**, independent of neuronal regulation, results in irregular, pulsating flow. Local chemical signals trigger such flow in response to changes in internal conditions, such as O_2, CO_2, H^+, and lactic acid levels. For example, during strenuous exercise, when O_2 levels decrease and CO_2, H^+, and lactic acid levels increase, the capillary beds in skeletal muscle become open, allowing increased local blood supply. In general, at rest, most capillary beds are closed but open only occasionally to allow supplies of O_2 and nutrients to the tissues.

2.7 Venules

A venule is a small vein formed of the fusion of multiple capillaries exiting from a capillary bed. They are approximately 8–100 μm in diameter. Multiple venules join to form veins. The walls of the venules are thin, consisting of the endothelium, a thin tunica media with few smooth muscle and elastic fibers, and an outer layer of connective tissue that consists of a very thin tunica externa (Fig. 2.6).

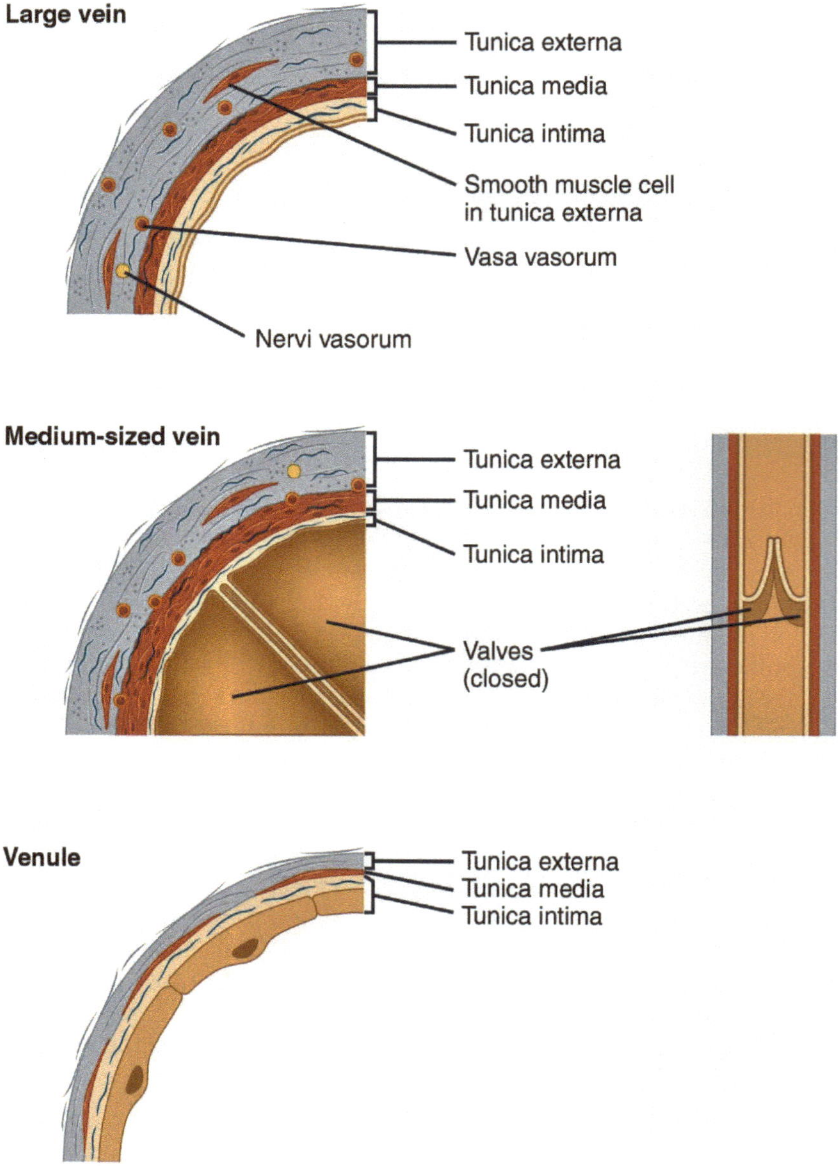

FIG. 2.6 Comparison of a large vein, medium-sized vein, and a small vein or a venule [6].

2.8 Veins

A vein is formed of the fusion of many venules. It conducts blood toward the heart. Compared to arteries, veins are thin walled with large and irregular lumens (see Fig. 2.6). Larger veins are often equipped with **valves** to promote the unidirectional flow of blood toward the heart and to prevent backflow of blood due to the pull of gravity and since they are low-pressure vessels.

Veins are considered as blood reservoirs since systemic veins contain approximately 64% of the blood volume at any given time. Their ability to hold a significant amount of blood is due to their high **capacitance**, which is their ability to distend to accommodate a large volume of blood, even at low pressure. The large lumens and relatively thin walls make veins far more distensible than arteries.

2.9 Blood flow

Approximately 5L of blood is ejected from the heart every minute. This blood needs to reach the tissues to allow the exchange of gases, nutrients, and wastes. It is achieved by having blood vessels connecting the heart to the tissues in two closed circuits, the systemic and pulmonary circuits. Blood must flow through the blood vessels to reach the tissues or be returned to the heart. It is initiated by the pumping of blood from the heart with high pressure. Ventricular contraction ejects blood into two major arteries, the aorta and the pulmonary artery. Blood in these vessels begins to flow in one direction only, and it always moves from regions of higher to lower pressure. Thus the key determinant for the blood to flow is blood pressure. With each stroke, the left ventricle ejects between 70 and 80mL of blood during systole. The initial momentum for the blood flow is due to the force with which the blood is pumped and the elastic recoil of the arteries releasing energy created by distention during systole.

Blood flow also depends on resistance within the vessel that opposes the flow of blood. This opposing force is due to the narrowing of the lumen further along the arterial branching, such as the arterioles. Arterioles are the major vessels contributing to the resistance within the vascular system. **Systemic vascular resistance** is a collective name for the resistance exerted by the vascular system, which is also called **peripheral vascular resistance** or **total peripheral resistance**. The blood pressure within the vessel must overcome resistance in order for the blood to flow. The relationship between blood flow, resistance, and blood pressure can be explained mathematically using a modified **Ohm's law** which deals with electricity, voltage, and resistance. The modified Ohm's law states that blood flow depends directly on blood pressure difference and inversely on peripheral vascular resistance. It can be mathematically shown as follows

$$Q = (P_1 - P_2)/R$$

(Q = flow, R = resistance, P_1 = pressure in one end, and P_2 = pressure in the other end, which is also represented as ΔP).

As discussed before, a large pressure drop occurs as blood moves through arterioles suggesting that arterioles present considerable resistance to flow due to their muscular walls with a narrow lumen. The three main factors that affect resistance are viscosity, vessel length, and vessel radius. The radius of the vessel is the most important of the three factors. The relationship between resistance and the radius is represented as $R = 1/r^4$, suggesting that resistance could increase 16 times when the radius is halved $(1/2^4)$.

2.10 Regulation of blood flow

Resistance and blood pressure play critical roles in the regulation of blood flow. There are many factors which, in turn, affect blood pressure and resistance. These include cardiac output, compliance, blood volume, blood viscosity, blood vessel length, and blood vessel diameter.

2.10.1 Cardiac output

Cardiac output determines the amount of blood to be carried by the blood vessels. In a normal situation, it is approximately 5 L. Any increase in cardiac output will, in turn, increase blood pressure and thus promote blood flow. Several factors increase cardiac output, including thyroid hormones, **hypercalcemia** (increased Ca^{2+} levels in the blood), and sympathetic activation. Similarly, any factor that decreases cardiac output will decrease blood pressure and blood flow. Those factors that decrease cardiac output include parasympathetic stimulation, elevated or decreased K^+ levels, **hypocalcemia** (decreased Ca^{2+} levels), **anoxia** (deficiency of O_2 at the tissues), and **acidosis** (lower than normal pH).

2.10.2 Compliance

Compliance is the ability to expand to accommodate increased content such as blood or air as water or air in a balloon. The greater the blood vessel's compliance, the more it can expand to accommodate blood without increased resistance or blood pressure. Veins are more compliant compared to arteries and hence hold more blood. Any stiffening of arteries would reduce compliance and increase resistance. Such abnormalities increase the workload of the heart.

2.10.3 Blood volume

Under normal circumstances, blood volume varies little. Any decrease in blood volume, **hypovolemia**, will decrease blood pressure and blood flow. Hypovolemia may be caused by blood loss due to injury, dehydration, vomiting, or certain medications used to treat hypertension. On the other hand, any increase in blood volume, **hypervolemia**, will increase blood pressure and blood flow. It may be caused by retention of water and sodium, as seen in patients with heart failure, certain types of kidney disease, **hyperaldosteronism** (overproduction of aldosterone), and glucocorticoid steroid treatments.

2.10.4 Blood viscosity

Viscosity refers to the thickness of fluids. Pure water, for example, is less viscous than blood. The viscosity of blood is due to blood cells, proteins, and other dissolved substances within the blood, which the pure water lacks. The viscosity of blood contributes to resistance—the greater the viscosity of blood, the slower the blood flows and vice versa. Since the vast majority of blood cells are red blood cells, any condition affecting **erythropoiesis** (a process of production of red blood cells), such as **polycythemia** (overproduction of red blood cells) or **anemia** (underproduction of red blood cells), can alter viscosity. Since the liver produces most plasma proteins, any condition affecting liver function can also change the viscosity and therefore decrease blood flow.

2.10.5 Blood vessel length

The length of the vessel is directly proportional to its resistance. The longer the blood vessel, the greater the resistance and lesser the flow, and the opposite is true for the shorter vessels. The blood vessel's length and its relation to blood flow can be compared to a garden hose. It would take longer to travel through a long hose compared to a shorter one.

An individual weighing 150 pounds has about 60,000 miles of blood vessels in the body [7]. Weight gain of approximately 10 pounds will add 2000–4000 miles of blood vessels, depending upon the tissue type gained. Adipose tissue, for example, has less vascular supply compared to skeletal muscle. Weight reduction to the optimum level, on the other hand, decreases the resistance of many miles of vessels, which, in turn, lessens the stress on the heart.

2.10.6 Blood vessel diameter

The diameter of blood vessels affects blood flow inversely. Wider the blood vessel, the faster the blood moves. The blood vessels' diameter varies throughout the body, not only based on the type of vessel but also in response to neural and chemical signals, regulating vasodilation and vasoconstriction. The contracted state of the smooth muscle is responsible for the vascular tone of the vessel and is the primary determinant of diameter, and thus of resistance and flow. The vessel diameter and resistance are inversely related. An increased diameter means less blood contacting the vessel wall and less friction and lower resistance, subsequently increasing flow. The opposite is true during decreased diameter.

The lumen diameter has a tremendous effect on resistance. A small change in diameter will result in a significant change in resistance. Recall that resistance is inversely proportional to the radius (one-half of the vessel's diameter) of the blood vessel, and if the radius is decreased by one-half of its original size, the resistance to flow will increase 16 times.

2.10.7 Blood pressure, cross-sectional area, vessel diameter, and velocity of blood flow

Fig. 2.7 compares vessel diameter, total cross-sectional area, mean blood pressure, and blood velocity through the systemic vessels. Part (A) summarizes the diameter of the vessels. The arteries start as large vessels and consistently decrease in size as they continue to branch. The smallest vessels being the capillaries begin to fuse to give rise to venules and veins. The size of the veins closest to the heart is the largest among all the vessels. As shown in parts (A) and (B), capillary beds have the greatest cross-sectional area with the highest number of branching among all types of vessels while being the smallest. Part (C) shows mean blood pressure in different blood vessel types. Blood pressure drops as blood travels from arteries to arterioles, capillaries, venules, and veins. The steepest drop is at the arterioles due to increased resistance to blood flow in these vessels. Part (D) shows the velocity (speed) of blood

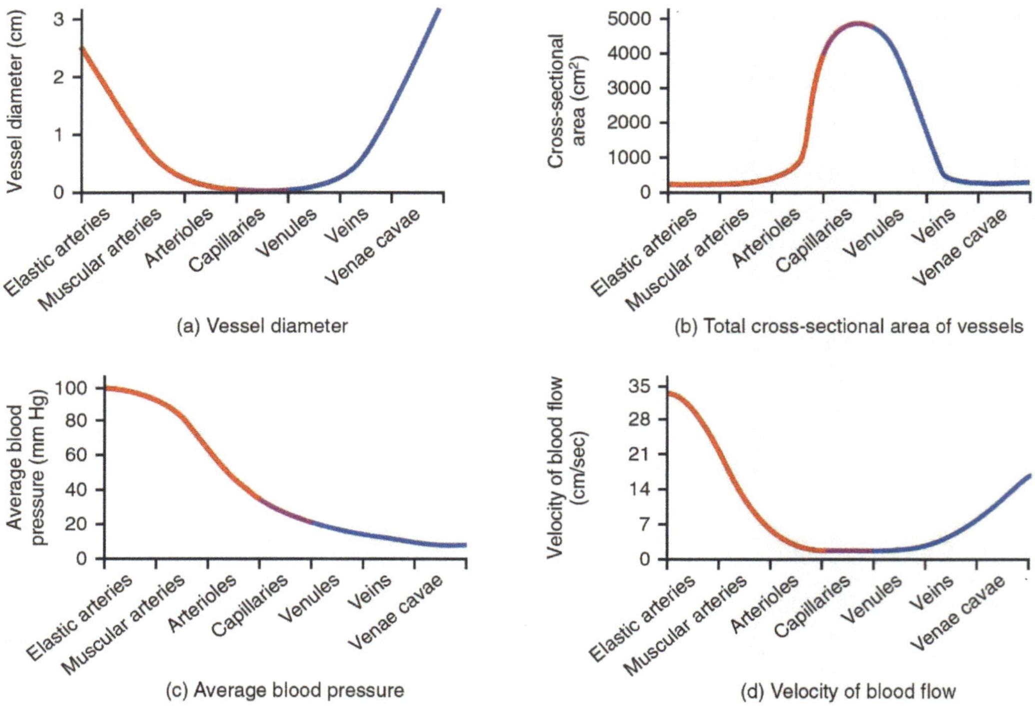

FIG. 2.7 The characteristics of blood vessels in the systemic circuit [8]. The relationships among blood vessels compared include (A) vessel diameter, (B) total cross-sectional area, (C) average blood pressure, and (D) velocity of blood flow.

flow, decreasing as it moves from arteries to arterioles to capillaries. The slowest flow of blood is at the capillaries to allow more time to exchange substances between the blood and the tissues. The rate of velocity increases as blood flows through the veins but not to the extent seen in large arteries due to decreasing blood pressure and increasing compliance in these vessels.

2.11 Capillary exchange

The cardiovascular system's primary purpose is to exchange gases, nutrients, wastes, and other substances between the cells, and the blood and capillaries play a crucial role in it. Small substances, such as gases, lipids, and lipid-soluble molecules, can diffuse directly through the endothelial cell membrane within the capillary wall. Glucose, amino acids, and some ions use transporters to move across the membrane by one of the many processes such as facilitated diffusion, active transport, cotransport, and counter transport mechanisms. Fenestrated capillaries permit the passage of substances through their fenestrae but are limited to the size of the fenestrae and the electrical charges of the transported substances. Blood cells and large plasma proteins can pass through the large spaces of the sinusoids. Endocytosis and exocytosis processes are utilized in the transportation of much larger structures, including pathogens. Water moves by osmosis.

2.11.1 Bulk flow

The bulk movement of fluids into and out of capillary beds is far more efficient than diffusion. The bulk exchange of fluids and solutes at most tissues occurs through the gaps between the endothelial cells called **intercellular clefts**. Blood plasma containing nutrients and gases moves out of the capillary at the arterial end in a process called **filtration**, whereas tissue fluid, rich in wastes, moves into the capillary at the venous end by **reabsorption**. A balance between two forces drives this form of transport, the **hydrostatic pressure**, which is the pressure of a fluid in a closed space, and the **osmotic pressure**, the pressure required to stop the movement of water from diffusing through the cell membrane by osmosis. Inside the capillary, the hydrostatic pressure is the same as blood pressure. The hydrostatic pressure within the capillary, called **capillary hydrostatic pressure** (CHP), is much higher than the tissue fluid's hydrostatic pressure (**interstitial fluid hydrostatic pressure**). The interstitial fluid hydrostatic pressure is less than CHP because the lymphatic system continuously drains the excess fluid from the interstitial space.

The higher capillary hydrostatic pressure drives the fluid out of the capillary (Fig. 2.8). Hydrostatic pressure is opposed by osmotic pressure. Osmotic pressure is also called blood **oncotic pressure** or blood **colloid oncotic pressure** (**BCOP**), which is generated mainly by the attraction of water toward an area of higher solute concentration. Since blood has plasma proteins and blood cells, it has a higher concentration of solutes than the tissue fluid, and hence the movement of water into the capillary by the process of osmosis.

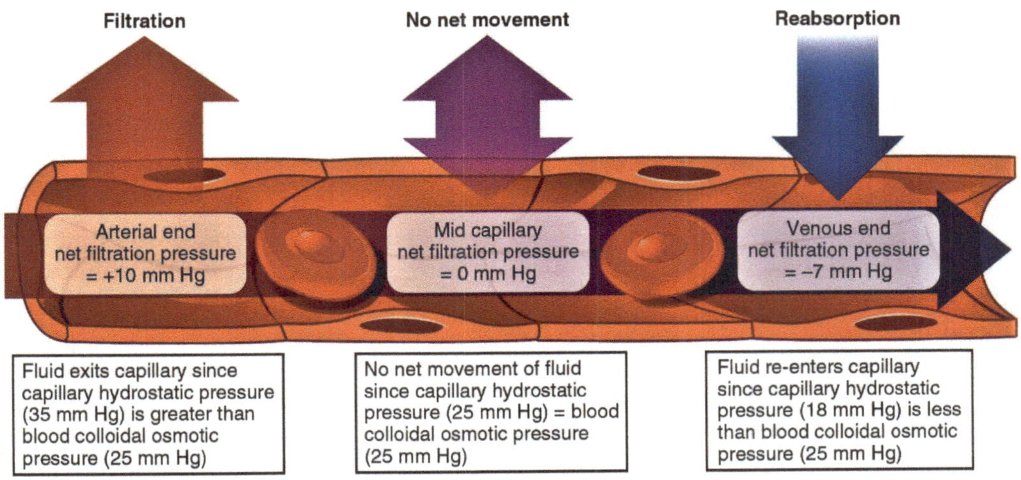

FIG. 2.8 Capillary exchange [9]. Net filtration occurs near the arterial end of the capillary since capillary hydrostatic pressure (CHP) is greater than blood colloidal osmotic pressure (BCOP). There is no net movement of fluid near the midpoint since CHP = BCOP. Net reabsorption occurs near the venous end since BCOP is greater than CHP.

Hydrostatic pressure at the arterial end is higher compared to the venous end since the blood pressure continues to drop from the arterial end of the capillary to the venous end. The hydrostatic pressure at the arterial end is approximately 35 mmHg, whereas the capillary hydrostatic pressure at the venous end is 18 mm Hg. However, the osmotic pressure stays the same throughout and is approximately 25 mm Hg. The **net filtration pressure** is the difference between the CHP and BCOP. It is greater at the arterial end (+10 mm Hg) than the venous end (−7 mm Hg). Therefore the balance shifts from the net outward flow at the arterial end to the net inward flow at the venous end, allowing more fluid to be filtered out than the fluid moving back in. Approximately 15% of the fluid is left behind in the interstitial space.

Excess fluid enters lymphatic capillaries, where the fluid gets its new name as lymph. Valves in the lymphatic capillaries prevent the backward flow of lymph, thus creating a unidirectional flow. Lymph is carried through the lymphatic system, which includes lymph nodes where filtration of the lymph takes place and ultimately lymph is drained into the subclavian vein. It is estimated that 24 L of filtrate are produced at the capillary beds per day, out of which 20.4 L are reabsorbed [10]. The remaining volume that was filtered but not reabsorbed is handled by the lymphatic system, which screens this fluid for debris and pathogens before returning it to the vascular system.

Any alterations in the dynamics of bulk flow could result in a fluid shift from one compartment to another. **Edema** is the buildup of fluid in the interstitial space, which could be caused by many situations. Liver disease such as cirrhosis may limit the liver from producing plasma proteins, which, in turn, would decrease BCOP. Similarly, kidney disease such as glomerulonephritis may cause injury to the glomerular capillaries allowing the escape of proteins from the blood, thus reducing BCOP. An increase in arterial blood pressure, such as in untreated hypertension, could increase the net filtration pressure and promote more filtration than reabsorption leading to edema. Damage to lymphatic capillaries may limit the transportation of lymph, where a buildup of tissue fluid could also cause edema.

2.12 Venous system

The function of the venous system is to return the blood back to the heart. As blood moves away from the heart, the arterial blood pressure decreases continuously and is approximately 18 mm Hg at the end of the capillary beds and the beginning of the venules and continues to fall (Fig. 2.7C), but the velocity of blood flow increases (Fig. 2.7D). Also, as the diameter increases from the smaller venules to larger veins, the cross-sectional area decreases (see Fig. 2.7A and B).

The blood pressure must be higher in the veins than the pressure within the atria for the venous return, and the atrial pressure during diastole is near zero and thus favors the return of blood to the heart. Despite the low pressure in the atria during diastole, the low blood pressure, large lumen, and the thin wall within the vein are the three characteristics that limit blood flow through the veins. Since a large portion of the blood resides in systemic veins (64%), any action that increases blood flow through the veins will increase venous return to the heart. Vasoconstriction due to sympathetic activation, for example, helps return the blood to the heart from the veins.

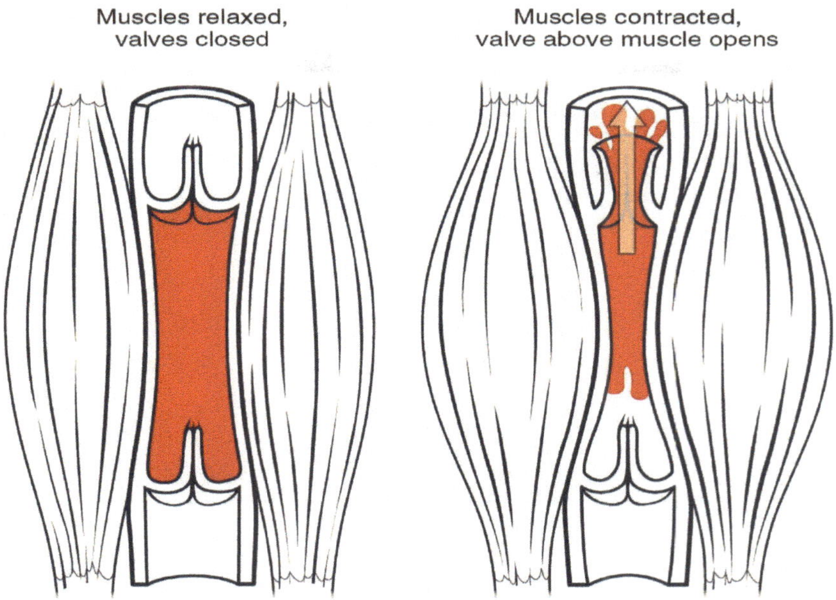

FIG. 2.9 Skeletal muscle pump [11]. The contraction of skeletal muscles surrounding a vein in the leg compresses the vein and increases the pressure in that area, allowing the valve above to force open and the blood to move upward toward the heart. The valve below stays closed to prevent the backward flow of blood.

Veins tend to expand and allow blood to pool, especially in the lower parts of the body, below the heart level. Certain mechanisms exist, however, to assist the flow of blood through the veins. Veins in the lower parts of the body are equipped with valves to prevent the backward flow of blood. In addition, skeletal muscle activity and the breathing mechanism play essential roles in returning blood to the heart through the veins and are therefore referred to as the **skeletal muscle pump** and the **respiratory pump**.

2.12.1 Skeletal muscle pump

Several skeletal muscles, especially in the lower parts of the body, assist in blood circulation through the veins to improve venous return. Contraction of the surrounding skeletal muscles can increase pressure within the veins, which, in turn, helps move blood along the veins, particularly against gravity. This mechanism is referred to as the **skeletal muscle pump**. As shown in Fig. 2.9, when leg muscles contract, such as during walking, they exert pressure on nearby veins. This elevated pressure causes blood to flow upward by opening valves above while closing the valves below, thus preventing backward blood flow.

2.12.2 Respiratory pump

Normal breathing helps facilitate venous return. During inspiration, the diaphragm moves into the abdominal cavity, expanding the thoracic cavity and decreasing the intrathoracic

TABLE 2.1 Systemic blood flow during rest, mild exercise, and maximal exercise in a healthy young individual [12,13].

Structure	Resting (mL/min)	Mild exercise (mL/min)	Percent change (%)	Maximal exercise (mL/min)	Percent change (%)
Skeletal muscle	1200	4500	375	12,500	1041
Heart	250	350	140	750	300
Brain	750	750	No change	750	No change
Integument	500	1500	300	1900	380
Kidney	1100	900	−81	600	−183
Gastrointestinal	1400	1100	−127	600	−233
Others (i.e., liver, spleen)	600	400	−150	400	−150
Total	5800	9500		17,500	

pressure while compressing the abdominal cavity and increasing the intra-abdominal pressure. As pressure within the thorax decreases, the thoracic veins' blood pressure decreases below the pressure in the abdominal veins, which causes blood to flow along its pressure gradient from veins in the abdominal cavity, where pressure is higher, into the thoracic cavity, where pressure is low. During expiration, when pressure increases within the thoracic cavity, pressure in the thoracic veins increases, increasing venous return into the heart while valves in the veins prevent blood from flowing backward from the thoracic to abdominal veins.

2.13 The role of exercise on systemic blood flow

Table 2.1 provides the distribution of systemic blood at rest and during mild and strenuous exercise. The percentage of blood reaching different systems varies considerably. The two organs that receive the highest percentage of blood compared to their size are the brain and the kidneys. The brain receives approximately 15% of the total blood supply, whereas the kidneys receive about 20%. The high metabolic rate that occurs in the brain is one reason for its rich blood supply. Similarly, the filtration of blood at the kidneys is the primary reason for its large blood supply.

Blood supply also increases as the body transitions from a resting state to light exercise and eventually to heavy exercise (see Table 2.1). During exercise, selective vasodilation occurs in the skeletal muscles, heart, lungs, and integument and vasoconstriction in the vessels supplying the kidneys and most digestive organs. The body distributes more blood to the skin during exercise to release excess heat generated by increased activity. Similarly, the blood supply to skeletal muscles, the heart, and the lungs also increases during exercise. One exception is the brain, which receives a constant blood supply independent of the level of activity.

Exercise promotes cardiovascular health in a number of ways. In addition to the improved blood supply, exercise increases the size and weight of the heart. While the heart's average weight for the nonathletic individual is approximately 300 g, the heart's mass will increase to 500 g in an athlete [13]. The increase in size makes the heart more robust and efficient at pumping blood. Venous return is also enhanced by both the skeletal muscle and respiratory pumps. Improved venous return increases preload and thus cardiac output. Generally, as little as 30 minutes of continuous exercise each day has beneficial effects and has been shown to lower the heart attack rate by nearly 50% [13].

References

[1] 2101 Blood Flow Through the Heart.jpg. Wikimedia Commons, the free media repository. Updated October 20, 2020. Accessed February 3, 2021. https://commons.wikimedia.org/w/index.php?title=File:2101_Blood_Flow_Through_the_Heart.jpg&oldid=495706270.

[2] 2102 Comparison of Artery and Vein.jpg. Wikimedia Commons, the free media repository. Updated October 5, 2020. Accessed February 2, 2021. https://commons.wikimedia.org/w/index.php?title=File:2102_Comparison_of_Artery_and_Vein.jpg&oldid=482213141.

[3] 2103 Muscular and Elastic Artery Arteriole.jpg. Wikimedia Commons, the free media repository. Updated October 18, 2020. Accessed February 3, 2021. https://commons.wikimedia.org/w/index.php?title=File:2103_Muscular_and_Elastic_Artery_Arteriole.jpg&oldid=493319728.

[4] 2104 Three Major Capillary Types.jpg. Wikimedia Commons, the free media repository. Updated October 18, 2020. Accessed February 3, 2021. https://openstax.org/books/anatomy-and-physiology/pages/20-1-structure-and-function-of-blood-vessels.

[5] 2105 Capillary Bed.jpg. Wikimedia Commons, the free media repository. Updated September 10, 2020. Accessed February 4, 2021. https://commons.wikimedia.org/w/index.php?title=File:2105_Capillary_Bed.jpg&oldid=454437322.

[6] 2106 Large Medium Vein Venule.jpg. Wikimedia Commons, the free media repository. Updated October 21, 2020. Accessed February 4, 2021. https://commons.wikimedia.org/w/index.php?title=File:2106_Large_Medium_Vein_Venule.jpg&oldid=496411739.

[7] OpenStax College, Anatomy & Physiology: Blood Flow, Blood Pressure, and Resistance, 2013. https://openstax.org/books/anatomy-and-physiology/pages/20-2-blood-flow-blood-pressure-and-resistance.

[8] Relationships among Vessels in the Systemic Circuit.jpg. Wikimedia Commons, the free media repository. Updated April 13, 2013. Accessed February 4, 2021. https://openstax.org/books/anatomy-and-physiology/pages/20-2-blood-flow-blood-pressure-and-resistance.

[9] 2108 Capillary ExchangeN.jp. Wikimedia Commons, the free media repository. Updated October 21, 2020. Accessed February 4, 2021. https://commons.wikimedia.org/w/index.php?title=File:2108_Capillary_ExchangeN.jpg&oldid=496698946.

[10] OpenStax College, Anatomy & Physiology: The Cardiovascular System: Blood Vessels and Circulation, 2013. https://courses.lumenlearning.com/suny-ap2/chapter/capillary-exchange/.

[11] 2114 Skeletal Muscle Vein Pump.jpg. Wikimedia Commons, the free media repository. Updated October 14, 2020. Accessed February 4, 2021. https://commons.wikimedia.org/w/index.php?title=File:2114_Skeletal_Muscle_Vein_Pump.jpg&oldid=489113800.

[12] L. Biga, S. Dawson, A. Harwell, et al., Homeostatic regulation of the vascular system, in: C. Commons (Ed.), Anatomy and Physiology, OpenStax CNX, 2013. chap The Cardiovascular System: Blood Vessels and Circulation https://open.oregonstate.education/aandp/chapter/20-4-homeostatic-regulation-of-the-vascular-system/.

[13] OpenStax College, Anatomy & Physiology: Homeostatic Regulation of the Vascular System, 2013. https://open.oregonstate.education/aandp/chapter/20-4-homeostatic-regulation-of-the-vascular-system/.

CHAPTER

3

Inflammation

OUTLINE

OBJECTIVES

- Define inflammation. Describe how inflammation is triggered.
- List the cell types associated with inflammation and the roles played by them.
- Name the chemical mediators associated with inflammation and the specific roles played by them.
- Identify the different forms of inflammation.
- Summarize the key characteristics of inflammation.
- Describe the cascade of events following acute inflammation.
- Name the most common antiinflammatory agents and their means of reducing inflammation.

Biology of Cardiovascular and Metabolic Diseases
https://doi.org/10.1016/B978-0-12-823421-1.00014-7

3.1 Introduction

Inflammation is a protective mechanism against injury or infection, including mechanical damage, oxygen deprivation (ischemia), genetic or immune defects, chemical agents, temperature extremes, or ionizing radiation. It helps eliminate the initial cause of injury, removes damaged cells and tissues from the original insult, and initiates tissue healing and repair. Restoration of the injured tissue to its original form and function depends on the extent of the damage. Without inflammation, an infection would not be controlled, and a wound would not heal, leading to detrimental consequences. Simultaneously, too much inflammation could have harmful implications such as the development of chronic conditions including atherosclerosis (build of plaque in the blood vessel wall) and cancer due to the production of harmful agents. Inflammation is a complex process observed in vascularized tissues and depends on the activation of many cellular and chemical components.

3.2 Types of inflammation

Inflammation can be classified as either **acute** or **chronic**. **Acute inflammation** is the initial response to harmful stimuli, where the effects of inflammation are visible within seconds to minutes and is achieved by the increased movements of plasma and leukocytes from the blood into the injured tissues. It continues until the immediate threat to the host is eliminated, which could last from several hours to 8–10 days. Prolonged inflammation, known as **chronic inflammation**, lasts 2 weeks or longer. Chronic inflammation leads to a gradual shift in the cell types present at the site of inflammation. It is characterized by a combination of destruction and healing of the tissue from the inflammatory process. An unsuccessful acute inflammatory response sometimes precedes chronic inflammation, or it can also occur as a distinct process without much previous acute inflammation. If inflammation stretches over for a longer period or inflammation targets tissues where it is not needed, it may cause harm to the local tissue. Chronic inflammation has been linked to heart disease, cancer, and type 2 diabetes, among others. When inflammation results in systemic changes, it is referred to as **acute phase response**, typically associated with fever and **leukocytosis** (increased number of leukocytes). In severe cases, systemic inflammation may result in **sepsis**, a bacterial infection affecting multiple organs.

3.3 Cellular components

Cells involved in inflammation include **leukocytes**, **platelets**, **mast cells**, **endothelial cells**, and **fibroblasts** (Table 3.1). Leukocytes are subdivided into **granular** and **agranular types** **where neutrophils**, **eosinophils**, and **basophils** make up granular leukocytes, and **monocytes** and **lymphocytes** are the two agranular leukocytes. Neutrophils are the most abundant leukocytes and are the predominant phagocytes of **early inflammation**. They are the first to arrive at the infection site by being attracted to **chemotactic agents** released by the damaged cells and activated phagocytes. Monocytes are the primary phagocytes in the **late**

TABLE 3.1 Cells involved in the process of inflammation [1,2].

Cell	Characteristic features and function(s)
Monocytes 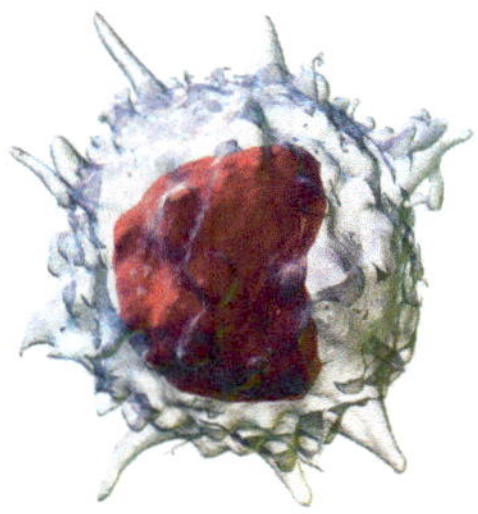	• Produced in the bone marrow and stored in the spleen (50%) or released into the blood (50%). • After circulating for 2–3 days, they may enter the tissue by being attracted to chemical substances released by the damaged cells at the site of injury or infection. • When a monocyte enters damaged tissue, it undergoes a series of changes to become a macrophage. • Combats certain infections and helps other leukocytes remove dead or damaged tissues.
Macrophages 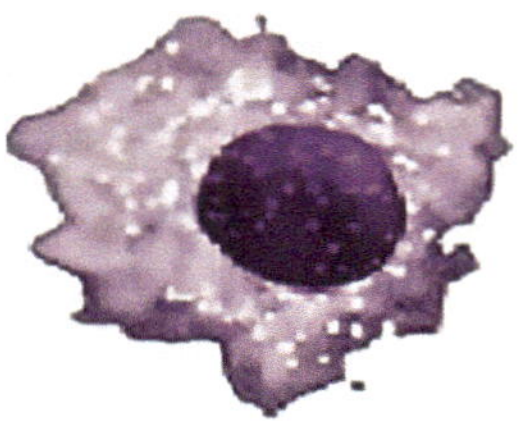	• Are monocytes migrated from the bloodstream into any tissue in the body. • Engulf and digest cellular debris and foreign substances in a process called phagocytosis (cell eating). • Cells involved in the build of plaque lesions of atherosclerosis.
Eosinophils 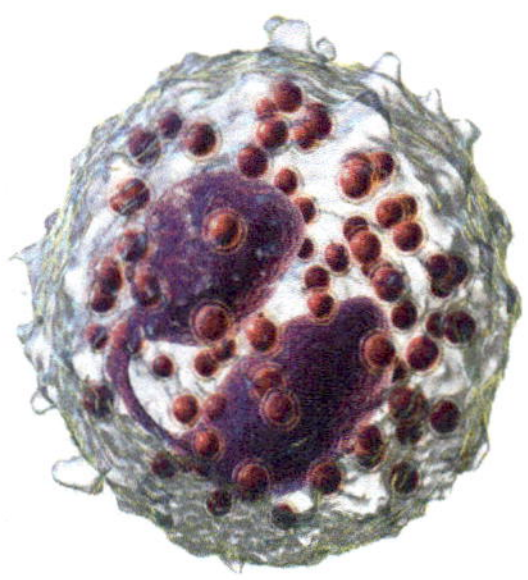	• Produced in the bone marrow. • Crucial for removing the parasitic infection. • Increased numbers during allergic reactions. • Capable of phagocytosis, antibacterial activity, and controlling inflammatory responses.
Basophils 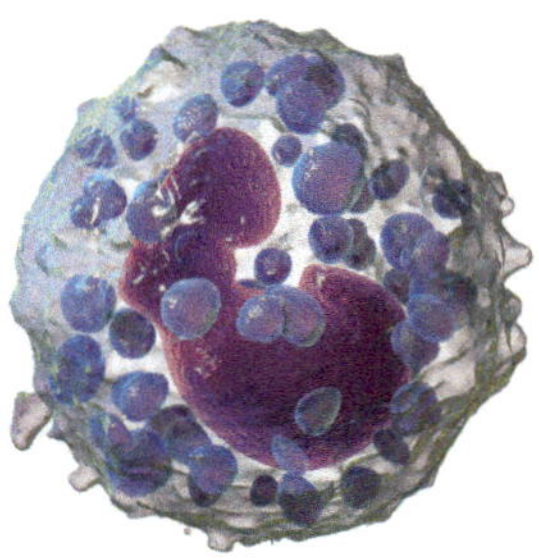	• Made in the bone marrow. • Respond by isolating and eliminating the allergen. • Release histamine, which contributes toward inflammation during an allergic reaction. • Plays an integral role in preventing blood clotting by containing heparin that acts as a natural blood thinner.

Continued

TABLE 3.1 Cells involved in the process of inflammation [1,2]—cont'd

Cell	Characteristic features and function(s)
Neutrophils 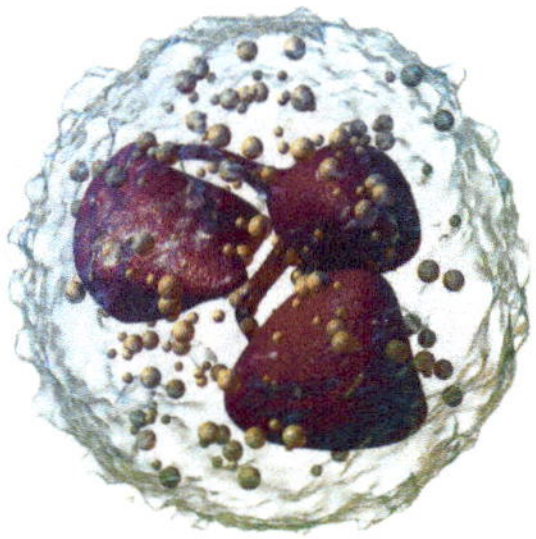	• Produced in the bone marrow. • First ones to appear at the site of bacterial infection. • Handle foreign invaders by phagocytosis.
Lymphocytes 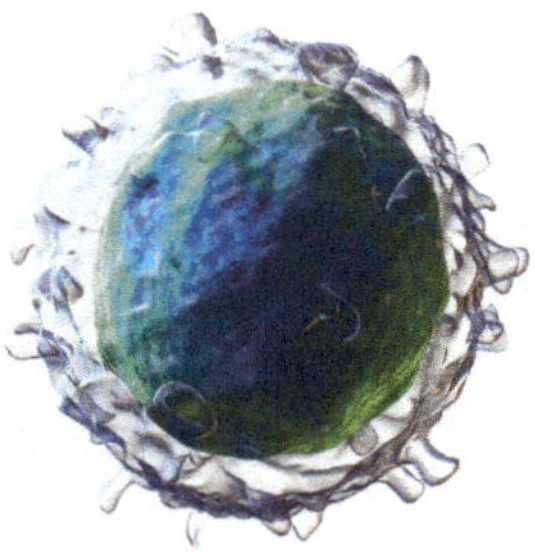	• Produced by the stem cells in the bone marrow. • Two main types—B and T lymphocytes. • Aid in host defense against pathogens. • Generate immune responses after exposure to new antigens and retain the memory of those specific antigens by way of B and T lymphocytes.
Endothelial cells 	• Line the vasculature and lymphatic system forming a semipermeable barrier between blood or lymph within vessels and the surrounding tissues. • Initiation and perpetuation of inflammation via angiogenesis (new blood vessel formation) and attraction of immune cells. • Control the type and number of immune cells that extravasate (move out of the vessel) into the interstitium. • Assist in the escape of leukocytes and plasma.
Mast cells 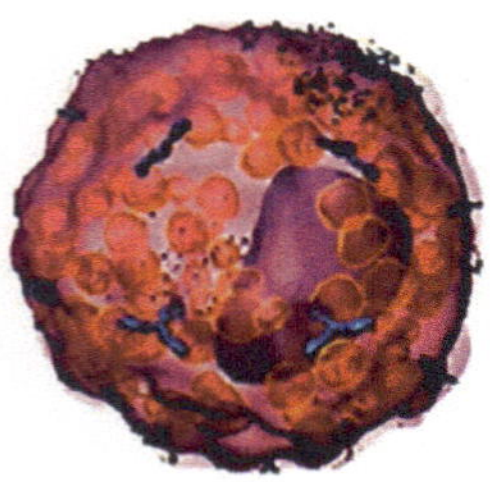	• Made in the bone marrow. • Interact directly with bacteria and play a vital role in host defense against pathogens. • Exert pro-inflammatory response by releasing histamine, leukotrienes, cytokines, chemokines, and neutral proteases.
Fibroblasts 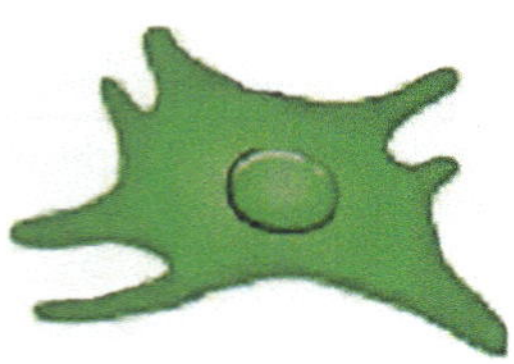	• Cells in the connective tissue. • Deposit fibrous matrices. • Fibroblast activation is mostly derived from macrophages. • Plays an essential role in tissue repair and wound healing.

inflammatory response. They are referred to as **macrophages** when they leave the general circulation and remain in the tissues. Monocytes and macrophages respond to various chemical stimuli at the site of infection or injury, phagocytose infected cells, remove cellular debris, and promote wound healing; thus, they prevent further damage. Eosinophils help defend against parasitic infection. Endothelial cells lining the blood vessels participate in inflammation in several ways. They allow the passage of leukocytes and plasma into the site of injury. **Mast cells** are found in the loose connective tissue close to blood vessels and are essential activators of inflammatory response. When stimulated, mast cells release their granules, consisting of many chemical agents, including **histamine**, chemotactic factors, **leukotrienes**, **cytokines**, **prostaglandins**, and growth factors within seconds.

3.4 Chemical mediators

A great variety of chemical substances are associated with inflammation. The sources of these agents include damaged cells and activated cells at the site of inflammation. The typical chemical agents released during inflammation include histamine, **bradykinins**, leukotrienes, and prostaglandins (Table 3.2). Histamine released by circulating mast cells promotes relaxation of arteriolar smooth muscle and vasodilation, which increases blood flow to the area of

TABLE 3.2 Chemical mediators of inflammation [3–11].

Chemical agent	Function(s)
Histamine 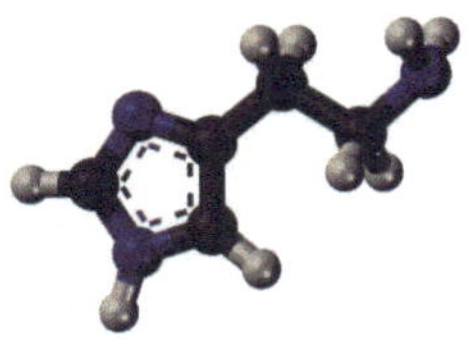	• Increases blood flow (vasodilation) to the inflammation site and vascular permeability to allow plasma, dissolved proteins, leukocytes, and other immune cells to enter inflamed tissue. • Critical role in allergic reactions.
Bradykinin 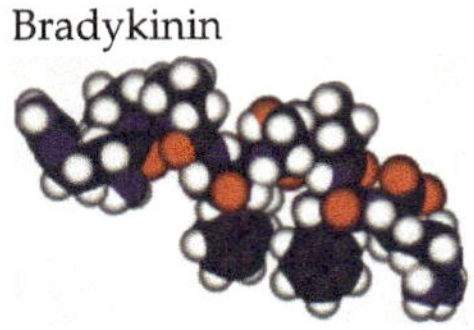	• Vasodilation, increased capillary permeability, and increased sensitivity to pain. • Mediators of acute inflammation. • Function in the transition to and maintenance of chronic inflammation.
Leukotrienes 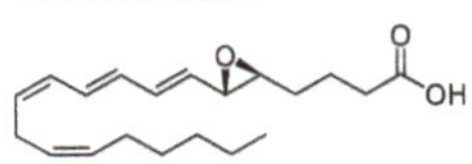	• Produced by leukocytes as the name suggests. • Associated with allergic reactions and act to sustain inflammatory reactions. • Helps recruit cells, such as neutrophils, to the tissue. • Aids in endothelial cell adherence and chemokine production by mast cells.
Prostaglandins 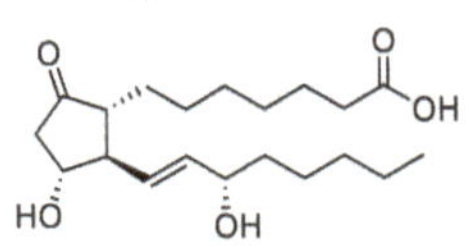	• Increases sensitivity to pain. • Increases blood flow and chemotaxis (chemical signals that attract leukocytes). • Contribute to chronic inflammation by inducing chemokines and recruiting inflammatory cells.

Continued

TABLE 3.2 Chemical mediators of inflammation [3–11]—cont'd

Chemical agent	Function(s)
Cytokines (chemokines) 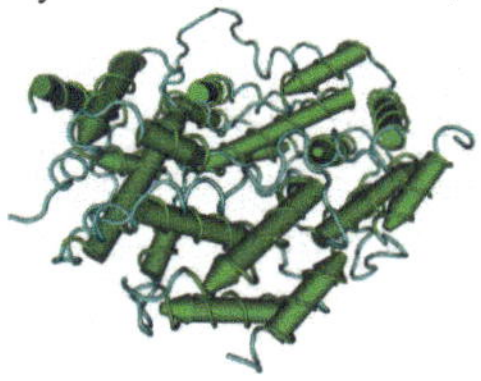	• Attracts leukocytes to tissues, fight infection, act as communication signals between cells. • Locates target cells and interacts with receptors on the target cells by binding to them, allowing cells to communicate with each other. The interaction triggers specific responses by the target cells. • Elevated cytokines, "cytokine storm," can have an adverse effect. • Some cytokines are pro-inflammatory (IL-1 and IL-6), whereas others reduce inflammation and promote healing (anti-inflammatory cytokines).
Chemokines 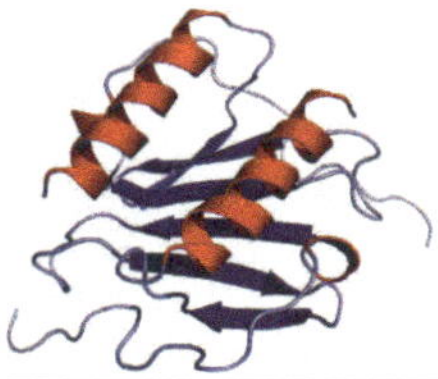	• A family of cytokines, or signaling molecules secreted by cells associated with inflammation. • Name is derived from their ability to induce directed chemotaxis (leukocyte movement in response to a chemical stimulus). • Attracts immune cells to the site of inflammation.
Interleukins (IL) 	• A type of cytokine. • Large subtypes or families of interleukins such as IL-1 and IL-2. • Provide protection against viral infection, mediate communication between cells, and regulate cell growth, differentiation, and motility.
Tumor necrosis factor-alpha (TNF-α) 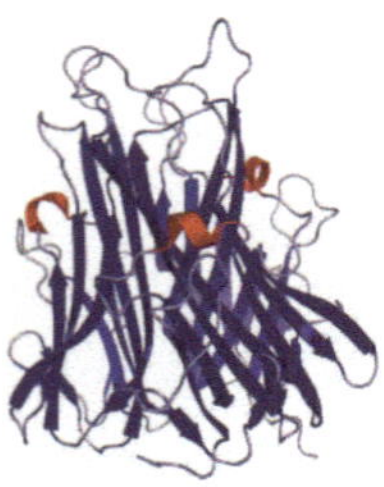	• A type of cytokine. • Inflammatory cytokine produced by macrophages. • Responsible for a range of signaling events within cells, leading to necrosis or apoptosis, the two types of cell death. • Responsible for the regulation of immune cells.
C-reactive protein (CRP) 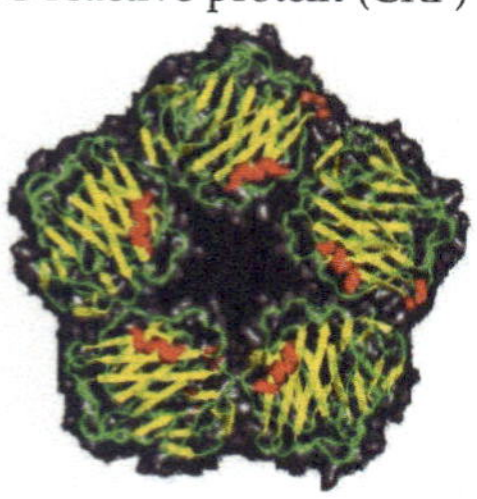	• Protein made in the liver and released in response to a rise in inflammation. • Associated with cardiovascular disease, coronary heart disease, diabetes, hypertension, and liver disease.

injury. Bradykinin is a protein that promotes vasodilation and increased capillary permeability as well as induces pain. Many cells secrete soluble substances called cytokines or **chemokines**. Most of these serve as local chemical agents, whereas the effects of others occur over long distances, such as induction of **fever**. **C-reactive protein** (CRP) is a blood-borne marker of inflammation and is associated with the activation of the **complement system**, a set of proteins involved in the body's defense. While some cytokines are pro-inflammatory inducing inflammatory responses, others are antiinflammatory, inhibiting the inflammatory response. Some of the cytokines bind to their receptors at their target cells and induce the release of other chemical agents. One such example is the binding of the cytokine **tumor necrosis factor-alpha (TNF-α)** to a cell resulting in the synthesis and release of **interleukin-1** (**IL-1**). The other major cytokines are **interferons** which protect against viral infection.

3.5 Characteristics of inflammation

Inflammation is associated with observable characteristics such as **warmth**, **redness**, **pain**, and **swelling** and are referred to as **cardinal signs**. Classic changes that occur in the smaller blood vessels such as arterioles and capillaries near the site of injury include vasodilation and increased permeability of the blood vessel allowing leukocytes and plasma to escape to the site of inflammation. The increased blood flow and the greater concentration of red blood cells at the injury site are the cause of warmth and redness. Increased permeability of the blood vessel wall allows leakage of plasma from the blood into the injury site, leading to local swelling (**edema**) in the surrounding tissue. Such buildup of fluid within the area of inflammation dilutes toxins produced by infected or dying cells. Pain associated with inflammation is often due to the buildup of fluid locally compressing surrounding tissues as well as the presence of specific chemical agents such as bradykinin and prostaglandins, which sensitize the processing of pain.

3.6 Acute inflammation

The sequence of events that occur during acute inflammation is summarized in Figs. 3.1 and 3.2. Local (resident) macrophages immediately begin to phagocytose foreign microbes, defending against infection during the first hour or so. Resident macrophages also secrete chemical mediators, which help recruit more leukocytes to the area. Almost immediately, arterioles and capillaries within the area dilate, increasing blood flow to the injury site. This localized vasodilation is mainly induced by histamine released from mast cells in the area of tissue damage. Histamine and other chemical agents act on endothelial cells to increase capillary permeability so that more leukocytes and plasma proteins, both of which are crucial to the defense, escape into the tissue of injury (Fig. 3.1). A buildup of plasma proteins in the interstitial fluid increases local **interstitial fluid colloid osmotic pressure**. Furthermore, the increased local blood flow elevates **capillary blood pressure**. An increase in these two pressures tends to move fluid out of the capillaries and decrease water reabsorption back into the capillaries. Such a shift in fluid balance results in localized **edema**.

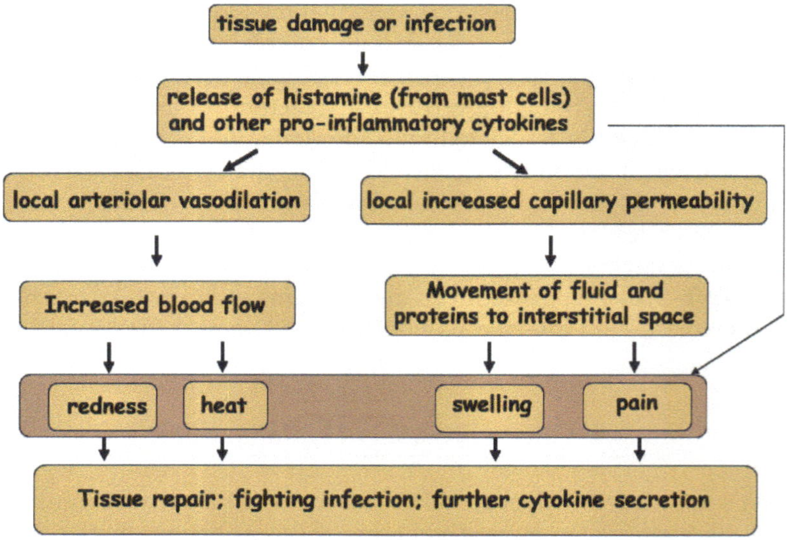

FIG. 3.1 Inflammatory response to injury or infection [12].

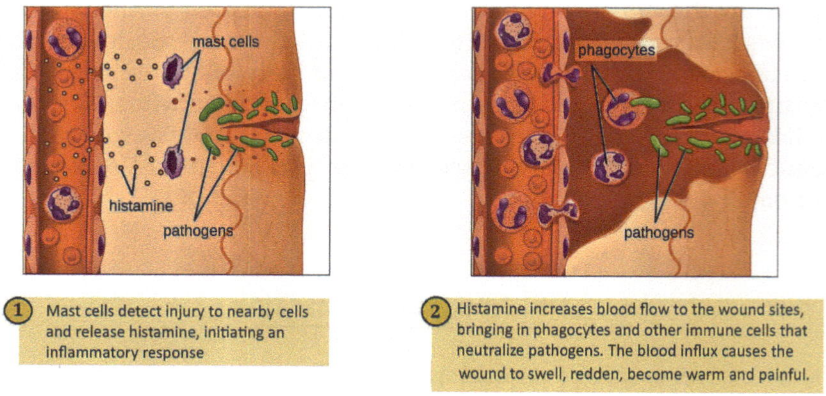

FIG. 3.2 Acute inflammatory response [13]. Watch this video for a summary of the process of inflammation https://www.youtube.com/watch?v=9bvMv5dQ7RU.

Chemical agents released during injury act on endothelial cells to increase the expression of receptors and adhesion molecules such as **selectins** which cause neutrophils and monocytes to slow their rate of movement as they roll along the interior of the blood vessel. Neutrophils and monocytes eventually adhere firmly to the endothelial lining through interaction with receptors such as **integrins**. The adhered leukocytes assume new forms to become thin and flat to be able to squeeze through narrow capillary spaces. After exiting, leukocytes move in the direction of chemicals released by activated resident leukocytes and other cells. Neutrophils arrive at the inflammatory scene earliest, within minutes, due to their large numbers, in general. During the next 8–12 h, monocytes follow neutrophils. As the body fights the pathogen, neutrophils, dead cells, lymph, and tissue fluids accumulate, resulting in pus formation. Macrophages will help remove the pus after a few days. Local chemical mediators now activate the recruited leukocytes and these activated leukocytes tend to remove infectious agents and cellular debris by phagocytosis.

The leaked plasma proteins include **clotting factors**. On exposure to **tissue thromboplastin** in the injured tissue, a plasma protein, **fibrinogen**, the final factor in the clotting system, is converted into **fibrin**. Fibrin forms clots in the spaces around the bacterial invaders and damaged cells. The walling off of the injured region from the surrounding tissues prevents or at least delays the spread of pathogens and their toxic products. Tissue repair of the injured areas follows. It can last a few days or longer, depending on the severity of the injured area.

3.7 Chronic inflammation

Chronic inflammation often begins as an asymptomatic, low-grade response. The fundamental causes of chronic inflammation include persistent infection, prolonged exposure to potentially toxic agents, and **autoimmunity**, where the body's immune system attacks its own structures. Unlike acute inflammation, chronic inflammation is characterized by the accumulation of macrophages and lymphocytes. It is also associated with repeated tissue destruction and attempts at healing and repair. The healing period begins during acute inflammation and may not be completed until after the injury is neutralized. If the damage is minor and destroyed tissues are capable of regeneration, it is possible to return injured tissues to their original forms and functions. On the other hand, if the damage is extensive and a resolution is not possible, the destroyed tissue is replaced by **scar tissue**. **Adipokines** are chemical agents released by the adipocytes (fat cells). Certain adipokines such as TNF-α and IL-6 contribute to inflammation in obese conditions, thus causing chronic low-grade systemic inflammation. Visceral fat, the fat around internal organs, is more likely to be chronically inflamed and is associated with an increased incidence of heart disease compared to subcutaneous fat, the fat that is found underneath the skin.

3.8 Resolution of inflammation

Poor resolution of inflammation has become a critical process in the development and continuation of many chronic inflammatory diseases. For the effective resolution of inflamed tissues to occur and to claim tissue homeostasis, specific cellular mechanisms are activated, which include the limitation or cessation of neutrophil recruitment, promotion of monocyte recruitment, induction of neutrophil apoptosis (cell death), and their subsequent enhancement and removal by phagocytes [14]. Together, these processes promote the termination of the inflammatory response and initiate tissue repair and healing. Therefore, the onset of inflammation is paralleled by a series of events aimed at balancing and controlling the

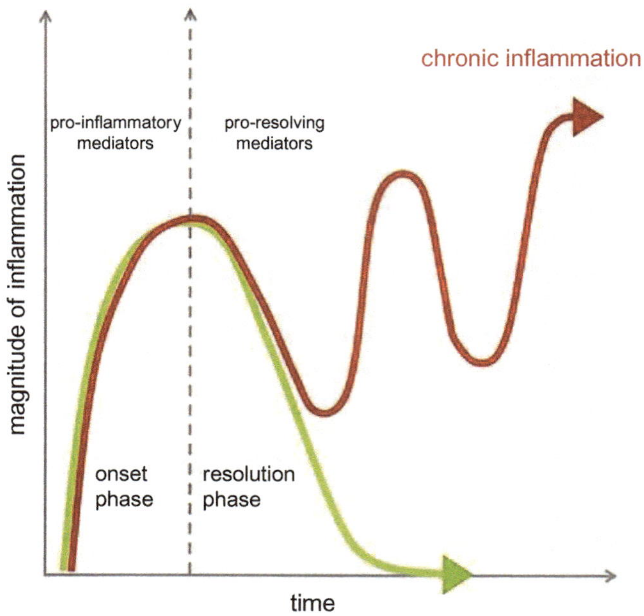

FIG. 3.3 Dynamics of the inflammatory response in chronic inflammation [14]. The acute inflammatory response is a highly coordinated sequence of events characterized by an onset phase coordinated by chemokines, cytokines, and pro-inflammatory mediators that is followed by an active resolution phase brought about by the engagement of specific cellular mechanisms under the control of several pro-resolving mediators to promote resolution of the tissue inflammation as well as healing and repair. A failure in pro-resolving pathways can extend in time the actions of pro-inflammatory mechanisms resulting in prolonged or chronic inflammation with recurrent exacerbations.

inflammatory response, ideally leading to inflammation resolution (Fig. 3.3). A failure to adequately activate pro-inflammatory mechanisms typically results in prolonged or chronic inflammation with recurrent injury and tissue damage.

3.9 Risk factors associated with chronic inflammation

Chronic inflammation is upregulated due to several key risk factors. These risk factors include a poor diet, sedentary lifestyle, and age, among others (Fig. 3.4). Age has been shown to increase inflammatory proteins and molecules due to the free radicals built up in the cells. A diet rich in saturated fat and refined sugars is associated with high amounts of pro-inflammatory proteins,which, in turn, also trigger chronic inflammation. Smoking has also been linked to increased inflammatory mediators.

3.10 Complications of chronic inflammation

Chronic inflammation is a significant contributor to many diseases that can lead to increases in morbidity. These include:

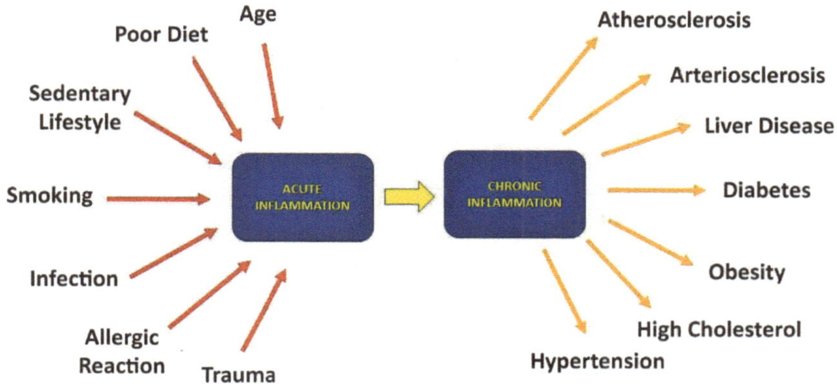

FIG. 3.4 Risk factors involved in acute inflammation may lead to chronic inflammation and one or more chronic conditions.

- *Cardiovascular disease* through atherosclerosis and coronary artery disease which can lead to myocardial infarction (heart attack) and stroke.
- *A liver disease* where a higher-than-average accumulation of fat in the liver can lead to chronic inflammation and scarring, disrupting normal metabolic processes.
- *Obesity* due to chronic inflammation of adipose tissue leads to the release of pro-inflammatory adipokines, insulin resistance, and poor processing of lipids including increased cholesterol levels.
- *Diabetes*, where the inflammation is associated with both type 1 and type 2 diabetes. Inflammation of pancreatic tissue and loss of beta cells of the islets of Langerhans would result in type 1 diabetes and chronic inflammation causing insulin resistance could lead to type 2 diabetes.
- *Hypercholesterolemia*, where chronic inflammation can lead to elevated levels of triglycerides and low-density lipoprotein ("bad cholesterol") while lowering the antiinflammatory, high-density lipoproteins ("good cholesterol").
- *Hypertension (high blood pressure)* where chronic inflammation can contribute to endothelial dysfunction, causing both microvascular and macrovascular complications.

3.11 Free radicals and inflammation

Free radicals, including reactive oxygen species (ROS), are O_2 molecules missing a paired electron (Fig. 3.5). Typically, "healthy" O_2 will have paired electrons, which keeps them stable. Normal cellular metabolism produces free radicals, and these play crucial roles in the activation of signaling pathways, which alter the intra- and extracellular metabolism [16]. However, an O_2 molecule that is missing an electron (e.g., no longer paired) makes them short-lived, highly reactive, and unstable. Free radicals will attack the closest stable molecule and "steal" its electron to have paired electrons again. Meanwhile, the molecule that lost its electron can become a free radical, which continues the chain reaction cascade causing damage to DNA, RNA, cellular lipids, proteins, carbohydrates, leading to the loss of specific cell

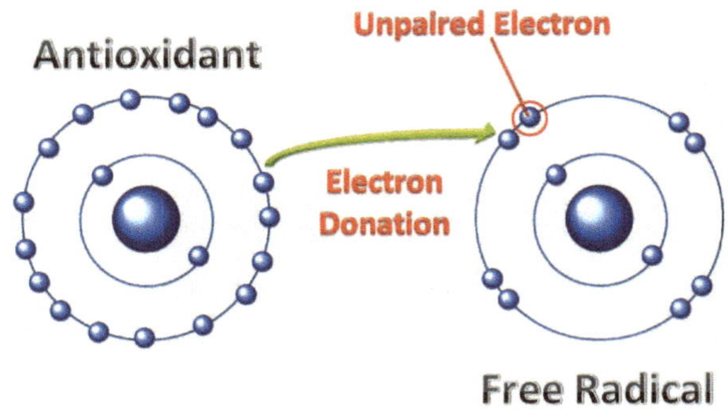

FIG. 3.5 An antioxidant donates an electron to a free radical which stabilizes the structure and prevents it from causing further damage [15].

structure and function and ultimately cell death. This damage to cellular components caused by free radicals is referred to as **oxidative stress**. Free radicals are elevated in diseased states such as obesity, diabetes, heart disease, inflammation, and are major contributors to the aging process. Free radicals in the body can also be formed by external factors such as pollution, UV exposure, and cigarette smoke.

3.12 Role of antioxidants

Antioxidants are compounds that inhibit oxidation or the production of free radicals. They act as "free radical scavengers" and hence prevent and repair the damage done by these free radicals by giving one of its electrons to the free radical, which stabilizes it (Fig. 3.5). Despite giving electrons to free radicals, they do not change into free radicals due to their structure and the high number of electrons.

Antioxidants often serve as antiinflammatory agents by minimizing the harmful effects of ROS. Vegetables are rich sources of antioxidants. Examples of antioxidants include beta-carotene (carrots), lutein (green leafy vegetables), lycopene (citrus fruits), and vitamins A (green leafy vegetables, carrots), C (citrus fruits), and E (green leafy vegetables). The body can produce some antioxidants, known as endogenous antioxidants which include **glutathione** and **superoxide dismutase**.

3.13 Antiinflammatory agents

Although inflammation is a protective process, excessive inflammation is harmful to the body due to exposure to a variety of chemical substances released by the injured and infectious cells and the deposition of debris and dead cells, which may cause local disruption of function. Furthermore, the repair process initiated as part of the inflammation process may limit the original function of an organ undergoing inflammation. Antiinflammatory agents help limit the harmful consequences of inflammation while maintaining the beneficial effects of the same.

The most commonly used antiinflammatory drugs are **nonsteroidal antiinflammatory drugs**, or **NSAIDs**, and **glucocorticoids**. **Aspirin** is an example of antiinflammatory drugs. It interferes with the inflammatory response by decreasing the production of prostaglandins. Although NSAIDs are beneficial, they can have undesirable side effects such as gastric ulceration. Glucocorticoids suppress almost every aspect of the inflammatory response. **Statins** also have antiinflammatory effects.

Natural antiinflammatory compounds often have fewer side effects. **Omega-3 essential fatty acids** are polyunsaturated and must be obtained in the diet. They are found in fish, green leafy vegetables, walnuts, flaxseed, and canola oil. Moreover, they are shown to decrease blood triglyceride concentrations, production of chemo-attractants, growth factors, adhesion molecules, and increase nitric oxide production. Thus Omega-3 fatty acids can help decrease blood pressure, thrombosis, cardiac arrhythmias, and the risk of developing atherosclerosis, cancer, and other conditions associated with inflammation. The Mediterranean-style diet consisting of whole grains, fish, olive oil, fresh fruits, and vegetables may significantly improve health and reduce cardiovascular disease and cancer risk. **Curcumin** (**turmeric**) and **green tea** are the other natural products with antiinflammatory benefits.

3.14 Diet and inflammation

The types and quantity of foods a person eats can significantly impact the level of inflammation and the risk of developing comorbidities. Diets that are high in refined starches, sugar, saturated and trans-fatty acids, and low in fiber increase the body's inflammatory processes, most likely by excessive production of pro-inflammatory cytokines [17]. Increases in circulating pro-inflammatory cytokines, recruitment of neutrophils, and oxidative stress can contribute to the generation of free radicals. To reduce the harmful effects of free radicals on inflammation, a diet high in fruits and vegetables and other plant-based, whole foods is healthy and lowers risks of certain diseases, and increases the level of antioxidants in the body. However, it is unclear whether this is because of the antioxidants, something else in the foods, or other factors. Examples of antioxidants include beta-carotene (carrots), lutein (green leafy vegetables), lycopene (citrus fruits), vitamins A (green leafy vegetables, carrots), C (citrus fruits), and E (green leafy vegetables).

3.15 Exercise and inflammation

Exercise reduces inflammation markers by decreasing adipocytokine production. It also reduces cytokine release from skeletal muscles, endothelial cells, and the immune system, such as TNF-α and C-reactive protein. Skeletal muscle plays a vital role in inflammatory signaling as cytokines, prostaglandins, and chemokines can be both released and taken up by muscle at rest and following exercise. Chronic inflammation compromises muscle mass and adaptability, but exercise training may exert an antiinflammatory effect. Lifelong exercise may positively impact muscle health by promoting antiinflammation in skeletal muscle [18].

References

[1] Blausen 0909 WhiteBloodCells.png. Wikimedia Commons, the free media repository. Updated October 25, 2020. Accessed February 9, 2021. https://commons.wikimedia.org/w/index.php?title=File:Blausen_0909_WhiteBloodCells.png&oldid=501275404.

[2] Hematopoiesis (human) diagram en.svg. Wikimedia Commons, the free media repository. Updated September 19, 2020. Accessed February 9, 2021. https://commons.wikimedia.org/w/index.php?title=File:Hematopoiesis_(human)_diagram_en.svg&oldid=464946037.

[3] Cytokine release.jpg. Wikimedia Commons, the free media repository. Updated October 10, 2020. Accessed February 9, 2021. https://commons.wikimedia.org/w/index.php?title=File:Cytokine_release.jpg&oldid=485571106.

[4] Computer-generated image of the structure for C-reactive protein.jpg. Wikimedia Commons, the free media repository. Updated November 1, 2020. Accessed February 10, 2021. https://commons.wikimedia.org/w/index.php?title=File:Computer-generated_image_of_the_structure_for_C-reactive_protein.jpg&oldid=508889628.

[5] TNFa Crystal Structure.rsh.png. Wikimedia Commons, the free media repository. Updated September 29, 2020. Accessed February 10, 2021. https://commons.wikimedia.org/w/index.php?title=File:TNFa_Crystal_Structure.rsh.png&oldid=475798465.

[6] 3LTQ.pdb.png. Wikimedia Commons, the free media repository. Updated September 13, 2020. Accessed February 10, 2021. https://commons.wikimedia.org/w/index.php?title=File:3LTQ.pdb.png&oldid=457656731.

[7] IL8 Solution Structure.rsh.png. Wikimedia Commons, the free media repository. Updated September 8, 2020. Accessed February 10, 2021. https://commons.wikimedia.org/w/index.php?title=File:IL8_Solution_Structure.rsh.png&oldid=452241071.

[8] Histamine 3D ball.png. Wikimedia Commons, the free media repository. Updated September 17, 2020. Accessed February 11, 2021. https://commons.wikimedia.org/w/index.php?title=File:Histamine_3D_ball.png&oldid=462416042.

[9] Bradykinin updated.png. Wikimedia Commons, the free media repository. Updated September 9, 2020. Accessed February 11, 2021. https://commons.wikimedia.org/w/index.php?title=File:Bradykinin_updated.png&oldid=453063797.

[10] Leukotriene A4.svg. Wikimedia Commons, the free media repository. Updated November 9, 2020. Accessed February 11, 2021. https://commons.wikimedia.org/w/index.php?title=File:Leukotriene_A4.svg&oldid=511604606.

[11] Prostaglandin E1.svg. Wikimedia Commons, the free media repository. Updated October 29, 2020. Accessed February 11, 2021. https://commons.wikimedia.org/w/index.php?title=File:Prostaglandin_E1.svg&oldid=506287882.

[12] Inflammatory response.png. Wikimedia Commons, the free media repository. Updated October 17, 2020. Accessed February 9, 2021. https://commons.wikimedia.org/w/index.php?title=File:Inflammatory_response.png&oldid=492439106.

[13] OSC Microbio 17 06 Erythema.jpg. Wikimedia Commons, the free media repository. Updated September 13, 2020. Accessed February 8, 2021. https://commons.wikimedia.org/w/index.php?title=File:OSC_Microbio_17_06_Erythema.jpg&oldid=457660920.

[14] C. Barnig, T. Bezema, P.C. Calder, et al., Activation of resolution pathways to prevent and fight chronic inflammation: lessons from asthma and inflammatory bowel disease. Review, Front. Immunol. 10 (2019) 1699, https://doi.org/10.3389/fimmu.2019.01699.

[15] Antioxidants Free radicals Lobo et al.,.png. Wikimedia Commons, the free media repository. Updated September 15, 2020. Accessed February 12, 2021. https://commons.wikimedia.org/w/index.php?title=File:Antioxidants_Free_radicals_Lobo_et_al.,.png&oldid=460276774.

[16] P. Arulselvan, M.T. Fard, W.S. Tan, et al., Role of antioxidants and natural products in inflammation, Oxid. Med. Cell. Longev. 2016 (2016), https://doi.org/10.1155/2016/5276130, 5276130.

[17] D. Giugliano, A. Ceriello, K. Esposito, The effects of diet on inflammation: emphasis on the metabolic syndrome. Review, J. Am. Coll. Cardiol. 48 (4) (2006) 677–685, https://doi.org/10.1016/j.jacc.2006.03.052.

[18] K.M. Lavin, R.K. Perkins, B. Jemiolo, U. Raue, S.W. Trappe, T.A. Trappe, Effects of aging and lifelong aerobic exercise on basal and exercise-induced inflammation, J. Appl. Physiol. 128 (1) (2020) 87–99.

CHAPTER

4

Obesity and metabolic syndrome

OUTLINE

OBJECTIVES

- Define body mass index. Explain how it can be calculated.
- Compare the body mass index of a normal person to that of an overweight and an obese individual.
- List the four pathways by which daily expenditure of energy takes place and the amount of energy each one of these consumes.
- Discuss in detail the causes of obesity.
- Explain how genetics is linked to obesity.
- List the different types of obesity and identify the type that leads to chronic diseases.
- Locate where leptin is made and summarize its role in the body. Describe its regulation of release and its mechanism of action.
- Locate where ghrelin is made and its role concerning obesity.

Biology of Cardiovascular and Metabolic Diseases
https://doi.org/10.1016/B978-0-12-823421-1.00001-9

- Identify the causes for leptin resistance.
- Relate obesity to insulin resistance.
- Summarize the strategies that are used in reversing obesity.
- Define metabolic syndrome.
- List the clusters of diseases that contribute to metabolic syndrome.
- Explain the underlying mechanism in the development of metabolic syndrome.
- Summarize the treatment choices for obesity and metabolic syndrome.

4.1 Introduction

Evolutionarily, the body tends to store fat and resist its loss. Any weight gain most often leads to **obesity**, a chronic disease characterized by an excess of adipose tissue in the body. It is the cause of a mismatch in energy balance, where calories taken in from food exceed those that are used, thus, resulting in an excess amount of fat in the bloodstream, referred to as **hyperlipidemia**. It affects people of all ages and ethnicities. In recent years, the onset of obesity is occurring at increasingly younger ages. The prevalence of overweight and obesity among children, adolescents, and adults has risen significantly over the past several decades. According to the *Center for Disease Control*, approximately 42.4% of the American adult population were reported to be obese in 2017–18 [1]. The health consequences of obesity are significant and lead to an increased risk of chronic diseases such as insulin resistance, type 2 diabetes, heart disease, and several types of cancers.

4.2 Body mass index as an index of fat accumulation

One of the most common markers for body fat content is the **body mass index** (BMI). The BMI can be calculated as weight expressed in kilograms, divided by the square of height, expressed in meters (kg/m^2):

$$\text{BMI} = (\text{weight in kg}/\text{height in m}^2) \text{ or } \left(\text{weight (lb)}/[\text{height (in.)}]^2 \times 703\right)$$

Normal BMI is between 18.5 and 24.9 kg/m^2, whereas BMI between 25 and 29.9 kg/m^2 is considered overweight, and BMI above 30 kg/m^2 is considered obese (Fig. 4.1). BMI, however, does not take into account large muscle mass, and therefore in athletes with large muscle mass, it may not be an accurate measure. BMI does not measure body fat directly but is correlated with more direct measures of body fat. The prevalence of obesity-related diseases is common at BMI values $>25\,kg/m^2$. For children, a healthy weight is usually when the BMI is at the 5th percentile up to the 85th percentile, overweight if their BMI is in the 85th percentile to less than the 95th percentile, and obese if their BMI is in the 95th percentile or above, based on growth charts for children who are the same age and sex [3]. Other alternative approaches to measuring fat accumulation are waist to hip circumference, skinfold thickness, bioelectrical impedance, and underwater weighing, but they are used less commonly.

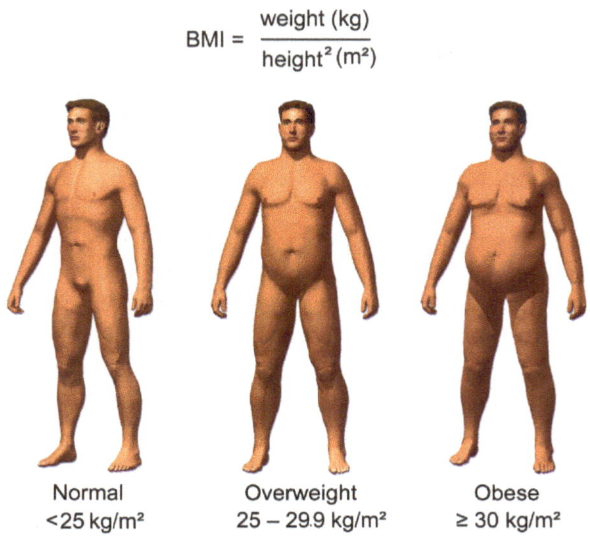

FIG. 4.1 Body mass index (BMI) is used to define normal weight, overweight, and obesity [2]. BMI is a measure of weight and height.

Energy balance is measured as the amounts of calories consumed (*energy in*) through eating and drinking compared to calories burned (*energy out*) through several energy using pathways. Daily energy expenditure is expended through four pathways (Fig. 4.2):

Resting metabolic rate (RMR), which is the energy expended to carry out bodily functions while the body is at rest, including the repair and maintenance of cells and tissues, the maintenance of resting heart rate, a constant body temperature, and many other physiological processes that are essential in maintaining homeostasis. It is the largest

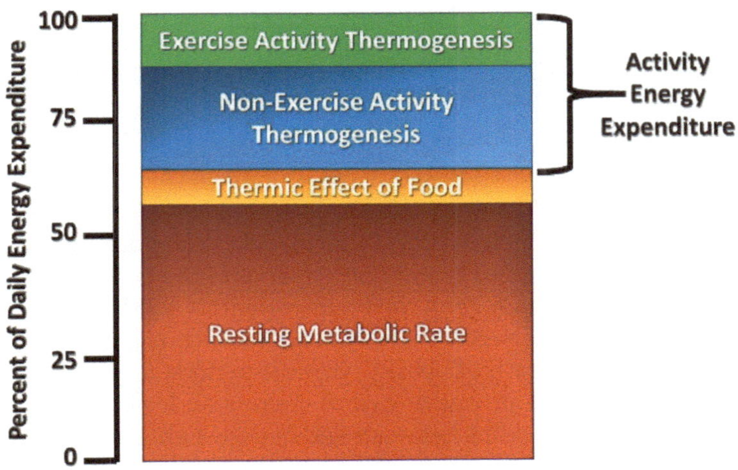

FIG. 4.2 Components of total daily energy expenditure.

energy expenditure component accounting for approximately 60%–70% of total daily energy.

The thermic effect of food (TEF) is the excess heat generated when food is consumed, digested, and absorbed. It accounts for approximately 10% of total energy expenditure. For example, a 400 kcal meal would result in the body using ~40 kcals to digest, absorb, and store the nutrients from the meal. Many factors affect TEF, such as age, meal size, meal frequency, and meal composition.

Nonexercise activity thermogenesis (NEAT) is the calories burned through movement outside of structured exercise. Examples of NEAT include cooking, cleaning, shopping, and even small movements such as fidgeting or playing a musical instrument. NEAT can vary substantially between people by up to 2000 kcal per day [4]. Individuals with obesity show low NEAT.

Exercise activity thermogenesis (EAT) is the energy expended from planned activities such as running, jogging, swimming, or weight training. It varies from as high as 40% in a highly active person to as low as 10% in a sedentary individual. Physical activity is the most important means by which energy is expended in the body as muscles use approximately 25%–30% of total energy per day. A single episode of strenuous exercise, for example, may increase basal energy expenditure for several hours after the physical activity is stopped. Exercise increases muscle mass and decreases body fat mass whereas inadequate physical activity results in the opposite.

Energy balance can fluctuate on a daily basis but over the long term if too many calories are consumed compared to what is burned, weight gain occurs (Fig. 4.3). Lack of exercise or increased intake of food leads to decreased use of energy and increased storage of fat in adipocytes (**adiposity**), liver, and other tissues and weight gain (**positive energy balance**) among overweight and obese individuals. Exercise promotes increased use of energy than normal resulting in weight loss (**negative energy balance**).

4.3 Causes of obesity

Obesity is caused by various behavioral, metabolic, genetic, and hormonal influences that affect energy balance.

4.3.1 Environmental factors

Lifestyle and environmental factors play a crucial role in the occurrence of obesity. The development of overweight and obesity is a consequence of the easy and inexpensive availability of high calorie yielding foods, combined with sedentary lifestyle changes, occurring in modern societies and affecting genetically predisposed subjects [5]. Childhood overnutrition seems to compound the occurrence of obesity. Critical periods *in utero* and early postnatal growth during which the development of neural circuits and metabolic pathways can be modified, resulting in elevated feeding behavior and body weight gain later in life. Enormous portion size, high-energy density, palatability, excessive amounts of refined starch, added sugars, high-fat content, and low levels of dietary fiber in fast food seem to be the reason for overweight and obesity.

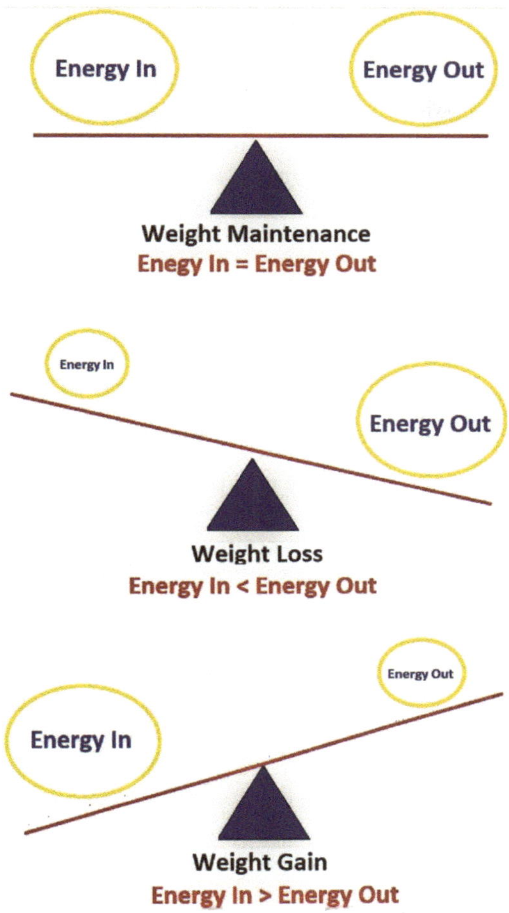

FIG. 4.3 Energy balance equation. If the *Energy In* is equal to the *Energy Out*, weight remains the same. If the *Energy In* is greater than the *Energy Out*, the weight will be gained. If the *Energy In* is lower than the *Energy Out*, the weight will be lost. The *Energy In* consists of calories from food or drink. The *Energy Out* is calories from resting energy expenditure, thermic effect of food, and physical activity.

4.3.2 Genetic factors

Although environmental factors play a significant role in the onset of obesity, there is a genetic correlation to body weight regulation. The interactions between environmental and genetic factors are likely to play an essential yet undefined role in the development of obesity. Studies using body mass index are a measure of obesity among monozygotic and dizygotic twins while keeping the environment constant to reveal that genetic component played an essential role [6]. Approximately 20%–25% of obesity cases are related to genetic factors and are carried over for generations. Obese parents are likely to give birth to children who become overweight, and their offspring are also often overweight.

Defects in one or more specific genes could cause alteration of pathways that regulate the feeding center (the **hypothalamus**), energy expenditure, and fat storage. There are many obesity genes identified. The human obese gene, the **Ob gene**, encodes and stimulates the production of leptin, which acts on the hypothalamus to promote (1) the production of chemicals that suppress appetite and (2) reduce the levels of specific chemical substances that trigger appetite. A malfunction of this gene is shown to affect the production of leptin, which disrupts the normal functioning of the hypothalamus in the regulation of feeding behavior. Thus any mutation in the gene that produces leptin or the genes that respond to leptin at the leptin-melanocortin circuit and **neuropeptide Y (NPY)** and **agouti-related peptide (AgRP)** pathway could bring about endless hunger and marked obesity. More common but less well understood are polygenic factors involving alterations in multiple genes that enhance the susceptibility to obesity.

4.4 Adipose tissue

Adipose tissue is a dynamic structure that contributes to a broad range of physiological processes. It is a specialized loose connective tissue consisting of mainly one type of cells, **adipocytes** (Fig. 4.4). The total number of adipocytes may increase under appropriate dietary conditions, including a high-fat diet. The increase in the number of adipocytes adds to body weight. The rate of formation of adipocytes is rapid at a young age compared to adulthood. The greater the rate of fat storage, the greater the number of fat cells added. For example, obese children have three times more adipocytes compared to normal lean children of the same age. Mature adipocytes cannot be eliminated except for surgical manipulation (**lipectomy**).

A central function of adipose tissue is energy storage. In times of excess caloric intake, the body stores energy in adipose tissue primarily in the form of triglycerides. Regulation of triglyceride synthesis and mobilization is essential for energy balance and regulation. During negative energy balance, when more calories are expended than are provided by ingested food, triglycerides are hydrolyzed to release free fatty acids and are oxidized to obtain energy.

Brown fat is a type of adipose tissue found mainly in young children and is associated with heat production (Fig. 4.5). On the other hand, white adipose tissue is the primary storage site for fat in adults. It is found in the subcutaneous layer of the skin, bone marrow, and around visceral organs and blood vessels. Its distribution is affected by sex, age, and genetic factors. The **beige adipocyte** is a type of fat cell described by the ability to produce heat and increase energy expenditure. Beige fat can be considered phenotypically as a fat cell that possesses characteristics between those of the **white fat cell,** an accumulator of energy, and the brown fat cell, which produces heat [9].

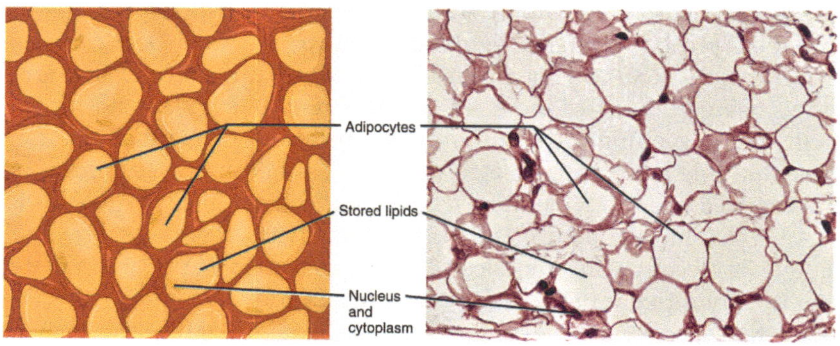

FIG. 4.4 Adipose tissue is made of adipocytes or fat cells [7] which store fat for energy purpose and provideinsulation. LM × 800.

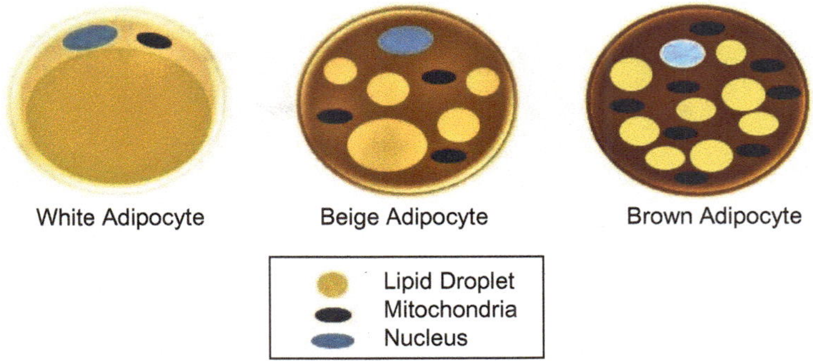

FIG. 4.5 Morphology of white, beige, and brown adipocytes [8].

4.4.1 Location of adipose tissue storage

Adipose tissue is located around internal organs (visceral fat), beneath the skin (subcutaneous fat), in bone marrow (yellow bone marrow), intermuscular (muscular system), and in the breast (breast tissue) (Fig. 4.6). Visceral fat or abdominal fat (also known as organ fat or intra-abdominal fat) is located inside the abdominal cavity, packed between the organs such as around the stomach, liver, intestines, and kidneys. Visceral fat is obtained from cross-sectional scanning of the abdomen at the lumbar Spine 4 and 5 (L4–L5) using magnetic resonance spectrometry (Fig. 4.7). An excess of visceral fat is known as central obesity, or "belly fat," in which the abdomen protrudes excessively. Excess visceral fat is also linked to type 2 diabetes [12], insulin resistance [13], cardiovascular diseases [14], and other obesity-related diseases. Men are more likely to have fat stored in the abdomen due to sex hormone differences compared to women. Estrogen hormone in the female tends to increase subcutaneous fat storage, especially in the buttocks, thighs, and hips [15]. Subcutaneous fat storage is not related to many classic obesity-related pathologies, and some evidence suggests it might be protective [16]. The World Health Organization has summarized characteristics of visceral adipose tissue that contribute to heart disease and possibly promote insulin resistance. They include increased fat cell number, higher blood flow rate per unit tissue mass, elevated cortisol, and androgen receptor numbers, and increased catecholamine-induced lipolysis.

4.5 Types of fat deposition

The body has a natural tendency to resist weight loss. For example, a loss of 10% body weight increases appetite significantly. Excessive accumulation of fat leading to obesity is instead a slow process. Most of the weight gain occurs between ages 27 and 44 years of age. Men begin at age 30, and women at age 27 start accumulating 0.2–0.8 kg (0.5–1.8 lbs) of body weight each year until age 60. Men having 25% or higher total body fat and women with 35% or greater total body fat are considered obese.

Android type of fat deposition, known as **central obesity**, is where fat deposition is in the visceral and abdominal regions and appears to have a greater health risk in developing hypertension, diabetes, osteoarthritis, pancreatitis, and many other conditions. **Gynoid** type of fat

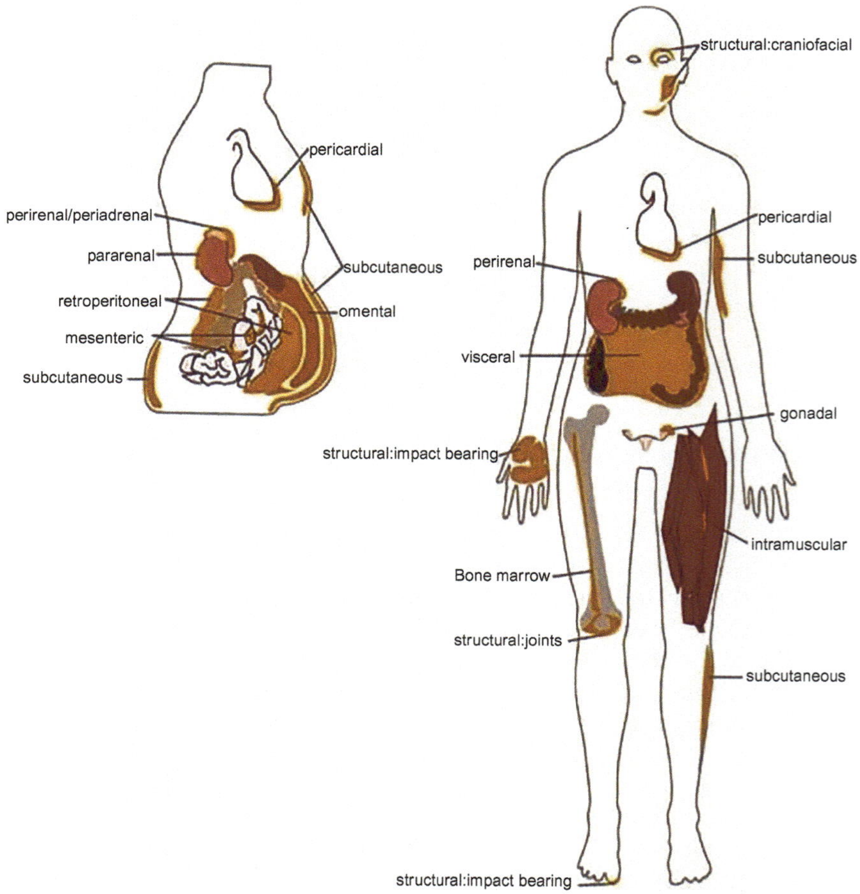

FIG. 4.6 A cross-sectional image of the abdomen obtained by magnetic resonance spectrometry by scanning the abdomen at the level of L4–L5 [10]. Computed tomography makes the measurement of visceral adiposity reasonably easy for the patient and the physician. Because of differences in the attenuation values of adipose, muscle, and bone tissue, it is straightforward to distinguish fat from muscle and bone on the image. The intra-abdominal (visceral adipose) tissue is highlighted in white. Studies have indicated that among equally overweight or obese individuals, an excess of intra-abdominal (visceral adipose) tissue is predictive of increased risk of the expected abnormalities of obesity [11].

deposition, on the other hand, occurs in the hips, thighs, and buttocks resulting in the pear-shaped adiposity. Gynoid type of fat distribution is commonly seen in women (Fig. 4.8). The exact mechanisms underlying the variability in adipose tissue distribution are largely unknown.

Interestingly, the location of the adipose tissue may be an important indicator of the insulin sensitivity of the tissue. Individuals who store their adipose tissue in the lower body (gynoid or pear shape) tend to be more insulin-sensitive as the adipose tissue acts like a sponge or reservoir that soaks up the excess free fatty acid and triglyceride, keeping them from accumulating in tissue outside of adipose tissue. In contrast, individuals with an apple or android shape contain their adipose tissue predominantly in the upper body. However, subcutaneous

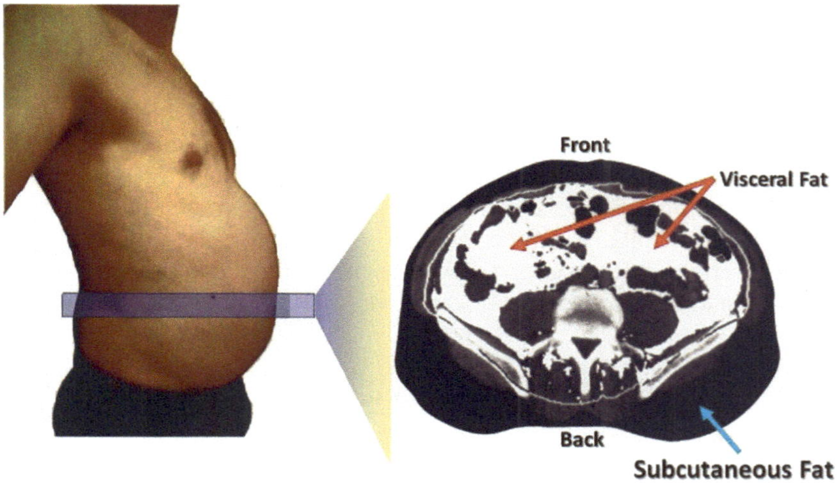

FIG. 4.7 A cross-sectional image of the abdomen obtained by magnetic resonance spectrometry by scanning the abdomen at the level of L4–L5 [10]. Computed tomography makes the measurement of visceral adiposity fairly easy for the patient and the physician. Because of differences in the attenuation values of adipose, muscle, and bone tissue, it is easy to distinguish fat from muscle and bone on the image. The intra-abdominal (visceral) adipose tissue is highlighted in white. Among equally overweight or obese individuals, an excess of intra-abdominal (visceral) adipose tissue is predictive of increased risk of the expected abnormalities of obesity.

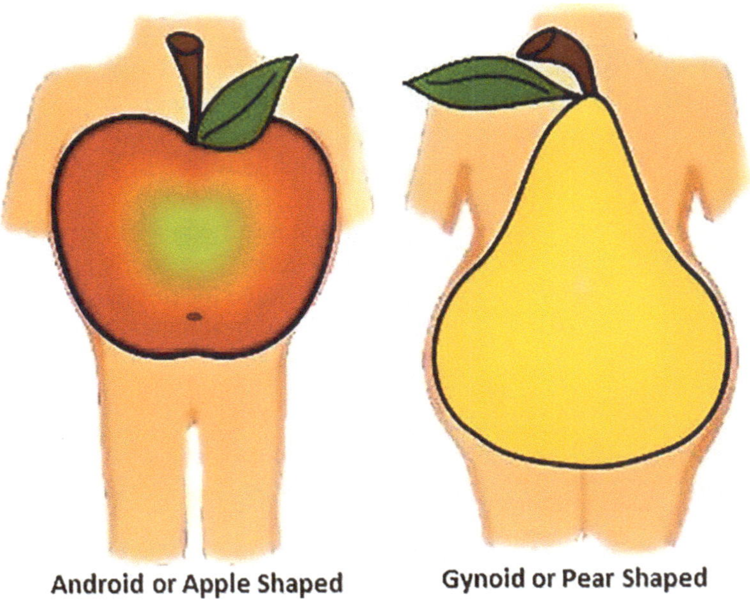

FIG. 4.8 Types of fat distribution. Android-shaped individuals have large waistlines that store body fat in the upper body (abdomen). Gynoid or pear-shaped individuals have smaller waists but store their body fat in the lower body parts such as the hips and thighs. Storage of body fat in the abdomen (apple-shaped) is associated with greater cardiovascular and metabolic disease risk than someone with body fat in the hips and thighs (pear-shaped).

adipose tissue in these individuals does not act like a sponge or reservoir. Instead, the excess free fatty acid and triglycerides are circulating and accumulate in unwanted places such as the liver, pancreas, skeletal muscle, and visceral adipose tissue, a phenomenon that has been described as an ectopic fat deposition. As a result, these individuals tend to be at a greater risk for insulin resistance and type 2 diabetes. Males tend to store more of their body fat in the waist than females, who store more of their excess adipose tissue in the hip and thighs. Therefore, subcutaneous adipose tissue located in the lower body might act as a metabolic sink, protecting various organs against the damage of excessive lipid accumulation and macrophage infiltration. The distribution of adipose tissue in the subcutaneous area may be one reason women have a lower incidence of cardiovascular and metabolic disease.

4.6 Leptin and ghrelin

Adipose tissue secretes various bioactive peptides, referred to as adipokines, that act both locally and systemically, including hormones. One essential hormone made by the adipocytes is **leptin**, which plays a significant role in regulating appetite and energy balance. Leptin is produced mostly in adipose tissue. High levels of adipose tissue would lead to higher levels of leptin. The brain reads leptin as a sign of fullness, so it "turns off" hunger. More leptin means decreased appetite. With this in mind, it would seem that individuals who are overweight or obese would feel less hungry and eat less, given the higher amount of excess adipose tissue, leading to weight loss. However, persistently elevated leptin levels cause the brain to become desensitized or resistant to the effects of leptin, and it stops responding, leading to increased hunger. The secretion of leptin and its regulation is a complex neurohumoral mechanism involving a vital brain structure called the **hypothalamus** and a variety of hormones (Fig. 4.9). The hypothalamus is the control center for hunger and satiety. However, the feeding process itself is modulated by the higher centers (e.g., amygdala and prefrontal cortex) and neurohumoral signals from the **gastrointestinal tract**. Both an increase in energy storage in the adipose tissue and when the individual is well fed, leptin is released by the adipocytes (Fig. 4.9).

Ghrelin is a hormone that communicates with the brain when hungry (Fig. 4.9). The primary function of ghrelin is to increase appetite. It is produced in the stomach and released into the bloodstream when the stomach is empty. It is called the hunger hormone; the emptier the stomach, the higher the levels of ghrelin. Circulating ghrelin levels increase during fasting and between meals. Specific neurogenic abnormalities such as lesions, tumors, neural pathways, or neurotransmitters affecting ventromedial nuclei of the hypothalamus have been shown to cause an animal to overeat.

Leptin can cross the blood-brain barrier and reach the hypothalamus, where it acts on two different hypothalamic pathways. It inhibits an **anabolic circuit** and at the same time stimulates a **catabolic circuit** to promote energy expenditure. The inhibition of food intake by leptin is via the inhibition of a sensory pathway situated in the hypothalamus involving the release of NPY and AgRP (Fig. 4.9). The opposite is true for ghrelin. The promotion of leptin's catabolic action causes a decrease in the energy stores and, therefore, weight loss and decreased production of leptin, an example of a negative feedback mechanism. Thus, leptin levels rapidly decline with caloric restriction and weight loss, leading to an increase in appetite and suppression of energy expenditure.

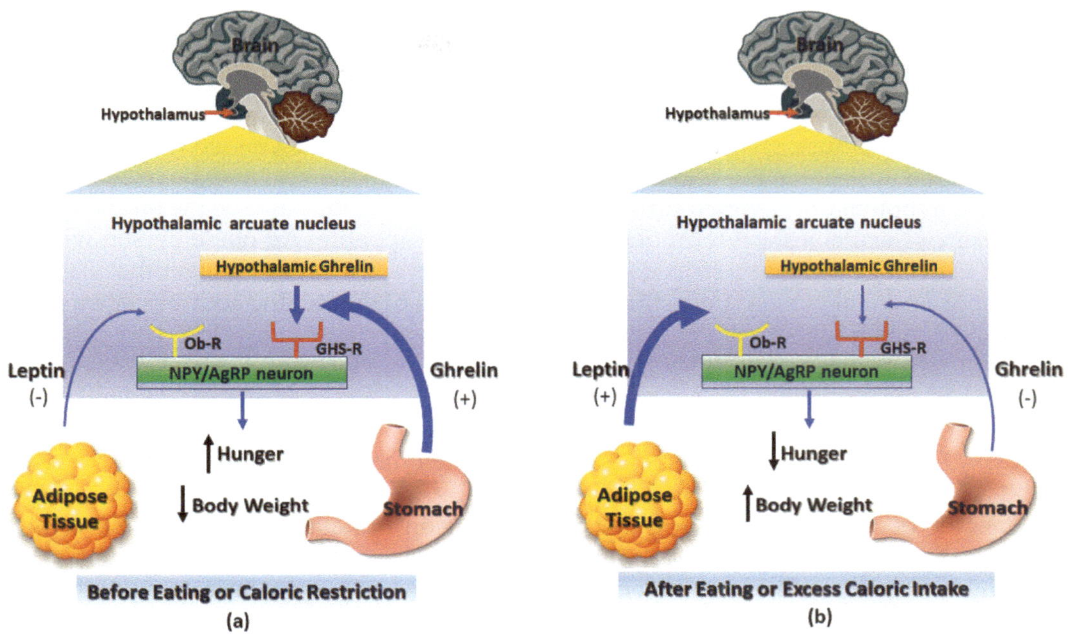

FIG. 4.9 A detailed view of neurohumoral involvement in the action of leptin and ghrelin (A) before eating or from caloric restriction, (B) after eating or excess caloric intake [17]. *AgRP*, agouti-related peptide; *GHS-R*, ghrelin hormone sensitive receptor; *NPY*, neuropeptide Y; *Ob-R*, leptin receptor. *Thick arrows* (+) = increased production, *Thin arrows* (−) = decreased production.

4.7 Development of resistance to hormones

Hormone resistance to feeding appears to play a role in the development and maintenance of obesity. Leptin resistance could be due to high levels of free fatty acids in the blood, which may, in turn, increase fat metabolites in the neurons and interfere with leptin signaling. Another hypothesis is the inability to cross the blood-brain barrier, possibly due to defective transport across endothelial cells. An obese person often develops leptin resistance, which results in increased levels of circulating leptin in these subjects. There is a four-fold increase in the circulating leptin levels in obese individuals than in lean subjects. The high levels of leptin in the blood are strongly associated with the combination of upper body obesity, glucose intolerance, hypertriglyceridemia, and hypertension. The high levels of leptin appear to activate the sympathetic system. Chronic activation of the sympathetic system by circulating leptin is linked to hypertension, hyperglycemia, glucose intolerance, and insulin resistance. Leptin resistance is shown to be reversible with diet modification and weight reduction.

Insulin secretion is directly proportional to the size of the fat mass. In the basal state and following meals, obese subjects secrete more insulin, although blood glucose is identical in obese and lean individuals. Leptin secretion is also in direct proportion to the size of the fat mass, although the amount of fat in the cell is not the exact stimulus for its secretion. Insulin has a very short **plasma half-life** of 2.5–3 min. Thus it provides a minute-to-minute update of ongoing metabolic activity via a signal that is directly proportional to total body fat. In contrast, leptin has a **half-life** of about 45 min and provides a more stable message to the brain about body fat stores.

The hypothalamic temperature may be another factor that controls appetite. Heat given off during digestion of food stimulates the satiety center, causing an increased energy expenditure. Increased appetite in the cold and decreased appetite in the heat may be related to such changes in hypothalamic temperature. Increased temperature may also be the mechanism by which exercise exerts a short-term depressant effect on appetite. An obese person who has reduced to average weight by strict dietary measures develops intense hunger that is greater than that of a normal person suggesting that the set point of the feeding control system in this individual is much higher than that of a nonobese person.

4.8 Metabolic syndrome

A cascade of disorders following chronic weight gain is **metabolic syndrome** (**MetS**) or **syndrome X**. Metabolic syndrome is a cluster of signs and symptoms such as abdominal obesity (large waist circumference), hypertension (high blood pressure), fasting hyperglycemia (high blood sugar), high blood triglycerides (fat in the blood), and decreased levels of high-density lipoproteins (HDLs, "good cholesterol"). According to the American Heart Association, determination of metabolic syndrome requires at least three of the five signs/symptoms: waist size of over 40″ for men or 35″ for women, blood pressure greater than 130/85 mmHg (prehypertensive), blood glucose above 110 mg/dL, low HDL (<50 mg/dL for women and 40 mg/dL for men), and increased triglycerides (greater than 150 mg/dL) (Fig. 4.10). **Hemoglobin A1C** (HbA1C), a measurement of blood glucose levels, is often greater than 6.5% compared to less than 5.7% in normal individuals. The viscosity of blood is expected to be thicker than usual due to increased lipids and glucose.

Individuals with metabolic syndrome have up to a five-fold increase in the risk of developing type 2 diabetes and are twice as likely to die from myocardial infarction or stroke than those without this syndrome. Individuals with this syndrome consume more food than the energy being spent. Thus, the utilization and storage of energy are different than in normal individuals. All of the metabolic syndrome features are closely related to the accumulation of excess fat in the abdominal cavity around the visceral organs. Subjects with metabolic syndrome are typically overweight with excess body fat around the waist, referred to as **central obesity** (Fig. 4.11) or visceral, male-pattern, or apple-shaped adiposity.

Metabolic syndrome is linked to the development of atherosclerosis, which could cause coronary artery disease, hypertension, and type 2 diabetes among other conditions. These individuals often have other conditions such as **hyperuricemia** (increased uric acid in the blood), **fatty liver disease**, **polycystic ovarian syndrome** (in women), and **erectile dysfunction** (in men). The rate of occurrence of metabolic syndrome increases with age. While approximately 34% of the adult US population appears to have metabolic syndrome, it increases to 60% in the age group above 50 years of age.

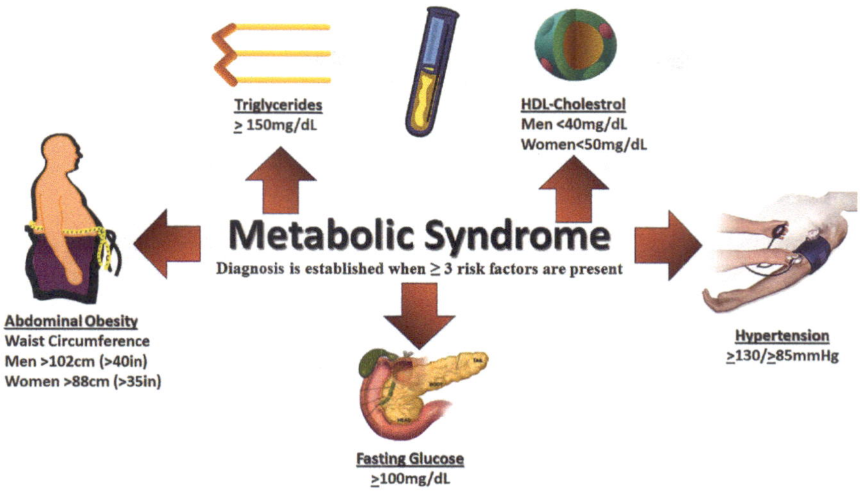

FIG. 4.10 Diagnosis of Metabolic Syndrome [18]. *HDL-C*, high-density lipoprotein ("good cholesterol"). Hypertension = high blood pressure, mg/dL = milligrams per deciliter of blood [19,20].

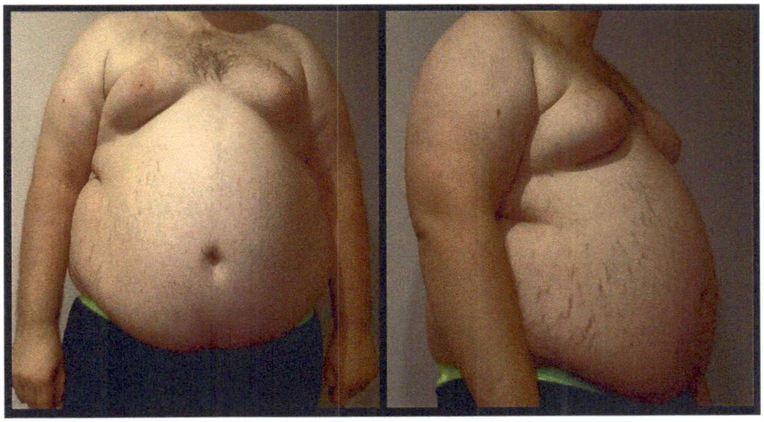

FIG. 4.11 Visceral or truncal obesity [21].

4.9 Pathogenesis of metabolic syndrome

The pathogenic mechanisms of the metabolic syndrome are complex and continue to be studied. Although the exact mechanisms of the complex pathways of metabolic syndrome are unknown, obesity, age, poor diet, and a sedentary lifestyle are key factors. Chronic stress appears to play a role by disrupting the hormonal balance of the hypothalamic-pituitary-adrenal axis. Increased circulating cortisol levels, a hormone secreted by the adrenal cortex, results in elevated glucose and insulin levels, which, in turn, leads to visceral adiposity, insulin resistance, dyslipidemia, and hypertension. Visceral adiposity and insulin resistance play large roles in most of the pathways involved in metabolic syndrome.

4.9.1 Visceral adipose tissue

Obesity is associated with large waist circumference and higher amounts of visceral adipose tissue (Fig. 4.12). The higher levels of visceral adipose tissue result in a high amount of free fatty acids (FFA) being secreted into the portal vein and are taken up by the liver. Increased abdominal fat and the subsequent increase in the release of FFAs into the hepatic portal circulation and, ultimately, its appearance in the liver may decrease the liver's ability to clear insulin, causing hyperinsulinemia, a higher than normal level of insulin. The liver can also become insulin resistant, leading to increased triglyceride synthesis, and become responsible for the increased low-density lipoprotein production (LDL) [25,26]. An increase in LDL and a reduction in HDL are indirect effects of insulin resistance caused by altered lipid metabolism in the liver. Visceral adipose tissue releases inflammatory cytokines that contribute to chronic noninfectious systemic inflammation. Even small increases in visceral fat can result in damage to the endothelium.

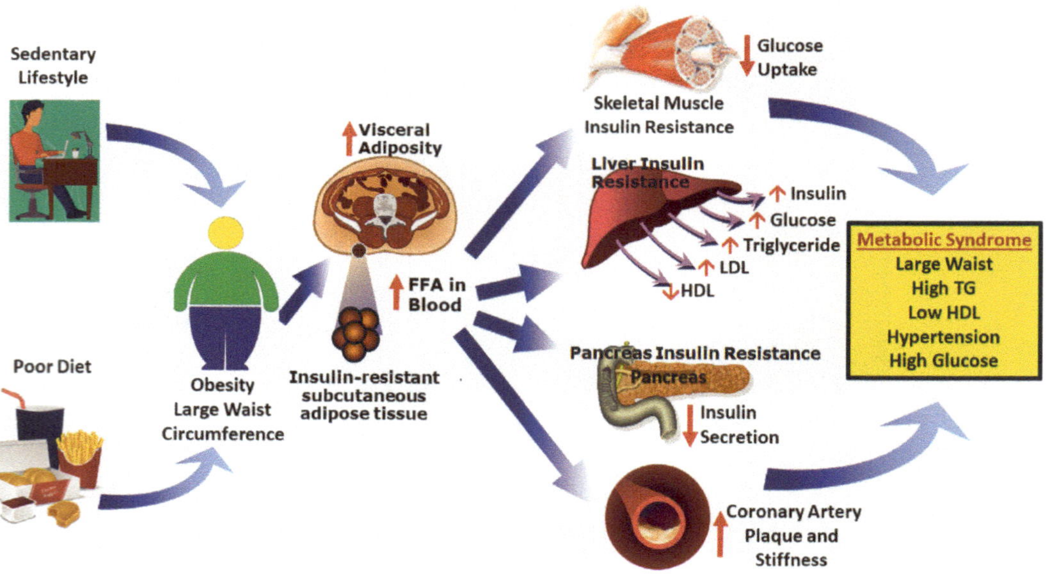

FIG. 4.12 Pathogenesis of the metabolic syndrome [22–24]. Large abdominal adiposity increases the accumulation of visceral adiposity and causes subcutaneous adipose tissue to become insulin resistant, releasing excess free fatty acids (FFA) into the blood. The excess FFA is taken up by the skeletal muscle, liver, and pancreas, causing them to become insulin resistant and plaque to build up in the coronary arteries. Collectively these events lead to the cluster of metabolic abnormalities associated with the metabolic syndrome, which includes central obesity (large waist circumference), atherogenic dyslipidemia (high TG, low HDL), hypertension (high blood pressure), and high blood glucose levels (hyperglycemia). *HDL*, high-density lipoprotein; *TG*, triglycerides.

4.9.2 Insulin resistance

A combination of physical inactivity and a high-fat, high-refined sugar diet is linked to **insulin resistance**. Insulin resistance and hyperinsulinemia may be a cause rather than a consequence of obesity. Insulin resistance appears to be associated with decreased lipoprotein lipase activity in the skeletal muscle, whereas high insulin levels inhibit hormone-sensitive lipase in the adipocytes. Insulin increases glucose uptake in the skeletal muscle and liver and inhibits lipolysis (fat breakdown) and liver gluconeogenesis (making energy from noncarbohydrate sources; Fig. 4.12). When adipose tissue becomes insulin resistant, the inhibitory effect of insulin on lipolysis is lifted. Thus, an increase in FFA in the blood prevents insulin action such as inhibition of fat breakdown [27]. As a result, adipose tissue will continue to release excess FFA into the blood.

The FFA circulate throughout the body and end up in skeletal muscle, liver, pancreas, and coronary arteries, among other structures. The increased levels of FFA will further disrupt the regular insulin and fatty acid metabolism in these tissues, exacerbating insulin resistance in the body. For example, the reduced glucose uptake in the skeletal muscle causes glucose to remain in the blood longer, leading to hyperglycemia (high blood glucose levels). The increased insulin resistance in the liver increases gluconeogenesis and lipogenesis and the secretion of glucose and FFA into the blood.

The pancreas will sense the rise in blood glucose and increase insulin secretion (hyperinsulinemia) to maintain **euglycemia** (normal blood glucose levels). However, over time, the pancreas is unable to compensate, and, therefore, insulin secretion decreases. FFA are also lipotoxic to beta cells of the pancreas, causing decreased insulin secretion [28].

Insulin resistance contributes to the development of hypertension due to loss of the vasodilator effect of insulin and vasoconstriction caused by an increase in sympathetic activation and sodium reabsorption in the kidneys [29]. Insulin resistance also causes the adipose tissue to release certain proteins that increase blood viscosity (thicker blood), induction of a prothrombotic state (increased blood clots that could contribute to artery blockage), and release of pro-inflammatory cytokines (contributes to increased plaque buildup) that contribute to increased risk of cardiovascular [25,30].

4.10 Treatment options for obesity and the metabolic syndrome

Treatments for obesity and metabolic syndrome include lifestyle changes, such as heart-healthy eating and increased physical activity, and Food and Drug Administration (FDA)-approved weight-loss medication. Sustained negative energy balance until the desired weight loss is achieved by reducing energy intake and increased energy expenditure. National Institutes of Health guidelines recommend a decrease in caloric intake of 500 kcal per day for overweight and moderately obese subjects to achieve a weight loss of approximately 1 lb per week. A reduction of 500–1000 kcal per day for a person with a basal metabolic rate greater than 35 kg/m^2 to lose 1–2 lb per week or 10% weight loss after 6 months.

Exercise and a healthy diet, including whole grains, fruits, and vegetables, and avoiding refined products have been shown to improve most factors associated with metabolic syndrome (i.e., levels of blood pressure, insulin, and triglycerides). Large quantities of "bulk" are recommended, which are nonnutritive cellulose substances to distend the stomach to appease hunger. Exercise improves insulin sensitivity by increasing glucose transporters, oxidative enzymes, and by decreasing abdominal fat. Adding resistance exercise to an endurance training program reduces insulin resistance and reduces body fat in people with hyperinsulinemia more than endurance training alone. Exercise promotes daily energy expenditure and reduces overweight situations. Forced exercise is often an essential part of treatment. Surgical procedures such as gastric bypass and gastric banding surgeries are prescribed to help obese and overweight conditions. A variety of drugs are available to lower overweight and obesity, including amphetamines (sympathomimetics) that directly inhibit the feeding centers in the brain, and *Orlistat*, a lipase inhibitor, that reduces the intestinal digestion of fat.

References

[1] C.M. Hales, M.D. Carroll, C.D. Fryar, C.L. Ogden, Prevalence of obesity and severe obesity among adults: United States, 2017–2018, NCHS Data Brief 360 (2020) 1–8.

[2] Obesity & BMI.png. Wikimedia Commons, the free media repository. Updated September 28, 2020. Accessed February 22, 2021. https://commons.wikimedia.org/w/index.php?title=File:Obesity_%26_BMI.png&oldid=474767494.

[3] BMI Percentile Calculator for Child and Teen. Updated June 25, 2020. Accessed February 18, 2021. https://www.cdc.gov/healthyweight/bmi/calculator.html.

[4] J.A. Levine, M.W. Vander Weg, J.O. Hill, R.C. Klesges, Non-exercise activity thermogenesis: the crouching tiger hidden dragon of societal weight gain. Research support, N.I.H., extramural review, Arterioscler. Thromb. Vasc. Biol. 26 (4) (2006) 729–736, https://doi.org/10.1161/01.ATV.0000205848.83210.73.

[5] P.L. Prieto-Hontoria, P. Perez-Matute, M. Fernandez-Galilea, M. Bustos, J.A. Martinez, M.J. Moreno-Aliaga, Role of obesity-associated dysfunctional adipose tissue in cancer: a molecular nutrition approach. Research support, non-U.S. Gov't review, Biochim. Biophys. Acta 1807 (6) (2011) 664–678, https://doi.org/10.1016/j.bbabio.2010.11.004.

[6] N.L. Segal, R. Feng, S.A. McGuire, D.B. Allison, S. Miller, Genetic and environmental contributions to body mass index: comparative analysis of monozygotic twins, dizygotic twins and same-age unrelated siblings. Twin study, Int. J. Obes. 33 (1) (2009) 37–41, https://doi.org/10.1038/ijo.2008.228.

[7] 409 Adipose Tissue.jpg. Wikimedia Commons, the free media repository. Updated October 7, 2020. Accessed February 17, 2021. https://commons.wikimedia.org/w/index.php?title=File:409_Adipose_Tissue.jpg&oldid=483721996.

[8] Adipocyte types.jpg. Wikimedia Commons, the free media repository. Updated April 30, 2020. Accessed February 17, 2021. https://commons.wikimedia.org/w/index.php?title=File:Adipocyte_types.jpg&oldid=415900513.

[9] F. Lizcano, D. Vargas, Biology of beige adipocyte and possible therapy for type 2 diabetes and obesity, Int. J. Endocrinol. 2016 (2016) 9542061, https://doi.org/10.1155/2016/9542061.

[10] Abdominal obesity in men.jpg. Wikimedia Commons, the free media repository. Updated May 28, 2020. Accessed February 23, 2021. https://commons.wikimedia.org/w/index.php?title=File:Abdominal_obesity_in_men.jpg&oldid=422266066.

[11] A. Cook, C. Cowan, Adipose, in: StemBook, 2009.

[12] C.T. Montague, S. O'Rahilly, The perils of portliness: causes and consequences of visceral adiposity. Review, Diabetes 49 (6) (2000) 883–888, https://doi.org/10.2337/diabetes.49.6.883.

[13] K. Janochova, M. Haluzik, M. Buzga, Visceral fat and insulin resistance - what we know? Review, Biomed. Pap. Med. Fac. Univ. Palacky Olomouc Czech Repub. 163 (1) (2019) 19–27, https://doi.org/10.5507/bp.2018.062.

[14] I.J. Neeland, R. Ross, J.P. Despres, et al., Visceral and ectopic fat, atherosclerosis, and cardiometabolic disease: a position statement. Research support, non-U.S. Gov't review, Lancet Diabetes Endocrinol. 7 (9) (2019) 715–725, https://doi.org/10.1016/S2213-8587(19)30084-1.

[15] L.M. Brown, D.J. Clegg, Central effects of estradiol in the regulation of food intake, body weight, and adiposity. Research support, N.I.H., extramural review, J. Steroid Biochem. Mol. Biol. 122 (1-3) (2010) 65–73, https://doi.org/10.1016/j.jsbmb.2009.12.005.

[16] S.A. Porter, J.M. Massaro, U. Hoffmann, R.S. Vasan, C.J. O'Donnel, C.S. Fox, Abdominal subcutaneous adipose tissue: a protective fat depot? Research support, N.I.H., extramural research support, non-U.S. Gov't, Diabetes Care 32 (6) (June 2009) 1068–1075, https://doi.org/10.2337/dc08-2280.

[17] M. Kojima, K. Kangawa, Drug insight: the functions of ghrelin and its potential as a multitherapeutic hormone. Research support, non-U.S. Gov't review, Nat. Clin. Pract. Endocrinol. Metab. 2 (2) (2006) 80–88, https://doi.org/10.1038/ncpendmet0080.

[18] Third report of the national cholesterol education program (NCEP) expert panel on detection, evaluation, and treatment of high blood cholesterol in adults (adult treatment panel III) final report. Guideline practice guideline, Circulation 106 (25) (2002) 3143–3421.

[19] Blood Pressure - Take Another Person.png. Wikimedia Commons, the free media repository. Updated October 24, 2020. Accessed February 25, 2021. https://commons.wikimedia.org/w/index.php?title=File:Blood_Pressure_-_Take_Another_Person.png&oldid=498986727.

[20] Blausen 0699 PancreasAnatomy2.png. Wikimedia Commons, the free media repository. Updated October 25, 2020. Accessed February 25, 2021. https://commons.wikimedia.org/w/index.php?title=File:Blausen_0699_PancreasAnatomy2.png&oldid=501280308.

[21] Obesity6.JPG. Wikimedia Commons, the free media repository. Updated September 6, 2020. Accessed February 24, 2021. https://commons.wikimedia.org/w/index.php?title=File:Obesity6.JPG&oldid=449844615.

[22] Fast-food-menu.svg. Wikimedia Commons, the free media repository. Updated September 30, 2020. Accessed February 25, 2021. https://commons.wikimedia.org/w/index.php?title=File:Fast-food-menu.svg&oldid=476400181.

[23] WHO EN HealthyAtHome-Physical-activity Be active at home during COVID-19 outbreak 3.png. Wikimedia Commons, the free media repository. Updated February 13, 2021. Accessed February 25, 2021. https://commons.wikimedia.org/w/index.php?title=File:WHO_EN_HealthyAtHome-Physical-activity_Be_active_at_home_during_COVID-19_outbreak_3.png&oldid=532504467.

[24] Q12174 noun 5364 ccAnaFelix obesity.svg. Wikimedia Commons, the free media repository. Updated September 12, 2020. Accessed February 25, 2021. https://commons.wikimedia.org/w/index.php?title=File:Q12174_noun_5364_ccAnaFelix_obesity.svg&oldid=456742154.

[25] Y. Rochlani, N.V. Pothineni, S. Kovelamudi, J.L. Mehta, Metabolic syndrome: pathophysiology, management, and modulation by natural compounds. Review, Ther. Adv. Cardiovasc. Dis. 11 (8) (2017) 215–225, https://doi.org/10.1177/1753944717711379.

[26] G.F. Lewis, G. Steiner, Acute effects of insulin in the control of VLDL production in humans. Implications for the insulin-resistant state. Review, Diabetes Care 19 (4) (1996) 390–393, https://doi.org/10.2337/diacare.19.4.390.

[27] G. Boden, G.I. Shulman, Free fatty acids in obesity and type 2 diabetes: defining their role in the development of insulin resistance and β-cell dysfunction, Eur. J. Clin. Investig. 32 (s3) (2002) 14–23, https://doi.org/10.1046/j.1365-2362.32.s3.3.x.

[28] J.E. Tooke, M.M. Hannemann, Adverse endothelial function and the insulin resistance syndrome, J. Intern. Med. 247 (4) (2000) 425–431, https://doi.org/10.1046/j.1365-2796.2000.00671.x.

[29] D. Tripathy, P. Mohanty, S. Dhindsa, et al., Elevation of free fatty acids induces inflammation and impairs vascular reactivity in healthy subjects, Diabetes 52 (12) (2003) 2882–2887, https://doi.org/10.2337/diabetes.52.12.2882.

[30] I. Juhan-Vague, M.-C. Alessi, A. Mavri, P.E. Morange, Plasminogen activator inhibitor-1, inflammation, obesity, insulin resistance and vascular risk, J. Thromb. Haemost. 1 (7) (2003) 1575–1579, https://doi.org/10.1046/j.1538-7836.2003.00279.x.

CHAPTER

5

Atherosclerosis

OUTLINE

OBJECTIVES

- Define arteriosclerosis and identify the causes for this condition.
- List modifiable and nonmodifiable risk factors for atherosclerosis.
- Describe the mechanisms involved in the pathogenesis of atherosclerosis.
- Explain how endothelial injury is linked to atherosclerosis.
- Summarize the process involved in the production of monocytes during inflammation.
- Explain the process of initiation of atherosclerosis in a coronary artery.
- Describe the mechanism by which monocytes are recruited into the area of inflammation.
- List the reasons for plaque rupture.

Biology of Cardiovascular and Metabolic Diseases
https://doi.org/10.1016/B978-0-12-823421-1.00002-0

- Explain the mechanisms by which high-density lipoproteins help reduce the development of atherosclerosis.
- Summarize the role of C-reactive protein in protecting the body against atherosclerosis.
- Define aneurysm and discuss the causes for this condition.
- Compare aneurysm and dissection.

5.1 Introduction

Arteriosclerosis is the hardening of an artery causing loss of compliance, the ability to stretch. Atherosclerosis, on the other hand, builds a coat of fat in the inner lining of the blood vessel wall and, in particular, arteries. Atherosclerosis is an inflammatory disease that originates within the endothelial lining and progresses to become a fibrotic plaque. Over time, the fat content of the plaque hardens and makes the vessel stiffer and the lumen size narrower, interfering with the normal blood supply to the tissues. Both atherosclerosis and arteriosclerosis could lead to severe complications such as stroke and myocardial infarction.

5.2 Risk factors shared by arteriosclerosis and atherosclerosis

The **nonmodifiable risk factors** include genetic abnormalities, family history, increasing age, and male gender. A significant risk factor for arteriosclerosis and atherosclerosis is advanced age, as the conditions tend to progress over time. Although the development of atherosclerotic plaque is a slow process, it clinically manifests in the middle age or later, thus increasing myocardial infarction incidence fivefold. Premenopausal women are relatively protected against atherosclerosis. However, after the age of menopause, the incidence of atherosclerosis-related diseases increases in women and at older ages exceeds that of men [1].

Hyperlipidemia, hypertension, smoking, diabetes, and inflammation are classified as **modifiable risk factors** for atherosclerosis. However, some of these conditions could be indirectly due to genetic abnormalities. Among several risk factors, hypertension is one of the significant risk factors for atherosclerosis since it causes degenerative changes in the walls of large and medium arteries that can lead to both aortic dissection and cerebrovascular hemorrhage. Smoking is a well-established modifiable risk factor and likely accounts for the increasing incidence and severity of atherosclerosis. Prolonged years of smoking one pack of cigarettes or more daily doubles the death rate from ischemic heart disease. Smoking cessation reduces that risk substantially. Type 2 diabetes mellitus markedly increases the risk of atherosclerosis. The incidence of myocardial infarction is twice as high in patients with diabetes than in those without. Inflammation is intimately associated with atherosclerotic plaque development and rupture.

While several circulating markers of inflammation correlate with ischemic heart disease risk, **C-reactive protein** (CRP) has emerged as one of the simplest to measure as a marker of inflammation. CRP is an acute-phase inflammatory agent, synthesized primarily by the liver, and its expression is increased by many inflammatory mediators. It binds to bacteria and activates the classical complement cascade, thus causing inflammation. **Hyperhomocysteinemia**, increased homocysteine levels in the blood, is another condition associated with premature vascular disease and atherosclerosis.

5.3 Pathogenesis of arteriosclerosis

Arteriosclerosis, the stiffening of the arteries, is due to the buildup of proteins in the arterial wall, which causes the wall to thicken and become less elastic. In a normal situation, compliance allows an artery to expand when blood is pumped through it from the heart and to recoil after the surge has passed, thus promoting blood flow. In arteriosclerosis, compliance is reduced while pressure and resistance within the vessel increases, which, in turn, results in conditions such as hypertension and coronary artery disease. Arteriosclerosis affects blood vessels of all sizes and may cause downstream ischemic injury. The buildup of protein in the blood vessel wall could be due to nonenzymatic glycosylation (glucose attachment) of proteins in the basement membrane (e.g., diabetes mellitus), leakage of proteins from the basement membrane into the vessel wall (e.g., diabetes mellitus), or high blood pressure pushing plasma proteins into the vessel wall (e.g., hypertension). Arteriosclerosis often begins with injury to the endothelium of an artery, which may be caused by irritation from high blood glucose, infection, tobacco use, hyperlipidemia, and other factors. Arterial walls that are constantly stressed by blood flowing with high pressure are also more likely to be injured. Atherosclerosis is a form of arteriosclerosis and is the leading cause of coronary artery disease and cerebrovascular diseases.

5.4 Atherosclerosis

Atherosclerosis is the progressive narrowing of the arteries due to plaque formation and is considered the major cause of cardiovascular diseases [2]. Tissue injury typically triggers an inflammatory response within the area, including its blood vessels. When inflammation spreads through the arterial wall, it weakens and scars the wall, leaving it stiff (sclerotic). As a result, compliance is reduced, and circulating triglycerides and cholesterol seep through the damaged lining of the blood vessel wall, where they are frequently joined by leukocytes, Ca^{2+}, and cellular debris. Eventually, this buildup, called **plaque**, can narrow the lumen of the arteries enough to impair blood flow (Figs. 5.1 and 5.2). Large elastic and muscular arteries and medium-sized muscular arteries are most commonly affected, especially those supplying the heart, brain, kidneys, and lower extremities. Myocardial infarction, stroke, aortic aneurysms, and peripheral vascular disease (gangrene of the legs) are the major consequences of atherosclerosis. At the early stages of stenosis, outward remodeling of the vessel tends to preserve the size of the lumen. However, there are limits on the remodeling extent, and gradually, the expanding atheroma impinges on the lumen, and blood flow is compromised. Although sudden plaque rupture is the most dangerous consequence, atherosclerosis also interferes with tissue function through chronically diminished arterial perfusion. The effects of vascular occlusion ultimately depend on arterial supply and the metabolic demand of the affected tissue.

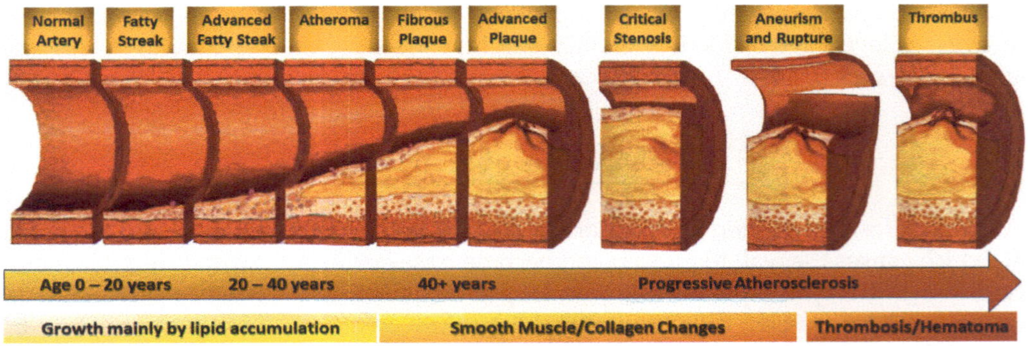

FIG. 5.1 Stages of plaque formation in a coronary artery [3].

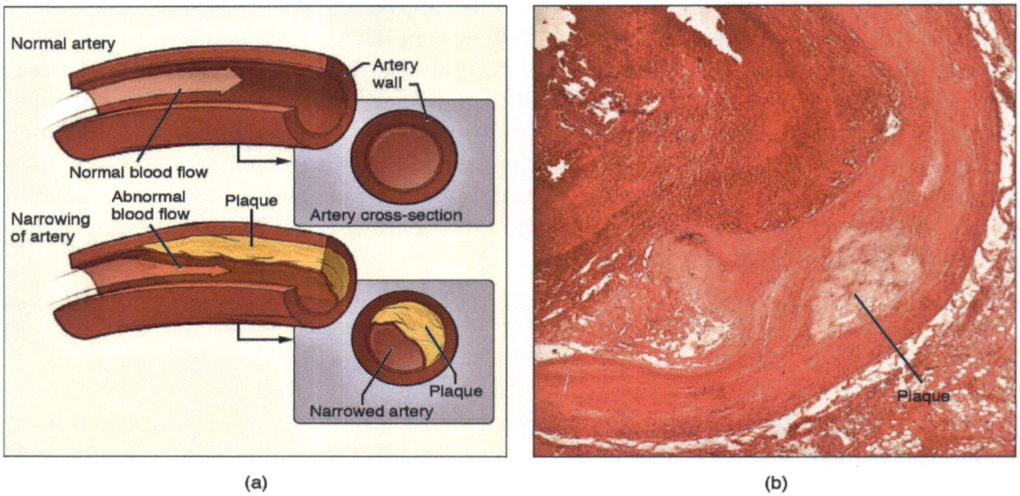

FIG. 5.2 A comparison of a normal artery to an artery with a plaque [4]. (A) Atherosclerosis can result from plaques formed by the buildup of fatty, calcified deposits in an artery. (B) Plaques can also take other forms, as shown in this micrograph of a coronary artery with a buildup of connective tissue within the artery wall. LM × 40. *Micrograph provided by the Regents of University of Michigan Medical School © 2012.*

5.5 Pathogenesis of atherosclerosis

The mechanisms responsible for the formation of a plaque are discussed in detail in the following sections.

5.5.1 Endothelial cell injury

Endothelial cells are very versatile. They can produce a variety of chemicals in response to alterations in metabolic activity. Endothelial cells could be activated by cytokines, lipid products, advanced glycation end-products (diabetes), hemodynamic stresses, hypoxia, viruses, complement components, and bacterial products, which may elicit inflammation and, in severe cases, septic shock. Activated endothelial cells express adhesion molecules and produce cytokines, growth factors, vasoactive peptides, procoagulant factors, anticoagulant factors, and several other agents. For example, endothelial cells release both **nitric oxide** (**NO**), a vasodilator, and **endothelin**, a vasoconstrictor. Besides, they release chemicals that maintain blood in a fluid state. They can also modulate the smooth muscle tone and its growth and metabolize hormones such as angiotensin. Normal endothelial cell function, therefore, is dependent on a delicate balance of these responses.

Atherosclerosis is considered to be a chronic inflammatory response of the arterial wall due to endothelial injury. Initially, it occurs in the intact endothelium. The intact but dysfunctional endothelial cells exhibit increased endothelial permeability, enhanced leukocyte adhesion, and altered gene expression. Endothelial injury and dysfunction may lead to increased vascular permeability, leukocyte adhesion, thrombosis, and atherosclerosis. Although many factors may be involved, the two most important causes of endothelial dysfunction are hemodynamic disturbances and hypercholesterolemia.

Hemodynamic disturbances: Blood flow with turbulence at the openings of branching vessels seems to be the cause of atherogenesis as plaques tend to occur at branching points and along the posterior wall of the abdominal aorta, where there are disturbed flow patterns.

Hypercholesterolemia: Since lipids are not water soluble, they are transported in the bloodstream bound to specific apoproteins. **Dyslipoproteinemias** are lipoprotein abnormalities, including increased low-density lipoprotein (LDL) and decreased high-density lipoprotein (HDL) levels, may result from mutations that lead to defects in apoproteins or lipoprotein receptors. They may also arise from other underlying disorders that affect circulating lipid levels, such as nephrotic syndrome (a kidney disease where there is more than a normal amount of protein excretion), alcoholism, hypothyroidism (decreased production of thyroid hormone than normal), and diabetes mellitus. All of these conditions are associated with an increased risk of atherosclerosis.

The mechanisms by which hyperlipidemia contributes to atherogenesis include the following:

- Chronic hyperlipidemia, particularly hypercholesterolemia, can directly impair endothelial cell function by increasing the production of local free radicals such as reactive oxygen species. Besides causing membrane and mitochondrial damage, free radicals accelerate NO decay, dampening its vasodilator activity.
- With chronic hyperlipidemia, lipoproteins accumulate within the intima, where they may aggregate or become oxidized by free radicals produced by inflammatory cells. **Macrophages** phagocytose such modified LDL through a variety of receptors. Chronic ingestion leads to the formation of lipid-filled macrophages called **foam cells**. Smooth muscle cells can similarly transform into lipid-laden foam cells ingesting modified lipids through LDL-receptors. Not only are the modified lipoproteins toxic to endothelial cells, smooth muscle cells, and macrophages, but their binding and uptake also stimulate the release of growth factors, cytokines, and chemokines that create a vicious cycle of monocyte recruitment and activation.

5.5.2 Role of macrophages

Macrophages play an essential role in the development of atherosclerosis. Monocytes are produced by the bone marrow from hematopoietic stem cell precursors called **monoblasts** and stored in the spleen (50%) or circulate in the blood (50%). Monocytes are attracted to a damaged site by chemical substances released by the damaged cells at the inflammation site. When a monocyte enters damaged tissue through the endothelium of a blood vessel, it undergoes a series of changes to become a macrophage (Fig. 5.3). Macrophages are the predominant cells involved in creating the progressive plaque lesions of atherosclerosis. They are known to fight off infection or injury by clearing pathogens and bring about resolution of inflammation. In response to an overabundance of pathogens or LDL, macrophages may contribute to the formation of atheroma (fatty buildup in a tissue) leading to atherosclerosis.

In response to inflammation, the bone marrow increases its production of a subset of monocytes referred to as **LY6C$^+$ monocytes** by the differentiation of the hematopoietic stem cells. LY6C$^+$ cells enter circulation from where they are recruited to the site of inflammation by chemoattractants such as monocytochemotatic protein-1 (MCP-1). MCP-1 is released by the endothelial cells of the inflamed tissue. Monocytes migrate into the blood vessel wall and differentiate to become macrophages.

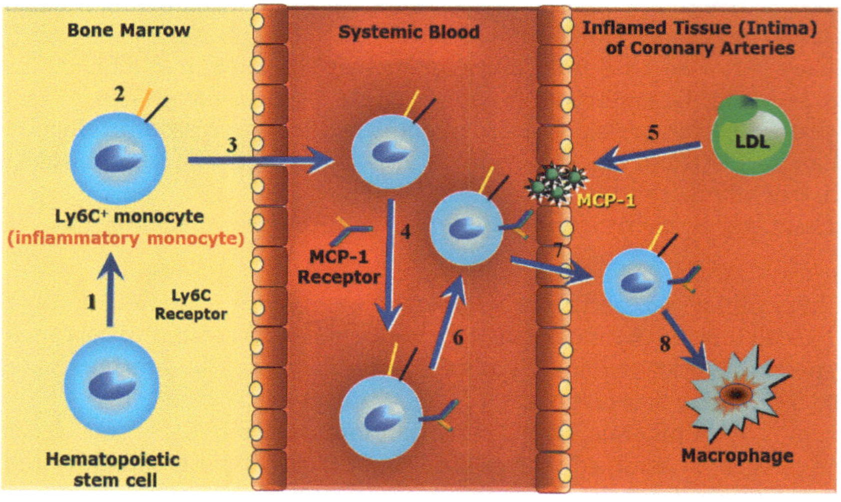

FIG. 5.3 Role of monocytes and macrophages in the early stages of atherosclerosis in the coronary artery [5]. (1) The bone marrow produces monocytes from hematopoietic stem cells. (2) A Ly6C receptor docks with the hematopoietic stem cells to form a Ly6C$^+$ inflammatory monocyte (a monocyte that responds to inflammation). (3) Ly6C$^+$ monocytes are released into the blood. (4) While circulating in the blood, a monocyte chemotactic protein-1 (MCP-1) receptor attaches to the Ly6C$^+$ monocyte. (5) In response to LDL infiltration into the intima, the endothelial cells of the inflamed tissue are activated to release MCP-1. (6) Ly6C$^+$ monocytes respond to the pro-inflammatory monocyte chemotactic protein-1 (MCP-1) by moving to the endothelial layer of the inflamed tissue. (7) The monocyte then moves across the endothelial layer. (Migration of the monocyte into the inflamed tissue will be described later.) (8) Monocyte differentiate into a macrophage.

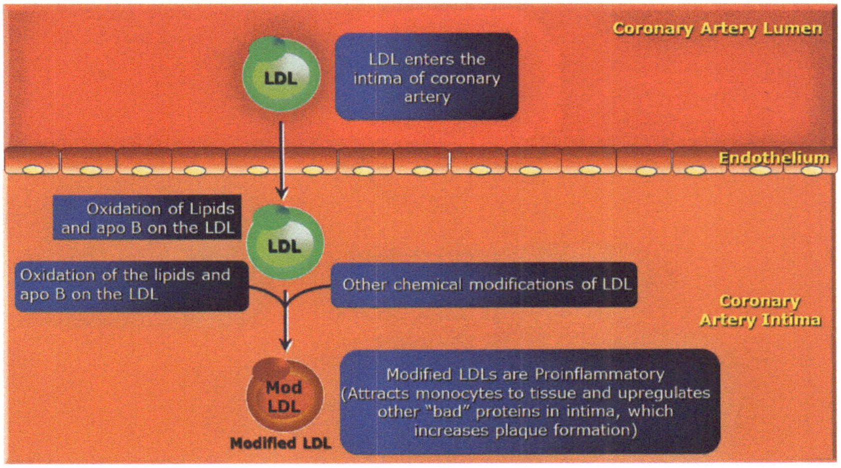

FIG. 5.4 Initial steps in the onset of atherosclerosis in a coronary artery. LDL cholesterol enters the intima of the coronary arteries where they are modified [6,7]. LDL in the blood enters the arterial wall by crossing the endothelial membrane. As LDL accumulates, it is subject to a variety of modifications. The best known of these is oxidation (loss of electrons) of the lipids and the apolipoprotein B (apoB). As modification on the apoB protein proceeds, it loses the affinity to the LDL receptor and, instead, has a high affinity (attraction) for binding to the scavenger receptors on a macrophage. Modified LDL particles are pro-inflammatory, leading to plaque formation in the coronary arteries.

5.5.3 The onset of atherosclerosis

Atherosclerosis can begin in several ways, but they all lead to the buildup of plaque. The following steps are commonly seen during the onset of atherosclerosis:

- LDL readily moves from the blood into the intima where it undergoes modifications to become proinflammatory (Fig. 5.4);
- LDL entry causes an increase in MCP-1 production at the endothelial layer (Fig. 5.5);
- Monocytes enter the intima from the blood by responding to MCP-1 (Figs. 5.6).

5.5.4 Plaque formation

LDLs are made in the liver and are released into the blood. Their production may be increased due to diets that consist of high saturated fats and refined sugar. As they circulate in the blood, they freely enter the tunica intima (Fig. 5.4). Upon their entry into the tissue, LDL is oxidized. Increased lipid oxidation results in the release of greater amounts of free radicals (Fig. 5.5). Additionally, as LDL is oxidized, it gets modified, and the modified LDL causes an increase in the production of MCP-1 by the endothelial cells. Accumulation of modified LDL in the intima and the cell injury associated with the release of free radicals will turn on an inflammatory process locally. As part of the inflammation process, monocytes are recruited into the area of inflammation.

When monocytes circulate in the blood, they usually bounce off the vessel wall and continue to cruise along. In atherosclerosis, the presence of MCP-1 attracts monocytes to the site of infection, and the expression of adhesion molecules on the endothelial surface slows the

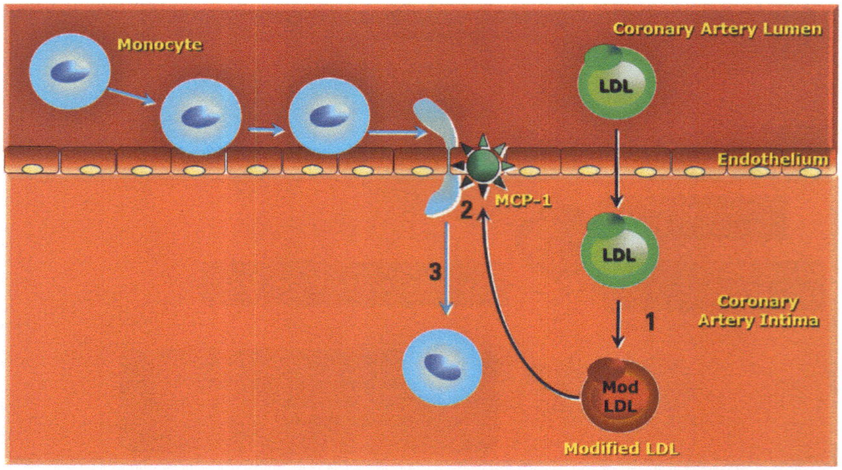

FIG. 5.5 Modified LDL is involved in many stages of the inflammatory process that leads to the development of atherosclerosis. (1) LDL enters tunica intima and is modified by undergoing oxidation. (2) Modified LDL activates endothelial cells to express monocyte chemotactic protein-1 (MCP-1) [8]. (3) MCP-1 attracts monocytes from the blood vessel lumen into the subendothelial space (intima), which is one of the very early stages in the inflammatory process leading to the development of atherosclerosis.

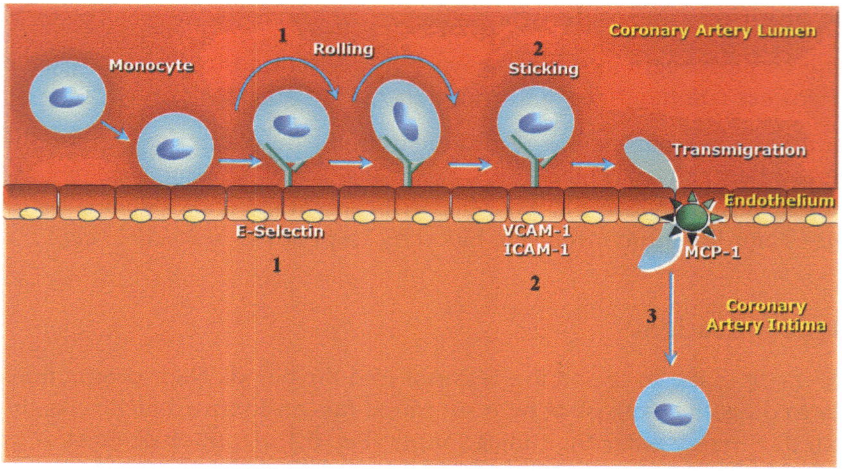

FIG. 5.6 Recruitment of monocytes [9] to the site of inflammation. (1) Monocytes, expressing E-selectin, receptors for the adhesion molecules, contact endothelial cells in response to monocyte chemotactic protein 1 (MCP-1) where the adhesion molecule such as vascular cell adhesion molecule-1 (VCAM-1) or intercellular adhesion molecule-1 (ICAM-1) slows the monocyte by inducing the rolling of the monocyte along the endothelial surface. (2) A firm attachment of monocyto to VCAM-1 or ICAM-1 occurs on the endothelial surface arresting their further rolling. (3) Once the monocyte is tightly bound, it assumes a flat and thin form and migrates into the intima where it will differentiate into a macrophage.

movement of monocytes. Adhesion molecules such as **vascular cell adhesion molecule-1 (VCAM-1)** and **intracellular adhesion molecule-1 (ICAM-1)** are involved in the recruitment of monocytes into the blood vessel wall. During inflammation, the expression of VCAM-1 or ICAM-1 increases on the surface of the endothelial cells. Monocytes, being attracted to the site of inflammation, expressing receptors for the adhesion molecules such as **E-selectin**, contact endothelial cells where the presence of adhesion molecules slow the monocytes by inducing rolling of these cells along the endothelial surface. A firm attachment to VCAM-1 or ICAM-1 occurs next, which interacts with integrins such as E-selectin, on the monocyte surface. Once the monocyte is tightly bound to the adhesion molecule, it migrates into the intima through the spaces between endothelial cells by rearranging its cytoskeleton and assuming a thin and flat form (Fig. 5.6). The intima now contains two critical components of initiation of atherosclerosis: modified LDL and monocytes.

The entry of monocytes and the modified LDLs into the intima play an essential role in the differentiation of monocytes into macrophages, a critical step in the inflammatory process to developing atherosclerosis. Macrophages release a variety of chemicals. Among these, tumor necrosis factor-α (TNF-α) and interleukin-1 (IL-1) activate endothelial cells to express more adhesion molecules (VCAM-1, ICAM-1) that bind monocytes, making them available for recruitment into the subendothelial space by MCP-1. The activated macrophages also express various scavenger receptors, several of which recognize the different modified LDL forms. Macrophages phagocytose LDL through the binding of LDL to scavenger receptors and are converted into the lipid-rich foam cells that are the hallmark of atherosclerosis. Accumulation of this modified LDL in the cell leads to the collection of cholesterol droplets in the macrophage and the formation of foam cells and later fatty streaks, which are the earliest stages of the formation of atheroma. Macrophages and foam cells produce cytokines and a variety of growth factors that lead to cell proliferation (enlargement of the fatty streak) and matrix production, as well as **metalloproteinases** that lead to matrix degeneration. Thus, macrophages and foam cells contribute to lesion growth and may contribute to instability and thrombotic events (Fig. 5.7).

Macrophages that are at the site of lesion are modified to become foam cells and begin to die, leading to a lipid core covered by a fibrous cap. Thrombus (blood clot) formation is another crucial step in the precipitation of unstable angina or myocardial infarction and occlusion during or following angioplasty. A thrombus may develop at only mild to moderate coronary stenosis (narrowing in a blood vessel). The majority of coronary events occur where there is less than 70% stenosis. Occlusive coronary thrombosis plays a role in over 80% of myocardial infarctions and about 95% of sudden death victims. Fig. 5.8 depicts the current understanding of the dynamics of the plaque's stability and formation of a thrombus. When a thrombus occupies over 75% of the lumen of an artery, blood flow to the tissue supplied is reduced enough to cause symptoms because of decreased O_2 (hypoxia) and accumulation of metabolic products such as lactic acid. More than 90% obstruction can result in **anoxia**, the complete deprivation of oxygen, and infarction, a mode of cell death.

The primary function of smooth muscle fibers in the tunica media is to regulate vasodilation and vasoconstriction. Endothelial cell dysfunction or loss due to vascular injury stimulates smooth muscle cell recruitment and proliferation, associated with matrix synthesis, thus causing thickening of the tunica intima (Fig. 5.8). Unlike smooth muscle fibers in the tunica media, the new smooth muscle fibers, found as part of the tunica intima, are motile and

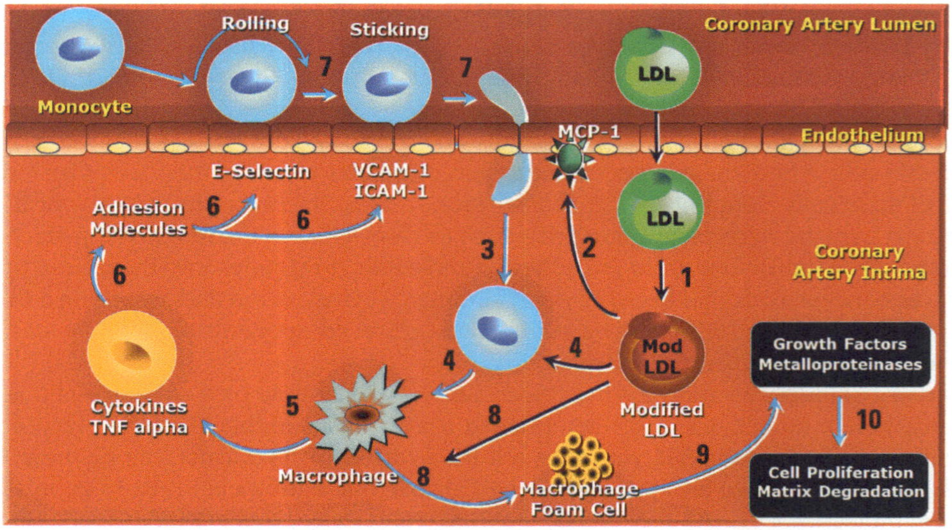

FIG. 5.7 Plaque development in atherosclerosis [6,10,11]. (1) LDL enters the intima and is modified via oxidation. (2) Modified LDL activates endothelial cells to express monocyte chemotactic protein-1 (MCP-1). (3) MCP-1 attracts monocytes from the vessel lumen and into the subendothelial space (intima). (4) Differentiation of monocytes into macrophages. (5) Macrophages release a variety of chemicals, including cytokines, tumor necrosis factor-α (TNF-α), and interleukin-1 (IL-1). (6) Cytokines activate endothelial cells to express adhesion molecules. (7) The adhesion molecules bind monocytes, making them available for recruitment into the subendothelial space by MCP-1. (8) Macrophages take up the modified LDL through their scavenger receptors, accumulate the lipid, and are converted into the lipid-rich foam cells that are the hallmark of atherosclerosis. Once the macrophage has taken up lipid and has been converted into a foam cell, all the cell types necessary for the inflammatory process that leads to atherosclerosis are present. (9) Macrophages and foam cells produce cytokines and a variety of growth factors that lead to cell proliferation (enlargement of the fatty streak), matrix production, and the release of metalloproteinases that cause matrix degeneration. (10) Thus, macrophages and foam cells both contribute to lesion growth and may contribute to instability and thrombotic events.

undergo cell division, acquiring the ability to produce biological agents. These new muscle fibers respond to growth factors and cytokines derived from platelets, endothelial cells, and macrophages. Over time, with the restoration of the endothelial layer, intimal smooth muscle fibers could return to the nonproliferative state. Smooth muscle cell proliferation and extracellular matrix deposition in the tunica intima contribute to the progressive growth of atherosclerotic lesions and convert a fatty streak into a mature **atheroma** (Fig. 5.8). The thickening of the tunica intima even after the healing of the vascular wall could impede blood flow. These growth factors also stimulate smooth muscle fibers to synthesize extracellular matrix, which helps stabilize atherosclerotic plaques. Thus, many of the atherosclerosis lesions are attributable to the chronic inflammatory reaction in the vessel wall. Activated inflammatory cells in an atheroma, on the other hand, may increase the breakdown of extracellular matrix components resulting in **unstable plaques**.

Plaque erosion or rupture is typically followed by partial or complete thrombosis resulting in acute tissue infarction such as myocardial or cerebral infarction. Plaques may rupture when they cannot withstand mechanical stresses. Both intrinsic factors such as plaque

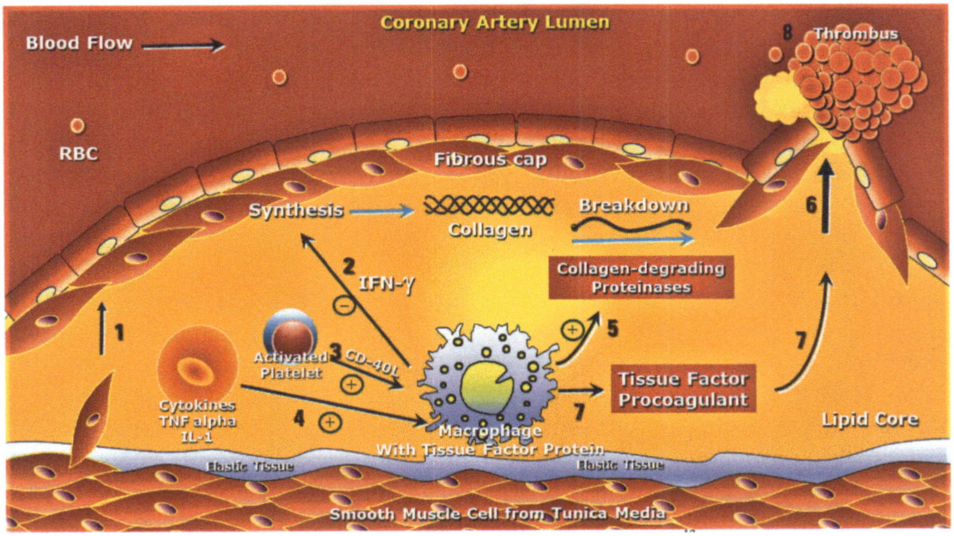

FIG. 5.8 Breakdown of the fibrous cap and thrombosis in the coronary arteries [12]. (1) Plaque accumulation in the intima sends a signal to the smooth muscle cells in the tunica media, causing the smooth muscle cells to multiply and move from the tunica media to the plaque's surface near the endothelial layer of the intima. (2) Inflammatory cells (macrophage with a protein called tissue factor) can send molecular messages (interferon-gamma) to the smooth muscle cells to inhibit the synthesis of new collagen from strengthening the plaque's fibrous cap. (3) At the same time, CD40 ligand (signal triggering molecule) on activated platelets triggers an inflammatory reaction of the macrophage. (4) Due to the buildup of plaque, cytokines (e.g., TNF-α, IL-1) are released by the lipids of the plaque and act on the macrophage. (5) Due to CD-40L and the cytokines, the macrophage releases proteolytic enzymes that degrade collagen and other structurally essential molecules of the plaque's fibrous cap. (6) As a result, when there is inflammation in the intima, the collagen responsible for the plaque's fibrous cap's integrity is under double attack, subject to both decreased synthesis and increased degradation. This sets the stage for plaque rupture through the fibrous cap into the lumen of the coronary artery. (7) Macrophage releases tissue factor, which is a potent procoagulant (associated with blood clotting). (8) After the rupture of plaque through the fibrous cap, tissue factor turns on a clotting cascade to form a thrombus (blood clot inside coronary artery). *RBC*, red blood cells; +, upregulated; −, downregulated.

composition and outside elements such as blood pressure, platelet reactivity, and blood vessel spasm could contribute to the rupture of a plaque. Plaques that contain large number of foam cells and lipids, and those in which the fibrous caps are thin or contain fewer smooth muscle fibers or have high number of inflammatory cells, are more likely to rupture. The fibrous cap undergoes constant remodeling which can either stabilize the plaque or render it more susceptible to rupture. Collagen is the major structural component of the fibrous cap and accounts for its mechanical strength and stability. Thus, the balance of collagen production and degradation affects the integrity of the fibrous cap. Collagen in atherosclerotic plaque is produced by smooth muscle cells, so that loss of these cells results in a less sturdy cap. Moreover, collagen turnover is controlled by enzymes elaborated mainly by macrophages and smooth muscle cells within the atheromatous plaque. Typically, plaque inflammation is associated with a net increase in collagen degradation and reduced collagen synthesis, thereby destabilizing the mechanical integrity of the fibrous cap. Inflammation induced by cholesterol deposits themselves may contribute to plaque destabilization.

A plaque can suddenly rupture, causing microscopic tears in the arterial wall, causing blood to leak into the tissue. Platelets are attracted to the site of injury and begin to form a blood clot. This clot can further obstruct the artery. If it occurs in a coronary or cerebral artery, it could cause a sudden heart attack or stroke. Alternatively, a plaque can break off and travel through the bloodstream as an **embolus** until it blocks a more distant, smaller artery. Even without total blockage, vessel narrowing can cause **ischemia** in the tissue "downstream" of the narrowed vessel. Ischemia, in turn, may result in **hypoxia**. Hypoxia involving the cardiac muscle or brain tissue can lead to cell death and severe brain or heart function impairment.

Sympathetic activation and the release of E and NE can increase systemic blood pressure or induce local vasoconstriction, increasing physical stress on a given plaque. Activation of the sympathetic system associated with wakening and rising can cause blood pressure spikes that have been linked to the apparent circadian periodicity for the onset of acute myocardial infarction peaking between 6 a.m. and noon. Similarly, intense emotional stress can also contribute to plaque disruption. Not all plaque ruptures result in occlusive thrombosis with catastrophic consequences. Plaque disruption and ensuing platelet aggregation and thrombosis are probably common, repetitive, and often clinically silent complications of atheroma. Healing of these subclinical plaque disruptions and resorption of their overlying thrombi is an essential mechanism in the growth of atherosclerotic lesions. Vasoconstriction at the sites of atheroma may also be stimulated by locally released platelet contents and endothelial cell dysfunction with the impaired secretion of endothelial-derived relaxing factors relative to contracting factors and mediators released from perivascular inflammatory cells.

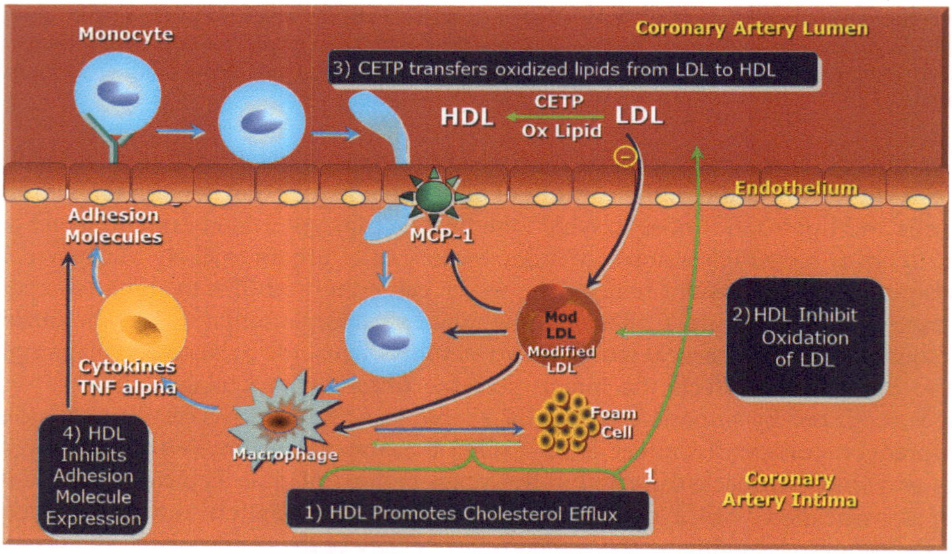

FIG. 5.9 Role of HDL in preventing plaque formation [13–15].

5.6 The role of high-density lipoprotein in the development of atherosclerosis

HDL is also called the "good cholesterol" due to its role in preventing atherogenesis (Fig. 5.9) through a variety of mechanisms. One of the best-known functions of HDL is the promotion of cholesterol efflux (moving out) from foam cells. The efflux of cholesterol from foam cells leads to a reduction in foam cell formation. Although macrophages may accumulate some lipids, they are not converted into foam cells in the presence of HDL [13]. As a result, the inflammatory process is prevented to a certain extent. One of the best-known mechanisms is the ability of HDL to inhibit the oxidation of LDL.

The importance of **paraoxonase** is now reasonably well established. Paraoxonase is transported in plasma as a component of HDL. It has been shown in vitro that paraoxonase inhibits the oxidative modification of LDL [16]. The working hypothesis of many groups is that paraoxonase in HDL may account for at least a proportion of the antioxidant properties of these lipoproteins, thus contributing to HDL's antiinflammatory effect. HDL has an antioxidant and antiinflammatory effect beyond transporting paraoxonase. **Cholesteryl ester transfer protein (CETP)** transfers oxidized lipids from LDL to HDL [14]. Once these oxidized lipids are in HDL, they are reduced by HDL apolipoproteins. It has also been shown that the liver takes up reduced lipids from HDL more rapidly than from LDL. Therefore, HDL reduces the oxidized lipids in LDL, a further antiinflammatory effect of HDL. The cytokine-mediated expression of adhesion molecules in endothelial cells is inhibited by HDL, which helps block a very early inflammatory stage in the development of atherosclerosis [14].

Treatment for atherosclerosis is mainly in the prevention or postponement of the development of the condition. It includes lifestyle changes, such as weight loss, regular exercise, and a diet low in sodium and saturated fats. Medications to lower cholesterol and blood pressure may be prescribed. Surgery is warranted if blockade of a specific vessel is diagnosed. In a procedure such as angioplasty, a catheter is inserted into the vessel at the point of occlusion, and a second catheter with a balloon-like tip is inflated to widen the opening. To prevent a subsequent collapse of the vessel, a small mesh called a **stent** is often inserted. In an **endarterectomy**, plaque is surgically removed from the walls of a vessel.

5.7 C-reactive protein (CRP), a biomarker of inflammation

Over the last decade, research has begun to unravel the association of the inflammatory cascade to the atherosclerotic process. Studies have examined various aspects of the immune system, including adhesion molecules (ICAM-1 and VCAM-1), E-selectin, and P-selectin, cytokines (interleukin-6, tumor necrosis factor-alpha), and other acute-phase reactants such as fibrinogen and serum amyloid A. CRP has emerged as the best clinical tool for cardiovascular risk detection of these inflammatory biomarkers.

CRP is a protein found in the blood, the levels of which rise in response to inflammation. The liver synthesizes it in response to factors released by adipocytes (Fig. 5.10). Elevated CRP correlates with an increased risk for cardiac events and mortality. It is measured in the blood

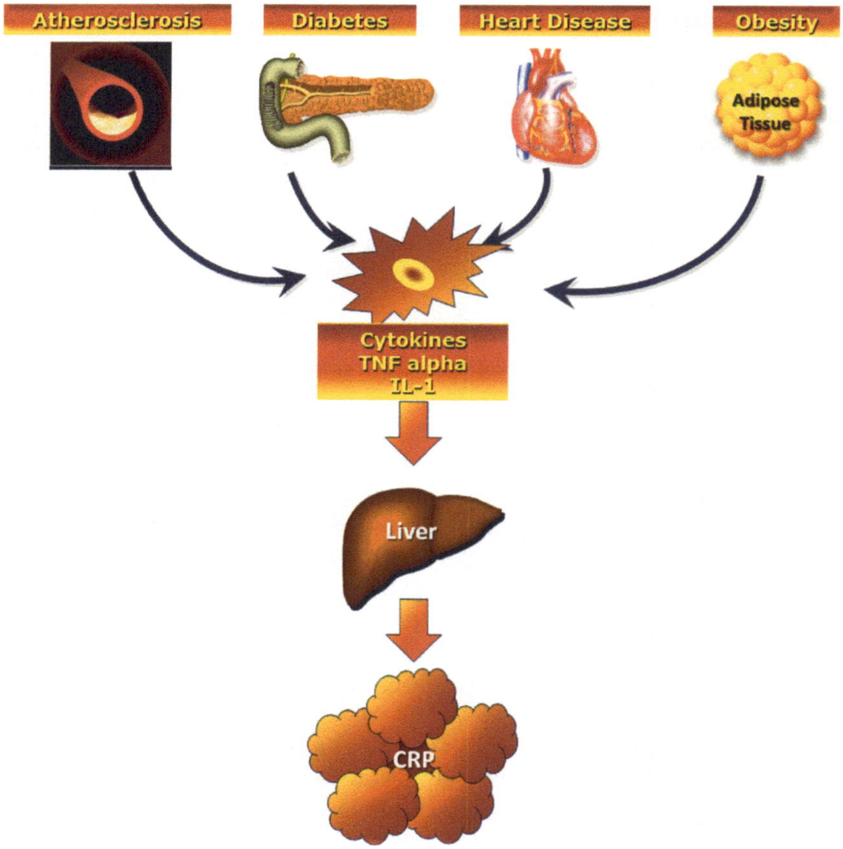

FIG. 5.10 Synthesis of C-reactive protein (CRP) [17]. Chronic inflammation caused by cardiovascular and metabolic diseases produces cytokines that are released into the blood. Cytokines will be taken up by the liver. In response to elevated inflammatory cytokines, the liver will synthesize and release CRP into the blood. Thus, CRP is an excellent marker of inflammation. Elevated CRP correlates with increased risk of cardiac events and mortality.

as highly sensitive-CRP (Hs-CRP) for better detection. Hs-CRP values range from 0.3 to 8.6 mg/L in healthy men and 0.2 to 9.1 mg/L in healthy women, not hormone replacement therapy. CRP proteins are defined as low risk: <1.0 mg/L, average risk: 1–3 mg/L, high risk: >3 mg/L.

It is common for individuals to have more than one uncontrolled cardiovascular disease risk factor (BMI >25 kg/m^2, current smoker, HDL <50, TG >150, glucose >110, blood pressure >130/85, LDL >70) and that the more risk factors an individual have, the higher the CRP levels. In contrast, the fewer the number of uncontrolled risk factors present, the lower the CRP level suggesting that if the number of uncontrolled risk factors can be reduced, the CRP level may also be reduced, thus emphasizing the potential importance of lifestyle modifications. Medications such as aspirin, statins, and thiazolidinediones tend to decrease CRP.

5.8 Aneurysms and dissection

An **aneurysm** is a localized abnormal enlargement of an artery. It may be congenital or acquired. **A true aneurysm** is when an aneurysm involves an intact arterial wall. In contrast, a **false aneurysm** or **pseudoaneurysm** is a defect in the vessel wall leading to an extravascular hematoma (a collection of blood due to a break in the wall), such as a leak at the sutured junction of a vascular graft. An **arterial dissection** is when blood enters a defect in the arterial wall and tunnels between its layers. Dissections are often but not always aneurysmal. Aneurysms and dissections may remain silent and rupture suddenly, often with catastrophic consequences.

There are different types of aneurysms based on their shapes and sizes. **Saccular** or **berry** aneurysms are round outpouching involving only a portion of the vessel wall and are most common in the brain. **Fusiform** aneurysms are circumferential dilations involving the entire artery and are often seen in a long vessel; they vary in diameter and length and affect extensive parts of the aortic arch, abdominal aorta, or iliac arteries (Fig. 5.11).

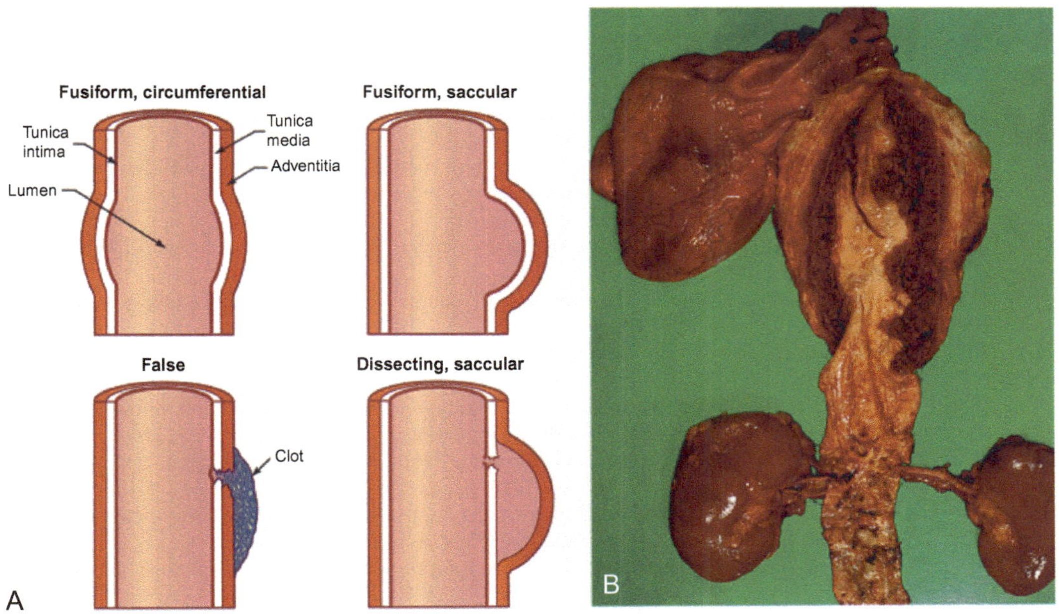

FIG. 5.11 Longitudinal sections of arteries showing different types of aneurysms or dissection. (A) Fusiform circumferential and fusiform saccular aneurysms are true aneurysms, caused by the weakening of the vessel wall. False and saccular aneurysms involve a break in the vessel wall, usually caused by trauma. (B) Dissecting aneurysm of the thoracic aorta. *(B) From I. Damjanov, J. Linder (Eds.), Anderson's pathology, 10th ed., St Louis, Mosby, 1996.*

5.9 Pathogenesis of aneurysms

Arterial walls constantly remodel by synthesizing, degrading, and repairing damage to their extracellular matrix constituents. Aneurysms may occur when the structure or function of the connective tissue within the vessel wall is compromised. Atherosclerosis and hypertension are the most common causes of an aneurysm. Other factors that weaken vessel walls and lead to aneurysms include trauma, vasculitis (inflammation of a blood vessel), congenital defects, and infections. In **Marfan syndrome**, a genetic abnormality, defective synthesis of the scaffolding protein fibrillin leads to weakening elastic tissue, leading to significant dilation in the aorta. Such aneurysms can rupture easily. Another example is the inflammation process, where the associated proteases alter the balance of collagen degradation and synthesis. Loss of smooth muscle cells or the synthesis of noncollagenous or nonelastic extracellular matrix in the place of healthy smooth muscle fibers could also weaken the vessel wall and lead to aneurysms.

References

[1] B.M. Leon, T.M. Maddox, Diabetes and cardiovascular disease: epidemiology, biological mechanisms, treatment recommendations and future research, World J. Diabetes 6 (13) (2015) 1246.

[2] S.S. Virani, A. Alonso, E.J. Benjamin, et al., Heart disease and stroke statistics-2020 update: a report from the American Heart Association, Circulation 141 (9) (2020) e139–e596, https://doi.org/10.1161/CIR.0000000000000757.

[3] Late complications of atherosclerosis.PNG. Wikimedia Commons, the free media repository. Updated September 24, 2020. Accessed March 1, 2021. https://commons.wikimedia.org/w/index.php?title=File:Late_complications_of_atherosclerosis.PNG&oldid=469694743.

[4] 2113ab Atherosclerosis.jpg. Wikimedia Commons, the free media repository. Updated October 21, 2020. Accessed March 1, 2021. https://commons.wikimedia.org/w/index.php?title=File:2113ab_Atherosclerosis.jpg&oldid=496698980.

[5] S. Gordon, P.R. Taylor, Monocyte and macrophage heterogeneity, Nat. Rev. Immunol. 5 (12) (2005) 953–964, https://doi.org/10.1038/nri1733.

[6] D. Steinberg, S. Parthasarathy, T.E. Carew, J.C. Khoo, J.L. Witztum, Beyond cholesterol, N. Engl. J. Med. 320 (14) (1989) 915–924.

[7] H. Itabe, T. Obama, R. Kato, The dynamics of oxidized LDL during atherogenesis, J. Lipids 2011 (2011) 418313, https://doi.org/10.1155/2011/418313.

[8] M. Navab, S.S. Imes, S.Y. Hama, et al., Monocyte transmigration induced by modification of low density lipoprotein in cocultures of human aortic wall cells is due to induction of monocyte chemotactic protein 1 synthesis and is abolished by high density lipoprotein, J. Clin. Invest. 88 (6) (1991) 2039–2046, https://doi.org/10.1172/jci115532.

[9] J. Mestas, K. Ley, Monocyte-endothelial cell interactions in the development of atherosclerosis, Trends Cardiovasc. Med. 18 (6) (2008) 228–232, https://doi.org/10.1016/j.tcm.2008.11.004.

[10] R. Ross, Atherosclerosis- -an inflammatory disease. Research support, U.S. Gov't, P.H.S. review, N. Engl. J. Med. 340 (2) (1999) 115–126, https://doi.org/10.1056/NEJM199901143400207.

[11] C.F. Nathan, Secretory products of macrophages, J. Clin. Invest. 79 (2) (1987) 319–326, https://doi.org/10.1172/jci112815.

[12] P. Libby, Molecular bases of the acute coronary syndromes, Circulation 91 (11) (1995) 2844–2850, https://doi.org/10.1161/01.cir.91.11.2844.

[13] A. Miyazaki, A.T.M.A. Rahim, T. Ohta, Y. Morino, S. Horiuchi, High density lipoprotein mediates selective reduction in cholesteryl esters from macrophage foam cells, Biochim. Biophys. Acta Lipids Lipid Metab. 1126 (1) (1992) 73–80, https://doi.org/10.1016/0005-2760(92)90219-L.

[14] G.W. Cockerill, K.A. Rye, J.R. Gamble, M.A. Vadas, P.J. Barter, High-density lipoproteins inhibit cytokine-induced expression of endothelial cell adhesion molecules. Research support, non-U.S. Gov't, Arterioscler. Thromb. Vasc. Biol. 15 (11) (1995) 1987–1994, https://doi.org/10.1161/01.atv.15.11.1987.
[15] J. Fan, T. Shimokama, S. Haraoka, O. Tokunaga, T. Watanabe, Monocyte-endothelial cell interactions in vitro, with reference to the influence of interleukin-1 and tumor necrosis factor, Biol. Cell. 79 (1) (1993) 17–26.
[16] M.I. Mackness, C. Abbott, S. Arrol, P.N. Durrington, The role of high-density lipoprotein and lipid-soluble antioxidant vitamins in inhibiting low-density lipoprotein oxidation, Biochem J. 294 (Pt 3) (1993) 829–834, https://doi.org/10.1042/bj2940829.
[17] G.M. Sorribes, Osteomyletis: A Review of a Diagnosis Based on CRP, Universitat de Barcelona, 2015.

CHAPTER

6

Coronary artery disease

OUTLINE

OBJECTIVES

- Describe coronary circulation.
- Identify any challenges associated with coronary circulation.
- Summarize the autoregulation of coronary circulation.
- Identify the risk factors for coronary artery disease.
- Explain the pathogenesis of coronary artery disease.

Biology of Cardiovascular and Metabolic Diseases
https://doi.org/10.1016/B978-0-12-823421-1.00006-8

- Define coronary ischemia, angina pectoris, and myocardial infarction.
- Compare ST-elevated myocardial infarction and subendocardial myocardial infarction.
- Describe the regulation of apoptosis of cardiomyocytes.
- Define cardiogenic shock. Identify the two forms of cardiogenic shock.
- Justify the remodeling process after MI.
- Identify the most common approaches to help patients with coronary artery disease.

6.1 Introduction

Heart disease is the leading cause of death in both men and women in the United States. More than 600,000 Americans die of heart disease every year, which accounts for approximately one in four deaths. The most common type of heart disease is **coronary artery disease** (CAD), the blockage of arteries that supply blood to the heart. Over 370,000 people die of this condition every year in this country alone.

6.2 Coronary arteries

Coronary arteries are the first branches of the aorta that supply oxygenated blood to the various parts of the heart. Based on the position of these branches, they are named **right** and **left coronary arteries** (Fig. 6.1). The right coronary artery supplies blood to the right atrium, portions of both right and left ventricles, and the conduction system of the heart. The left coronary artery carries blood to the left atrium and left ventricle, and interventricular septum.

The right coronary artery gives off branches to one or more **marginal arteries** below the right atrium to supply blood to the superficial portions of the right ventricle. The right coronary artery also gives rise to the **posterior interventricular artery** or **posterior descending artery** on the posterior surface of the heart and branches off further to supply the interventricular septum and portions of both ventricles. **The Circumflex artery**, a branch of the left coronary artery, fuses with the small branches of the right coronary artery. **The anterior interventricular artery** or **left anterior descending artery** (LAD) is the second major branch of the left coronary artery.

6.3 Coronary veins

Coronary veins return deoxygenated blood back to the heart. Smaller veins fuse to become a large, thin-walled vein, **coronary sinus**, on the posterior surface of the heart emptying directly into the right atrium. The veins are typically situated parallel to the arteries (see Fig. 6.1). The **great cardiac vein** receives several major branches, including the **posterior cardiac vein**, the **middle cardiac vein**, and the **small cardiac vein**. The posterior cardiac vein receives blood from the regions supplied by the marginal artery. The middle cardiac vein collects blood from the areas supplied by the posterior interventricular artery. The small cardiac

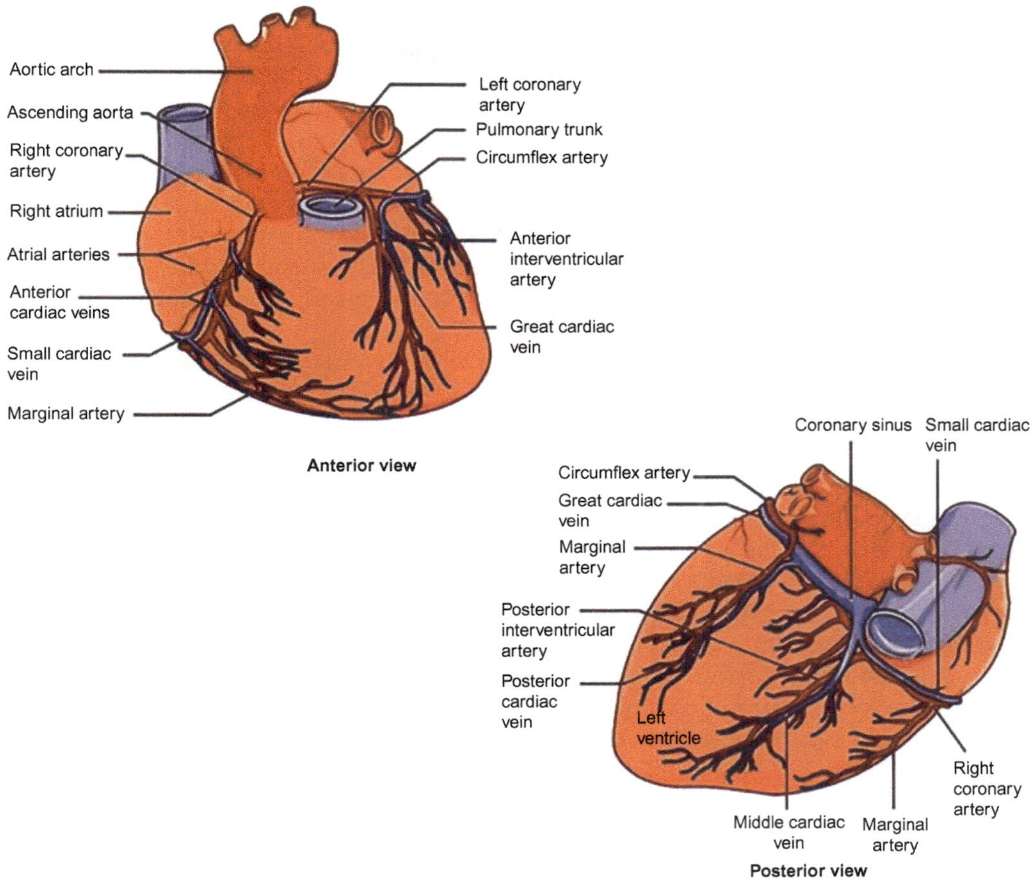

FIG. 6.1 Coronary circulation. The anterior and posterior views of the heart show prominent coronary surface vessels.

vein drains the blood from the posterior surfaces of the right atrium and ventricle. The **anterior cardiac veins** drain the anterior surface of the right ventricle. The great cardiac vein joins the coronary sinus on the posterior surface.

6.4 Phasic variations in coronary blood flow

Coronary blood flow exhibits phasic variations. During systole, as the myocardial tissue pressure rises, the arteries penetrating the myocardium are compressed and momentarily shut off. Thus the left coronary artery flow is sharply reduced and even flows backward for a very short interval. At the end of systole, the coronary flow rises to a maximum and then declines slowly during diastole. About 75% of the total coronary flow occurs during diastole for this reason. Because the branches of the coronary arteries traverse the heart from the epicardium to the endocardium, the vessels supplying the subendocardial portion of the left ventricle are most likely to be compressed by the myocardium during systole. Since the coronary arteries are perfused predominantly during diastole, the flow in these penetrating arteries is determined by the pressure gradient between the main epicardial arteries (aortic diastolic pressure) and the pressure in the subendocardium (left ventricular diastolic pressure), as well as by the duration of diastole. Therefore, when aortic end-diastolic pressure is decreased or left ventricular diastolic pressure increases, perfusion through the myocardium is diminished and this is especially true in patients with advanced CAD. Consequently, the subendocardial region of the left ventricle is more prone to ischemic damage and is the most common site for myocardial infarction.

During systole, the high pressure is transmitted to the myocardial tissues, which then compress the coronary vessels. This compression is more marked in the vessels near the ventricular space (the subendocardial vessels). The subepicardial vessels are partially protected from compression because of their location and the ventricular muscle thickness. Therefore, flow through these vessels is partially maintained during systole.

Another factor contributing to the increased incidence of ischemia in the subendocardial regions of the left ventricle is the earlier exhaustion of autoregulation in the subendocardial vessels. When coronary blood pressure is decreased, the subendocardial vessels dilate. This vasodilation leads to a further fall in **perfusion pressure** which causes reduced flow in the subendocardial muscle. In contrast to flow in the left ventricle, the flow in the right coronary artery is not disturbed during systole. The pressure is low in the right ventricular cavity. Therefore, flow occurs during diastole as well as during systole.

The ratio of blood capillaries to myocardial fibers is 1:1, and this arrangement is also true of skeletal muscle fibers. When cardiac muscle hypertrophies, the number of muscle fibers will not change, but the muscle fiber size increases. During cardiac muscle hypertrophy, the ratio of blood capillary to the muscle fiber remains the same, unlike in the skeletal muscle fiber, where the number of capillaries increases. Thus, each capillary must supply a greater bulk of muscle, and the distance that the nutrients have to travel to reach the center of the muscle is increased. These two factors limit the amount of cardiac hypertrophy that can occur without the development of ischemia. The myocardium, therefore, is at a severe disadvantage in its blood supply. The additional challenges with coronary circulation include the lack of functional anastomoses between the two major coronary arteries and the O_2 extraction from the coronary circulation is very high, and therefore, the coronary venous blood is less than 25% saturated with O_2. Moreover, coronary arteries are highly vulnerable and among the worst affected by degenerative atherosclerotic changes, which are usually widespread.

6.5 Regulation of coronary circulation

The primary resistance to coronary blood flow occurs at the smaller arterioles, which lie within the myocardium. The response of the precapillary arterioles to the metabolic needs of the myocardium is the most critical factor influencing coronary flow. The larger vessels found on the epicardial surface usually offer very little resistance to coronary flow. Neural stimuli such as sympathetic activation can, however, induce these large vessels to constrict markedly, thereby causing an increase in coronary vascular resistance.

The metabolism of the myocardium is predominantly aerobic. Thus, the coronary flow is directly proportional to the amount of work that the myocardium is performing. The heart extracts a large amount of O_2 from the blood at rest. O_2 consumption can be increased only by raising its flow by vasodilation. Thus, when myocardial metabolism is increased (or coronary flow is diminished), the local concentration of metabolites rises. These metabolites relax the smooth muscles of the precapillary vessels diminishing the coronary vascular resistance with a consequent increase in coronary blood flow. Several factors and substances have been suggested as mediators that adjust the metabolic activity of the myocardium to that of coronary vascular resistance. These include a diminution in the partial pressure of O_2, and increases in CO_2, H^+ or K^+ concentration, lactic acid, prostaglandin, adenine nucleotides, and adenosine.

Activation of the sympathetic system and the release of norepinephrine and epinephrine increase the heart rate and blood pressure and increase myocardial metabolism with a consequent increase in myocardial blood flow. In the normal range, an increase in blood pressure will increase the coronary flow for a short time, after which it returns to its basal value because of the increased tone of the coronary vessels (**autoregulation**). A fall in blood pressure leads to a decrease in O_2 concentration in the myocardium, with the ensuing accumulation of metabolites producing coronary vasodilation. With severe hypotension, however, this vasodilation will be insufficient to restore coronary flow. Thus, myocardial contractions will be impaired leading to a vicious cycle that can end in death, if not treated adequately. The increase in O_2 consumption associated with the increase in heart rate will dilate the coronary arteries, leading to increased coronary blood flow. If the coronary vessels cannot dilate, the increase in heart rate may be deleterious.

6.6 Coronary artery disease (CAD)

CAD is a vascular disorder involving coronary arteries that become clogged due to the buildup of cholesterol and other materials referred to as **plaque** on their inner walls (Fig. 6.2). This process of the buildup of plaque within the blood vessel is called

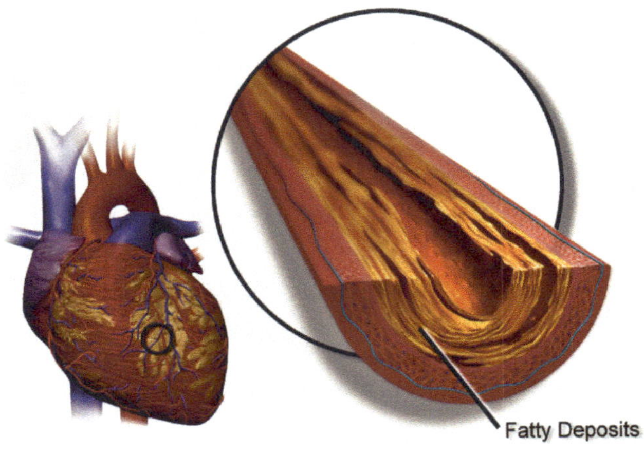

FIG. 6.2 Plaque buildup within the coronary artery [1].

atherosclerosis (see Chapter 5). Besides atherosclerosis, other causes of CAD include increased resistance in the coronary arteries, greater demand for blood supply due to increased ventricular volume, left ventricular hypertrophy, higher heart rate, valvular disease, shock, or anemia.

Coronary arteries are the most vulnerable arteries to develop atherosclerosis. The plaque first grows within the walls of the coronary arteries, which narrows the lumen of the blood vessels and, therefore, reduces blood supply to the tissue downstream. The plaque in the coronary artery brings about an imbalance between the coronary supply of blood and myocardial demand for O_2 and nutrients. Less than normal blood supply to the cardiac muscle is referred to as myocardial **ischemia** and is reversible in the early stages. The narrow vessel may be able to keep up with limited blood supply in the resting conditions but may not maintain blood supply when the demand for O_2 increases in situations such as exercise. A person having limited blood supply to the heart may become breathless, fatigued, and feel chest pain while climbing stairs, for example, because there is not enough blood getting through the narrow vessel. This phenomenon is predictable and reproducible and is referred to as **stable angina** (angina in Greek means "to strangle").

Over time, the plaque covering may rupture and expose fatty tissue and other dead material into the blood. Circulating platelets are attracted to the exposed plaque and are activated by the plaque material, leading to the production of a blood clot over the original plaque (thrombosis). Ischemia may be associated with pain in the left shoulder and left arm and is called **angina pectoris**. Within 10s of coronary occlusion, myocardial ischemia can lead to a hypoxic condition, i.e., less O_2 available to the tissue. At this time, aerobic respiration is replaced by anaerobic pathways and the buildup of lactic acid. Cardiac myocytes are viable for approximately 20min in anaerobic conditions. If the condition is not reversed within this window of time, these cells may die (Fig. 6.3).

If the decreased blood supply to the tissue is not corrected, the condition that follows is **infarction.** Infarction is an irreversible condition. Extensive myocardial infarction eventually will impair the pumping ability of the heart, referred to as **heart failure**.

6.7 Risk factors for CAD

The risk factors for CAD can be grouped as **nonmodifiable** and **modifiable risk factors**. Nonmodifiable risk factors include advanced age, family history, male gender, or women after menopause. Modifiable risk factors include dyslipidemia, hypertension, endothelial injury, smoking, high LDL levels, low HDL levels, insulin resistance, diabetes mellitus, thrombosis, hyperhomocysteinemia (higher than normal levels of homocysteine in the blood), obesity, sedentary lifestyle, and atherogenic diet, although some of these conditions

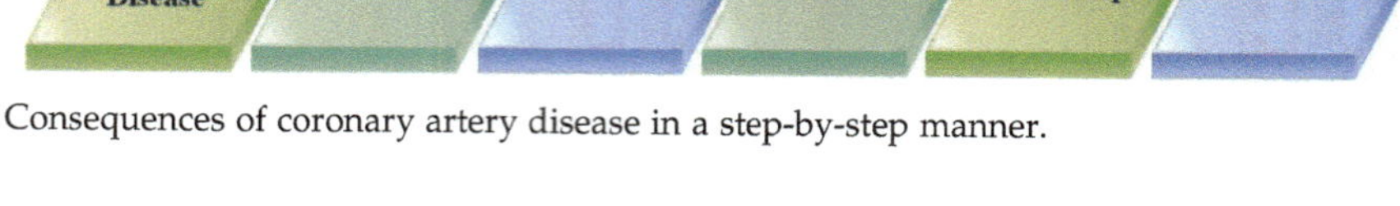

FIG. 6.3 Consequences of coronary artery disease in a step-by-step manner.

could be due to underlying genetic conditions. Several newly identified risk factors include markers of inflammation such as C-reactive protein, infection, periodontal disease, air pollution, and coronary artery calcification.

6.8 Myocardial infarction (MI)

MI is the loss of myocardial tissue caused by irreversible damage to the cardiac muscle fibers due to prolonged ischemia and hypoxia. An MI often occurs when the buildup of plaque occludes a coronary artery depriving of blood supply to cardiac muscle tissue (Fig. 6.4). Another cause of MI is when a portion of an unstable plaque travels through the coronary artery and lodges in one of the smaller branches. Severe stress may induce spasms of the smooth muscle in the vessel walls and limit blood supply to the muscle which could lead to an MI.

An **acute MI** is often due to the sudden occurrence of ischemia causing myocardial injury and subsequent infarction. It occurs most commonly in persons older than 65 years of age. In acute MI, there is sudden pain beneath the sternum (**angina pectoris**), often radiating down the left arm in males, whereas female symptoms vary. Other signs and symptoms may include difficulty breathing, shortness of breath (dyspnea), irregular heartbeat (palpitations), nausea and vomiting, sweating (diaphoresis), anxiety, and fainting (syncope). The American Heart Association estimates that about 40% of those with MI die within 1 year following this event.

An MI is confirmed by examining the patient's ECG, which often reveals alterations in the ST and Q segments. Some MI classification schemes are **ST-elevated MI (STEMI)** or **transmural infarction** and **nonelevated MI (non-STEMI** or **NSTEMI)** or **subendocardial MI**. Individuals with STEMI are at the highest risk for complications and, thus, require immediate intervention. Non-STEMI typically involves smaller infarctions which result in ST-segment depression and inversion of T waves (see Fig. 6.5B).

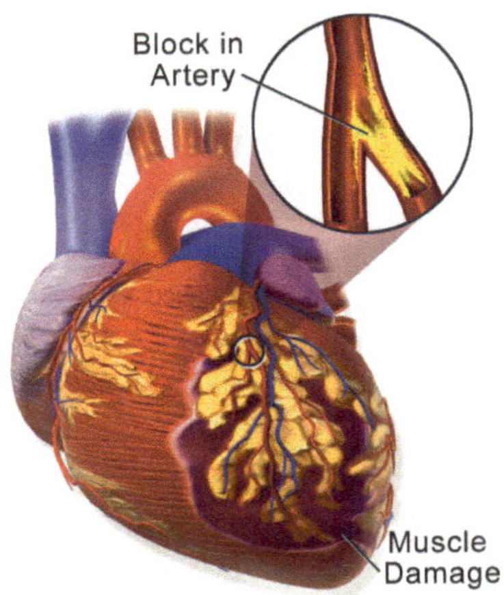

FIG. 6.4 Heart with myocardial infarction [2].

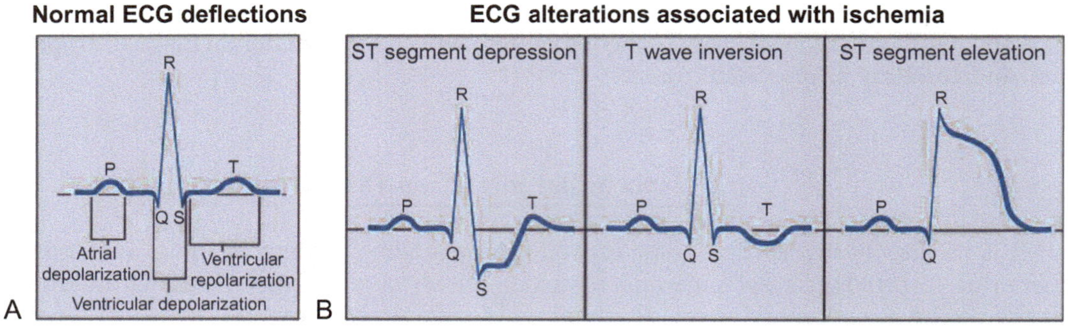

FIG. 6.5 A normal ECG (A) is compared with (B) ECG alterations associated with ischemia.

6.9 Pathogenesis of myocardial infarction

If blood flow is not rapidly restored during myocardial ischemia, cardiomyocyte death (apoptosis) may follow. Apoptosis is a programmed cell death process characterized by cell shrinkage, membrane blebbing (bulging outward), DNA fragmentation, and formation of apoptotic bodies [3,4]. Apoptosis may be initiated by several extrinsic or intrinsic signals, but the ultimate result is the same in most cells. Death receptors that belong to tumor necrosis factor (TNF)-alpha receptor superfamily play a critical role in initiating the process. Activation of TNF-alpha subfamily receptors leads to the formation of a **death-inducing signaling complex (DISC)**, consisting of **caspase 8** and adaptor proteins, such as **Fas-associated protein with death domain (FADD)**. Mitochondria play a significant role by providing energy for apoptosis and releasing important proapoptotic agents into the cytoplasm. They also serve as the primary site of action of the apoptosis regulatory proteins and a major source of reactive oxygen species (ROS) due to endoplasmic reticulum stress, all of which have been implicated in cellular damage and death [3,4]. The final steps of apoptotic death are highly conserved and likely mediated by a similar set of caspases, a family of enzymes that break down cell proteins (Fig. 6.6).

6.10 Cardiogenic shock

Cardiogenic shock is a state in which a weakened heart cannot pump enough blood to meet the body's needs—systolic or diastolic myocardial dysfunction results in a reduction in cardiac output and often pulmonary congestion (fluid in the lungs). Systemic and coronary hypoperfusion occur, resulting in progressive ischemia. Although many compensatory mechanisms are activated to support the circulation, these compensatory mechanisms may become maladaptive and produce a worsening of hemodynamics. After MI, the release of inflammatory cytokines may lead to **inducible nitric oxide** expression, excess nitric oxide, and inappropriate vasodilation, which causes further reduction in systemic and coronary perfusion. A vicious spiral of progressive myocardial dysfunction occurs that ultimately results in death, if it is not interrupted (Fig. 6.7) [8].

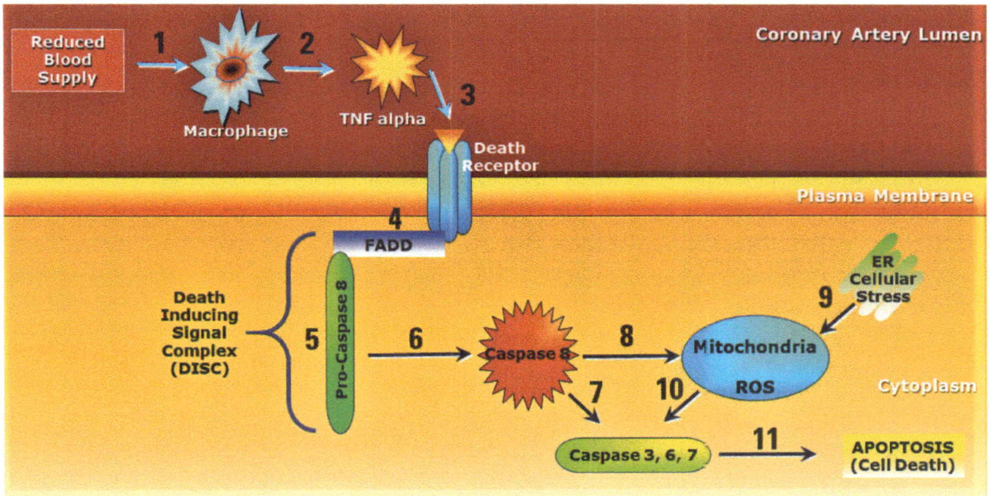

FIG. 6.6 Regulation of cardiomyocyte apoptosis (programmed cell death) [3,4]. Apoptosis pathways include those initiated by death receptor ligation, mitochondrial stress (e.g., oxidants), or endoplasmic reticulum (ER) stress (e.g., depletion of ER Ca^{2+}). (1) Limited blood supply to the myocardium and subsequent injury to the tissue triggers inflammation which increases macrophage recruitment to the heart. (2) Macrophages release tumor necrosis factor (TNF)-alpha. (3) TNF-alpha binds to the transmembrane death receptor. (4) Activation of Fas-associated death domain protein (FADD). (5) Activation of pro-caspase 8. Together, FADD and pro-caspase 8 form a signaling platform termed the death-inducing signaling complex (DISC). (6) Caspase 8 activation. (7) Direct or (8) indirect (via mitochondria) activation of caspases 3, 6, and 7. (9) Endoplasmic reticular (ER) or other types of cellular stress trigger apoptosis. (10) Release of reactive oxygen species (ROS) from the mitochondria as well as caspase 8 activate caspases 3, 6, and 7. (11) Apoptosis is induced by the activation of caspases 3, 6, and 7.

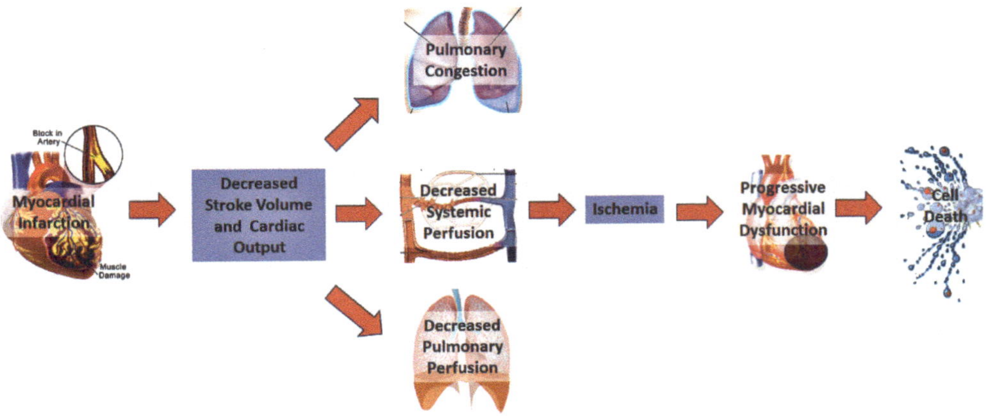

FIG. 6.7 Pathophysiology of cardiogenic shock due to myocardial infarction [4–7]. During cardiogenic shock, cardiac output, a function of both stroke volume and heart rate, is compromised. Decreased cardiac output results in decreased blood pressure and decreased pulmonary and systemic perfusion. The heart muscle itself receives inadequate blood leading to an ischemic state. Because impaired tissue perfusion weakens the heart and impairs its ability to pump blood forward, the ventricle does not fully eject its blood volume at systole. As a result, fluid backs up in the lungs. Over time, the lack of adequate coronary artery perfusion leads to progressive myocardial dysfunction and eventually cell death.

6.11 Cardiac remodeling after myocardial infarction

A series of changes occur in the myocardium soon after MI which is organized into four phases: (1) death of cardiomyocytes, (2) acute inflammation and tissue digestion, (3) formation of new tissue, and (4) scar tissue formation. Death of cardiomyocytes from apoptosis starts approximately 1 h after the onset of myocardial infarction. From about 4h to 2days, acute inflammation persists. During this phase, there is an influx of neutrophils and other inflammatory cells. Macrophages are formed that phagocytize and remove dead cardiomyocytes in the infarcted area. On day 2 and lasting up to 14days, new tissue formation occurs due to the presence of fibroblasts, macrophages, myofibroblasts, new blood vessels, and extracellular proteins. Finally, between 14days and 2 months after MI, scar tissue is formed in the infarcted area (Fig. 6.8). The scar tissue consists of cross-linked, collagen fibers. The time intervals for each phase can vary based on the degree of infarction and the individual's overall health status [9].

6.12 Revascularization of the heart

Clinical procedures may be used to restore myocardial blood flow with the specific intent of symptom relief and improved morbidity and mortality. Over the past decade, cardiac rehabilitation referral patterns for patients with coronary artery revascularization have evolved

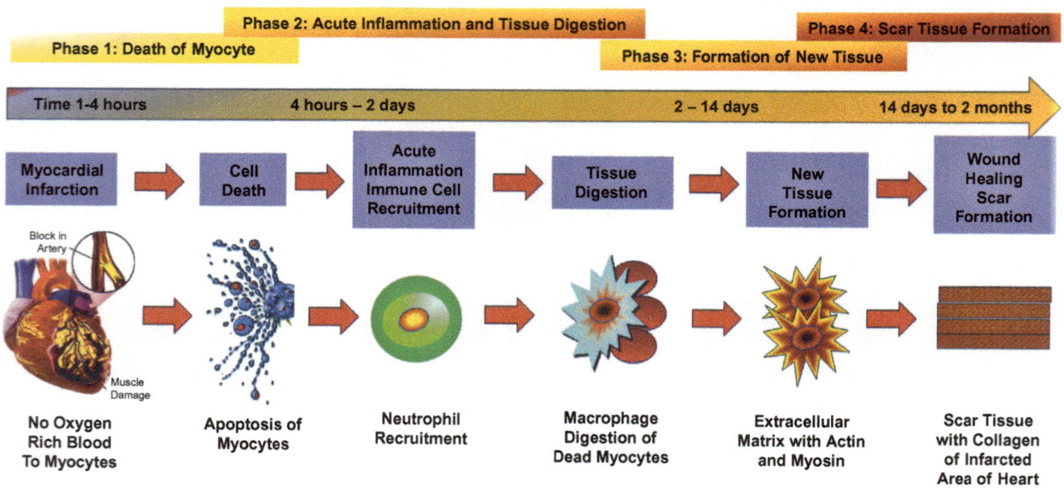

FIG. 6.8 Phases of cardiac healing and remodeling after myocardial infarction [9]. Phase 1: Death of cardiomyocytes starts at ~1h after coronary artery occlusion and can either be the result of apoptosis or necrosis. Phase 2: Between 4h and 2 days after MI, acute inflammation. Inflammatory cells (neutrophils and monocytes) move from the blood into the myocardium. Macrophages phagocytize and remove dead cardiomyocytes in the infarcted area. Phase 3: Inflammation causes granulation tissue formation (collagen-rich tissue) between 2 and 14days after the myocardial infarction. Phase 4: Granulation tissue (fibrous tissue) matures to become scar tissue between 14days and 2 months after MI.

beyond coronary artery bypass surgery. Because of advances in invasive coronary technology, cardiac rehabilitation programs must also prepare for a growing number of patients who have experienced interventions, including **percutaneous transluminal coronary angioplasty** (aka angioplasty; **PTCA**) alone and, more frequently, in combination with **stent** therapy, which involves the placement of a mesh tube along the artery wall to prevent occlusion.

Angioplasty and **stenting:** Angioplasty is a procedure in which the occlusion is mechanically widened with a balloon. PTCA is a nonsurgical procedure that relieves obstruction of the coronary arteries, allowing more blood and O_2 to be delivered to the cardiac muscle. PTCA is now referred to as **percutaneous coronary intervention**, or PCI, as this term includes balloons, stents, and atherectomy devices. A catheter with an expandable tip is inserted into a superficial vessel, usually in the leg, and then directed to the occlusion site. Once the tip is at the point of occlusion, the balloon is inflated to compress the plaque and to widen the vessel to increase blood flow (Fig. 6.9). A stent is the placement of a mesh at the site of occlusion to reinforce the weakened and damaged walls (Fig. 6.10) (watch this YouTube video: http://youtu.be/I45kJJoCa6s). When successful, PCI can relieve chest pain of angina, improve the prognosis of individuals with unstable angina, and minimize or prevent a heart attack without having the patient undergo open heart **coronary artery bypass graft** (**CABG**) surgery.

About 35%–40% of patients who have balloon angioplasty are at risk of more blockages in the treated area called **restenosis**, usually within 6 months after balloon angioplasty. Arteries that have stents can close, as well. Restenosis occurs in about 20% of patients with stents.

CABG is a surgical procedure to redirect blood around a section of a blocked branch of a coronary artery (Fig. 6.11) to restore normal blood flow. A piece of a healthy blood vessel from elsewhere in the body, such as a saphenous vein from a leg or mammary artery in the chest is used. During the procedure, one end of the graft is attached above the blockage, and the other end is sewn below the blockage. Blood bypasses the blockage by going through the new graft. It is the most common type of open-heart surgery in the United States, with more than 500,000 surgeries performed each year [13] (watch this YouTube video: https://youtu.be/aVPayssQbNk).

6.13 Myocardial reperfusion injury

When oxygen-rich blood returns to the area of reperfusion in the heart following myocardial infarction, neutrophils produce reactive oxygen species (ROS), oxygen with unpaired electrons that are highly damaging to cells. ROS, in turn, causes damages to the cardiomyocyte cell membrane by lipid peroxidation, which is the process by which free radicals (molecules with unpaired electrons) "steal" electrons from the lipids in cell membranes, resulting in cell damage to the sarcolemmal membrane. The damage to the sarcolemmal membrane causes the release of Ca^{2+} into the already Ca^{2+}-overloaded cardiomyocyte. In turn, a vicious cycle occurs in which the Ca^{2+} overload causes the sarcoplasmic reticulum to release more Ca^{2+} into the cardiomyocyte. The Ca^{2+}overload combined with the available ATP in the myofibrils of the cardiomyocyte causes hypercontracture (extreme cell shortening of the cardiomyocytes) in the first few minutes of reperfusion which leads to lethal reperfusion injury (death of the cardiomyocyte).

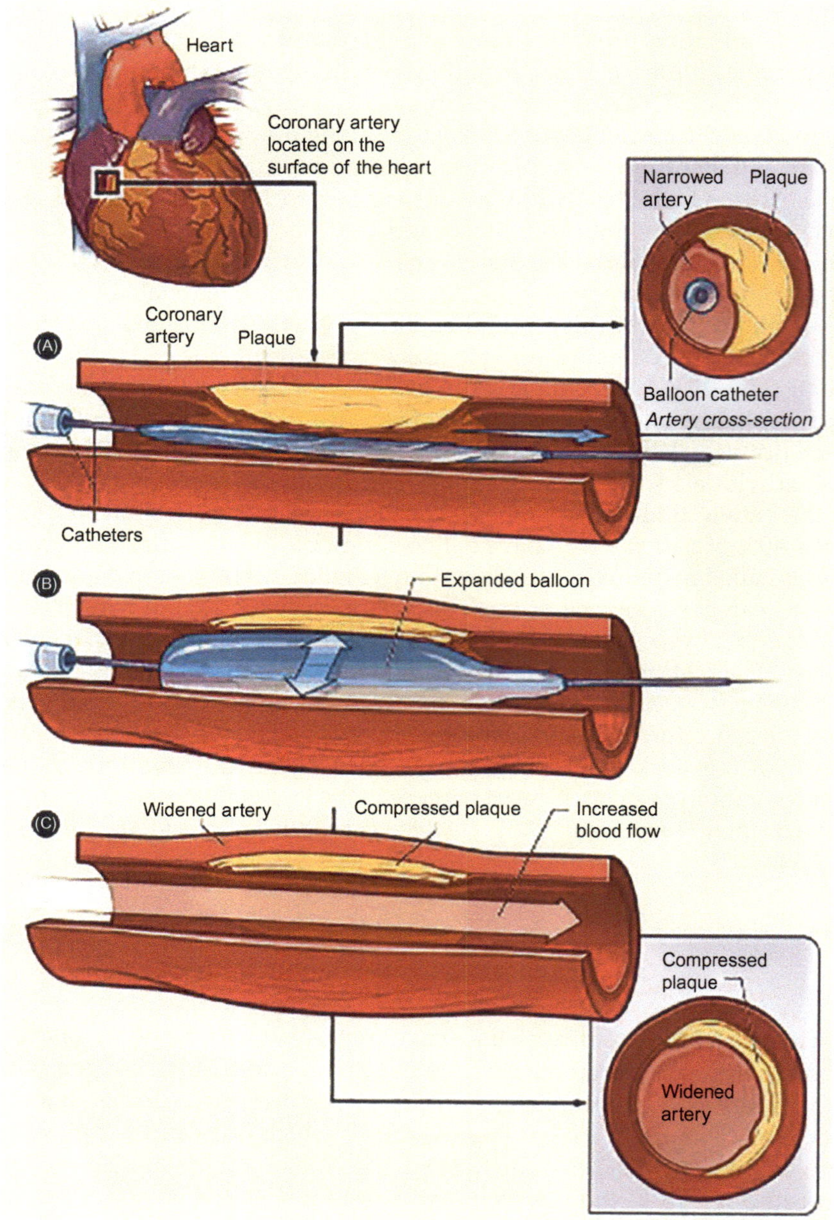

FIG. 6.9 Percutaneous interventions (balloon angioplasty) [10]. The illustration shows a cross-section of a coronary artery with plaque buildup. (A) Insertion of the deflated balloon catheter into the narrowed coronary artery. (B) The balloon is inflated. (C) Compression of the plaque.

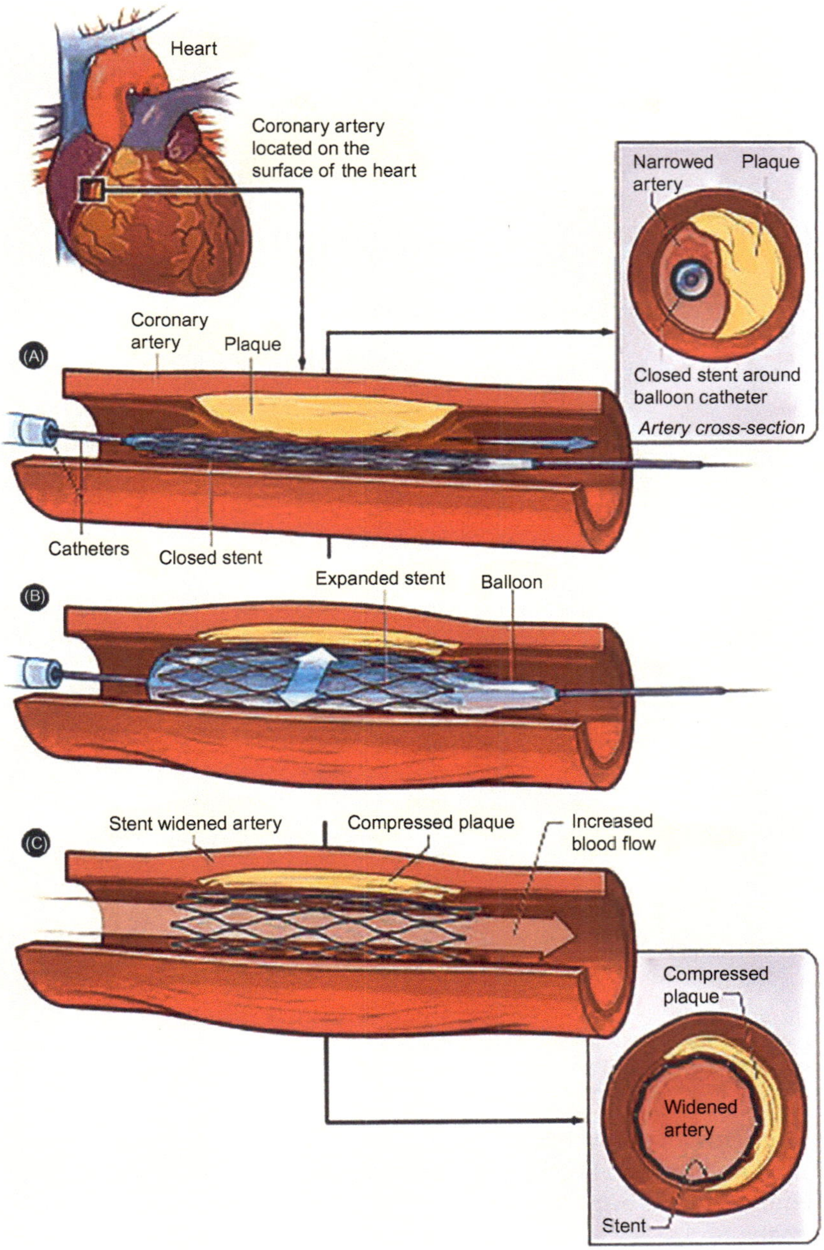

FIG. 6.10 Balloon angioplasty with stent. (A) Insertion of the deflated balloon catheter and closed stent into the narrowed coronary artery. (B) The balloon is inflated, expanding the stent. (C) Compression of the plaque and stent-widened artery. Open source image from the National Institutes of Health, United States.

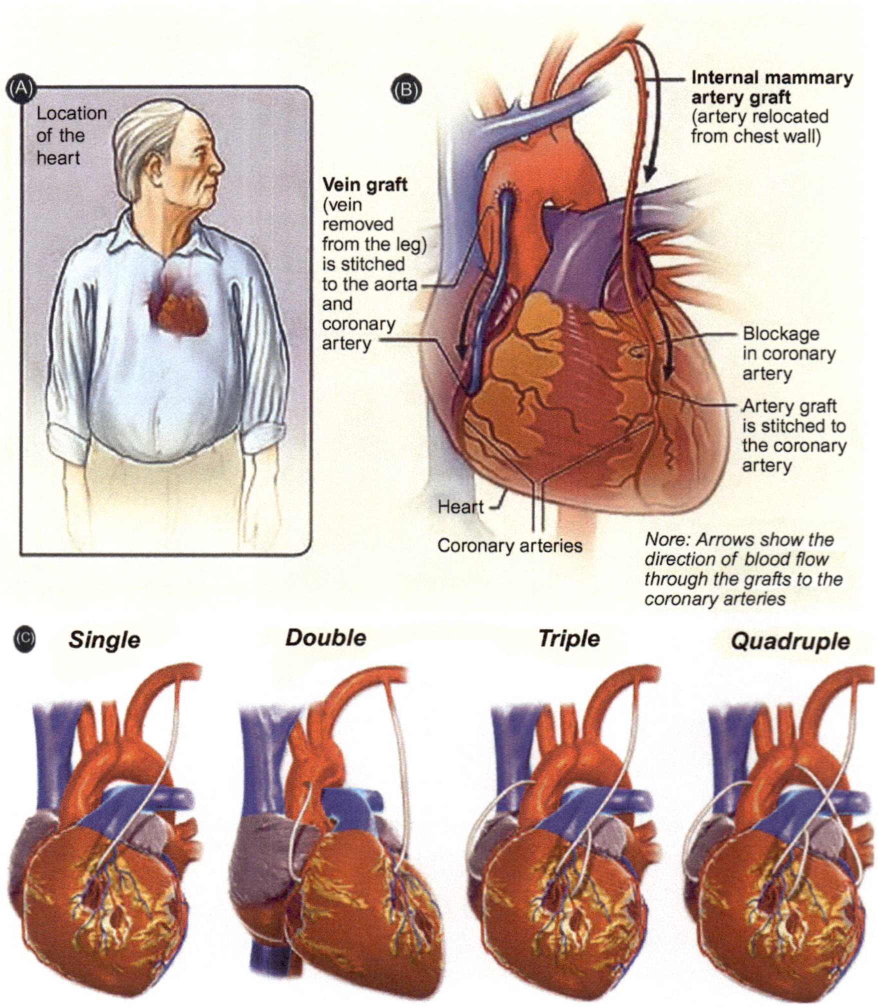

FIG. 6.11 Coronary artery bypass graft (CABG) [11,12]. (A) Location of the heart in the chest cavity. (B) Overview of coronary artery bypass graft. (C) Illustration depicting single, double, triple, and quadruple bypass.

The lethal reperfusion injury results in the death of cardiac myocytes that were viable immediately before myocardial reperfusion. Cell swelling, contracture of myofibrils, disruption of the sarcolemma, and the appearance of intramitochondrial calcium phosphate particles are some of the changes observed during reperfusion injury. Four stages of cardiac dysfunction are observed. The first type is **myocardial stunning**, a term denoting mechanical dysfunction that persists after reperfusion despite the absence of irreversible damage and restoration of normal or near-normal coronary flow. The myocardium usually recovers from this reversible form of injury after several days or weeks. Significant reductions in myocardial infarct size are

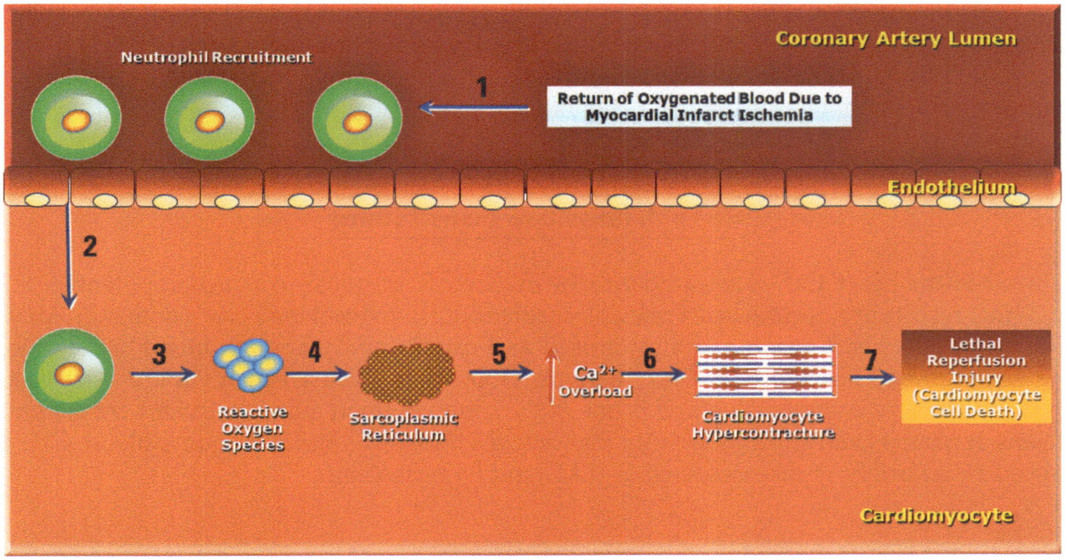

FIG. 6.12 Lethal reperfusion injury after myocardial ischemia [14,15]. (1) Oxygen-rich blood returns to the area of reperfusion in the heart following myocardial infarction. (2) Neutrophils are recruited into the myocardium. (3) Neutrophils produce reactive oxygen species (ROS, oxygen with unpaired electrons that are highly damaging to cells). (4) ROS damages the cardiomyocyte cell membrane by lipid peroxidation, which is the process in which free radicals (molecules with unpaired electrons) "steal" electrons from the lipids in cell membranes, resulting in cell damage to the sarcolemmal membrane. (5) The damage to the sarcolemmal membrane causes the release of Ca^{2+} into the already Ca^{2+}-overloaded cardiomyocyte. In turn, a vicious cycle occurs in which the Ca^{2+} overload causes the sarcoplasmic reticulum to release more Ca^{2+} into the cardiomyocyte. (6) The Ca^{2+} overload combined with the available ATP in the myofibrils of the cardiomyocyte causes hypercontracture (extreme cell shortening of the cardiomyocytes) in the first few minutes of reperfusion. (7) The hypercontracture of the myofibrils leads to lethal reperfusion injury (death of the cardiomyocyte).

obtained by early and successful myocardial reperfusion after a sustained episode of acute myocardial ischemia. Lethal reperfusion injury is a paradoxical form of myocardial injury caused by the restoration of coronary blood flow after an ischemic episode (Fig. 6.12) and must be avoided by the administration of a cardioprotective intervention at the beginning of myocardial reperfusion.

6.14 Exercise and prevention of myocardial injury

Regular exercise is an important lifestyle intervention in preventing and treating cardiovascular disease (CVD) and reperfusion injury and increase survival chances after MI. Exercise-induced beneficial effects on the myocardium are due to the decrease in several risk factors associated with cardiovascular disease (CVD), such as high cholesterol, hypertension, and obesity. Furthermore, it has been demonstrated that exercise can reproduce the "ischemic preconditioning" (IP), which refers to the capacity of short periods of ischemia to render the myocardium more resistant to subsequent ischemic insult and to limit infarct size during prolonged ischemia. A healthy lifestyle that incorporates good nutrition, weight management, and physical activity can play a significant role in avoiding CVD. Aerobic exercise programs are implemented for cardiac patients.

6.15 Congestive heart failure

In a healthy heart, Ca^{2+} is released into the myocyte in response to an action potential (Fig. 6.13) that results in a normal heart rate and cardiac output to meet the body's demands. However, when the heart becomes compromised and cannot pump enough blood, a condition called congestive heart failure (CHF) may develop. CHF is caused by a variety of diseases such as CAD and hypertension and is one of the most common ailments treated by the physicians. It results from any abnormality that reduces the heart's ability to pump blood. CHF is caused by a variety of diseases such as CAD and hypertension.

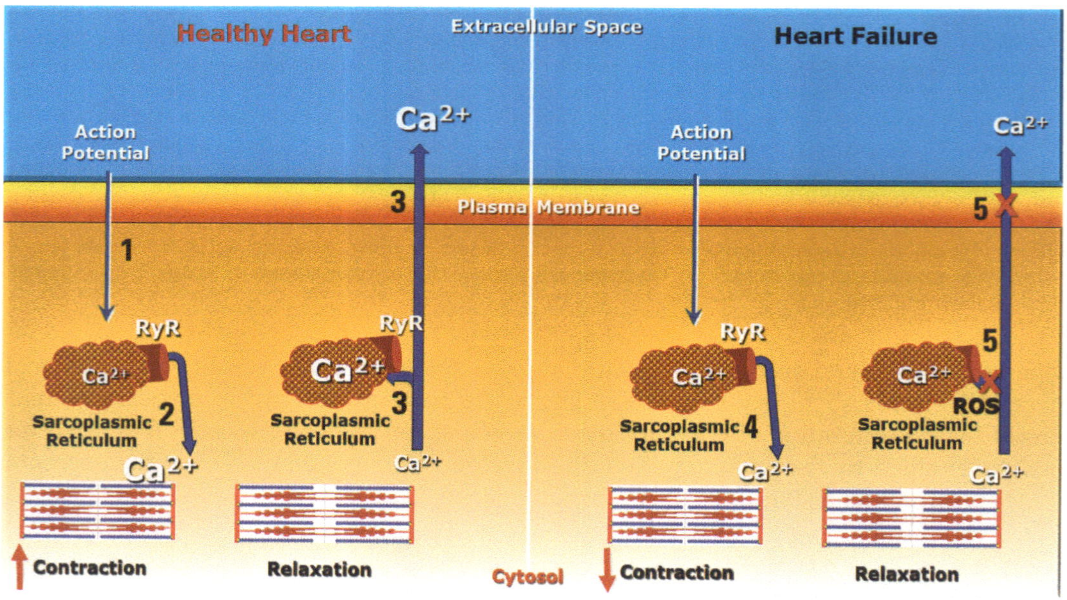

FIG. 6.13 Major contributors to an abnormal contraction in heart failure [9]. (1) Action potential activates Ryanodine receptors (RyR) on the sarcoplasmic reticulum to release Ca^{2+} into the cytosol. (2) Ca^{2+} initiates contraction within the myocardium leading to forceful ejection of blood from the heart. (3) During relaxation of the heart, Ca^{2+} is taken back into the sarcoplasmic reticulum and released into the extracellular fluid. (4) In a failing heart, Ca^{2+} released from the sarcoplasmic reticulum leads to contraction of the heart but rather weak. Therefore, less than normal amount of blood is ejected from the heart. (5) During relaxation, Ca^{2+} remains in the cytosol, with less taken up into the sarcoplasmic reticulum due to damaged ryanodine receptors (RyR) from the action of reactive oxygen species (ROS). Less Ca^{2+} exits the cytosol due to damaged channels in the plasma membrane. As a result, steps 4 and 5 continue in a vicious cycle leading to weak contractions of the heart.

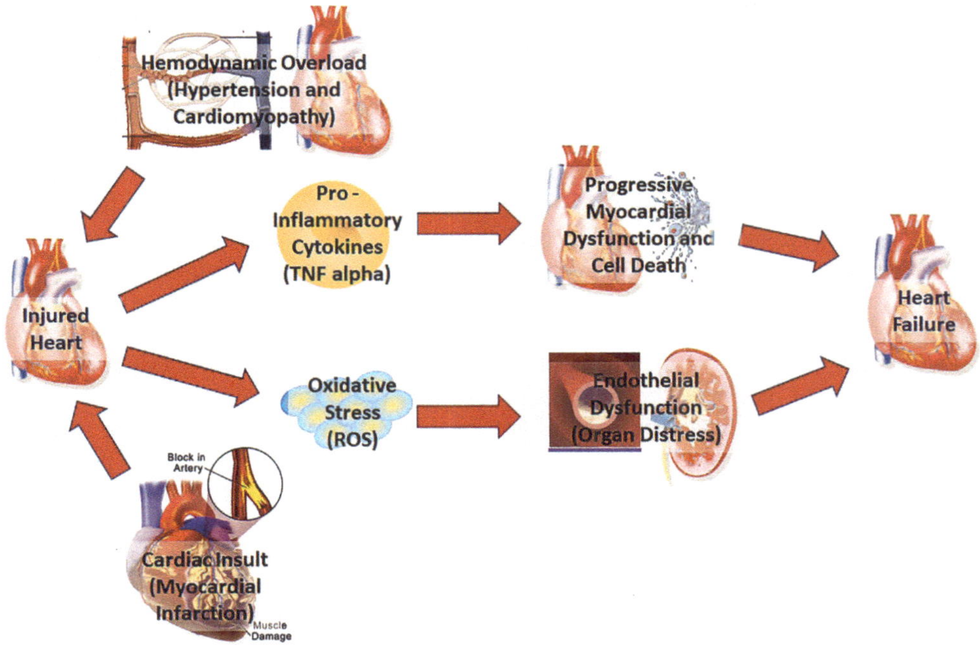

FIG. 6.14 Progression of heart failure [16]. The progression of heart failure is due to multiple events: hemodynamic overload (blood circulation stress) or cardiac insult leading to injury to the heart. In turn, the pro-inflammatory cytokines, tumor necrosis factor-α (TNF-α), and reactive oxygen species (ROS) increase oxidative stress. Together, these changes contribute to progressive myocardial and endothelial dysfunction leading to cell death and organ distress. The heart attempts to compensate by working harder through increased heart rate and contractility, but over time, the heart fails.

Heart failure may occur suddenly or over a period of time. When heart function deteriorates over the years, one or more conditions may exist. The strength of muscle contractions may be reduced, and the heart chambers' ability to fill with blood may be limited, resulting in less blood to pump (Fig. 6.14). Conversely, the pumping chambers may enlarge and fill with too much blood when the heart muscle is not strong enough to pump out all the blood it receives.

6.16 The progression of heart failure

The progression of CHF involves loss of contractility of the heart. The failing heart causes leakage of Ca^{2+} from the sarcoplasmic reticulum through dysfunctional **ryanodine receptors,** which regulate the passage of Ca^{2+} from the sarcoplasmic reticulum to the cytoplasm. The failing heart could also have impaired Ca^{2+}-ATPase pumps causing less Ca^{2+} to be pumped back into the sarcoplasmic reticulum and more to remain in the cytoplasm. The result is a reduced transient Ca^{2+}concentration (change in Ca^{2+} from the relaxed state to the contracted state). The more significant the change in Ca^{2+}concentration from rest to contraction, the better the contraction of the heart is. Due to leakage of Ca^{2+}, there is less of a change from rest to contraction, and therefore, a decrease in contractility. Collectively, the reduced transient Ca^{2+} concentrations and diastolic Ca^{2+} cause a reduction in contractile force of the heart (the heart will pump less blood with each beat) and an increase in heart arrhythmias (irregular heartbeats). Intracellular Na^+ concentration is also increased in failing myocytes adding to the issues with arrhythmias and a reduction in the amounts of mitochondrial Ca^{2+}. The increased Na^+ reduces the antioxidant capacity to heal the myocyte and enhances mitochondrial production of ROS. An increase in ROS could be the cause of malfunctioning of ryanodine receptors on the sarcoplasmic reticulum causing more Ca^{2+} to be leaked into the cytoplasm, decreasing the pumping power of the heart and triggering arrhythmias (abnormal heartbeat). Beta-blockers are used to correct this problem by improving the heart's ability to relax and slow the heart rate. Over time, beta-blockers improve the heart's pumping ability.

Heart failure is a chronic condition that gets worse with time (Fig. 6.14). The stages range from "high risk of developing heart failure" to "advanced heart failure," and provide treatment plans. As the condition gets worse, the heart muscle pumps less blood to the organs, and an individual moves toward the next stage of heart failure. The stages classified by the American Heart Association (AHA) and American College of Cardiology (ACC) [17] are different than the New York Heart Association (NYHA) clinical classifications of heart failure that rank patients as class I–II–III–IV, according to the degree of symptoms or functional limits [18]. Treatment at each stage of heart failure may involve changes in medications, lifestyle, and cardiac devices.

NYHA Classification of Heart failure includes the following four classes:

Class 1: No limitations in physical activity.
Class 2: Mild limitations in physical activity.
Class 3: Moderate limitations in physical activity.
Class 4: Symptoms occur at rest and any physical activity is not possible without symptoms.

ACC Heart Failure Stages are listed as follows:

Stage A: Patients at high risk for heart failure but have no symptoms or structural heart disease.
Stage B: Patients have structural heart disease but are asymptomatic.
Stage C: Patients have structural heart disease plus symptoms.
Stage D: Patients have refractory heart failure that requires modified interventions.

6.17 Common blood tests to diagnose MI

Standard blood tests indicating an MI include (1) elevated levels of **creatine kinase MB** (an enzyme that breaks down the conversion of creatine to phosphocreatine); (2) **troponin** (a protein involved in muscle contraction), both of which are released by damaged cardiac muscle fibers. Measurement of (3) **lactate dehydrogenase-1** (**LDH-1**) helps confirm tissue damage as it is found mainly in the heart.

6.18 Treatment options for MI

Immediate treatments for MI include administering supplemental O_2, aspirin, and nitroglycerine administered sublingually (under the tongue) to facilitate its absorption. Aspirin prevents aggregation of platelets which is the initial process in thrombosis. Nitroglycerine is a vasodilator that relaxes the smooth muscle in the tunica media of coronary vessels among others.

Long-term treatment includes injections of thrombolytic agents such as streptokinase to dissolve the clot, balloon angioplasty, stents to open blocked vessels, and bypass surgery to allow blood to pass around the site of the blockage. If the damage is extensive, coronary artery replacement with a heart transplant or ventricular assist device, a mechanical device that supplements the pumping activity of the heart, may be employed.

References

[1] Blausen 0257 CoronaryArtery Plaque.png. Wikimedia Commons, the free media repository. Updated October 25, 2020. Accessed March 10, 2021. https://commons.wikimedia.org/w/index.php?title=File:Blausen_0257_CoronaryArtery_Plaque.png&oldid=501291774.

[2] Blausen 0463 HeartAttack.png. Wikimedia Commons, the free media repository. Updated September 21, 2020. Accessed March 10, 2021. https://commons.wikimedia.org/w/index.php?title=File:Blausen_0463_HeartAttack.png&oldid=466659174.

[3] C. Gill, R. Mestril, A. Samali, Losing heart: the role of apoptosis in heart disease—a novel therapeutic target? FASEB J. 16 (2) (2002) 135–146, https://doi.org/10.1096/fj.01-0629com.

[4] M.A. Fortuno, N. Lopez, A. Gonzalez, J. Diez, Involvement of cardiomyocyte survival-apoptosis balance in hypertensive cardiac remodeling. Report, Expert. Rev. Cardiovasc. Ther. 1 (2) (2003) 293.

[5] 2105 Capillary Bed.jpg. Wikimedia Commons, the free media repository. Updated September 10, 2020. Accessed February 4, 2021. https://commons.wikimedia.org/w/index.php?title=File:2105_Capillary_Bed.jpg&oldid=454437322.

[6] The human lungs.png. Wikimedia Commons, the free media repository. Updated October 18, 2020. Accessed March 12, 2021. https://commons.wikimedia.org/w/index.php?title=File:The_human_lungs.png&oldid=493873391.

[7] Blausen 0993 PleuralEffusion.png. Wikimedia Commons, the free media repository. Updated October 25, 2020. Accessed March 13, 2021. https://commons.wikimedia.org/w/index.php?title=File:Blausen_0993_PleuralEffusion.png&oldid=501275294.

[8] S.M. Hollenberg, C.J. Kavinsky, J.E. Parrillo, Cardiogenic shock, Ann. Intern. Med. 131 (1) (1999) 47–59.

[9] Y. Matsui, J. Morimoto, T. Uede, Role of matricellular proteins in cardiac tissue remodeling after myocardial infarction, World J. Biol. Chem. 1 (5) (2010) 69.

[10] PTCA NIH.gif. Wikimedia Commons, the free media repository. Updated October 10, 2020. Accessed March 15, 2021. https://commons.wikimedia.org/w/index.php?title=File:PTCA_NIH.gif&oldid=486341667.

[11] Coronary artery bypass grafting.jpg. Wikimedia Commons, the free media repository. Updated March 2, 2021. Accessed March 15, 2021. https://commons.wikimedia.org/w/index.php?title=File:Coronary_artery_bypass_grafting.jpg&oldid=538394685.

[12] Blausen 0152 CABG All.png. Wikimedia Commons, the free media repository. Updated October 25, 2020. Accessed March 15, 2021. https://commons.wikimedia.org/w/index.php?title=File:Blausen_0152_CABG_All.png&oldid=501294332.

[13] L. Melly, G. Torregrossa, T. Lee, J.-L. Jansens, J.D. Puskas, Fifty years of coronary artery bypass grafting, J. Thorac. Dis. 10 (3) (2018) 1960–1967, https://doi.org/10.21037/jtd.2018.02.43.

[14] D.M. Yellon, D.J. Hausenloy, Myocardial reperfusion injury, N. Engl. J. Med. 357 (11) (2007) 1121–1135, https://doi.org/10.1056/NEJMra071667.

[15] D.J. Hausenloy, D.M. Yellon, Myocardial ischemia-reperfusion injury: a neglected therapeutic target, J. Clin. Invest. 123 (1) (2013) 92–100, https://doi.org/10.1172/jci62874.
[16] P. Pacher, J.S. Beckman, L. Liaudet, Nitric oxide and peroxynitrite in health and disease, Physiol. Rev. 87 (1) (2007) 315–424, https://doi.org/10.1152/physrev.00029.2006.
[17] S.A. Hunt, D.W. Baker, M.H. Chin, et al., ACC/AHA guidelines for the evaluation and management of chronic heart failure in the adult: executive summary a report of the American College of Cardiology/American Heart Association task force on practice guidelines (committee to revise the 1995 guidelines for the evaluation and management of heart failure): developed in collaboration with the International Society for Heart and Lung Transplantation; endorsed by the Heart Failure Society of America. Guideline practice guideline, Circulation 104 (24) (2001) 2996–3007, https://doi.org/10.1161/hc4901.102568.
[18] A. Malik, D. Brito, L. Chhabra, Congestive heart failure, in: StatPearls [Internet], StatPearls Publishing, 2021. Updated February 11, 2021 https://www.ncbi.nlm.nih.gov/books/NBK430873/.

CHAPTER

7

Renal physiology

OUTLINE

OBJECTIVES

- State the main functions of the kidneys.
- Draw the parts of a nephron while tracing the flow of the filtrate.

Biology of Cardiovascular and Metabolic Diseases
https://doi.org/10.1016/B978-0-12-823421-1.00005-6

- Identify how the blood vessel arrangement within the kidneys is unique.
- Locate the juxtaglomerular apparatus within the nephron and mention the role played by this structure.
- Define filtration, reabsorption, and secretion as parts of the process of urine formation.
- Explain the filtration mechanism in the nephron, including the forces involved in the process.
- Define glomerular filtration rate and summarize the mechanisms that exist to regulate it.
- Describe the mechanism for the reabsorption of salts and water in the proximal tubule.
- Differentiate between transcellular and paracellular transport mechanisms.
- Define transport maximum using glucose as an example .
- Compare the percentage of water reabsorbed at different segments of the tubule.
- Summarize the role of hormones in the regulation of fluid and electrolyte levels in the body by acting at the nephron.
- Explain how the renin-angiotensin-aldosterone system regulates electrolyte and water levels in the body.

7.1 Introduction

Kidneys are a pair of bean-shaped structures with several vital functions in the body. They play a significant role in filtering the blood and getting rid of metabolic waste products, such as urea, uric acid, creatinine, bilirubin, H^+, and foreign substances such as drugs and toxins, pesticides, and food additives. Kidneys are essential in the regulation of blood pH and the concentrations of solutes. They not only contribute to regulating blood volume and blood pressure but also serve an endocrine function by producing **erythropoietin** (EPO) to stimulate red blood cell production, **renin** in regulating blood volume, blood pressure, and solute concentrations and help synthesize **vitamin D**. The urinary system, controlled by the autonomic nervous system, stores urine until a convenient time for disposal and provides the anatomical structures to transport it to the outside of the body.

7.2 External anatomy of the kidney

The kidneys are paired bean-shaped organs that are approximately the size of a fist and possess a reddish brown color. They are located behind the peritoneal cavity close to the posterior abdominal wall. The position of the left kidney is at the level of T12 to L3 vertebrae, whereas the right kidney is situated slightly lower due to the proximity of the liver (Fig. 7.1). They are covered by a fibrous capsule that helps to hold their shape and protect them. This capsule is lined by a shock-absorbing layer of adipose tissue which, in turn, is encompassed by a tough renal fascia.

7.3 Internal anatomy

Each kidney is organized into an outer **cortex** and an inner **medulla** (Fig. 7.2). The **renal columns** are connective tissue extensions that project from the cortex to the medulla and

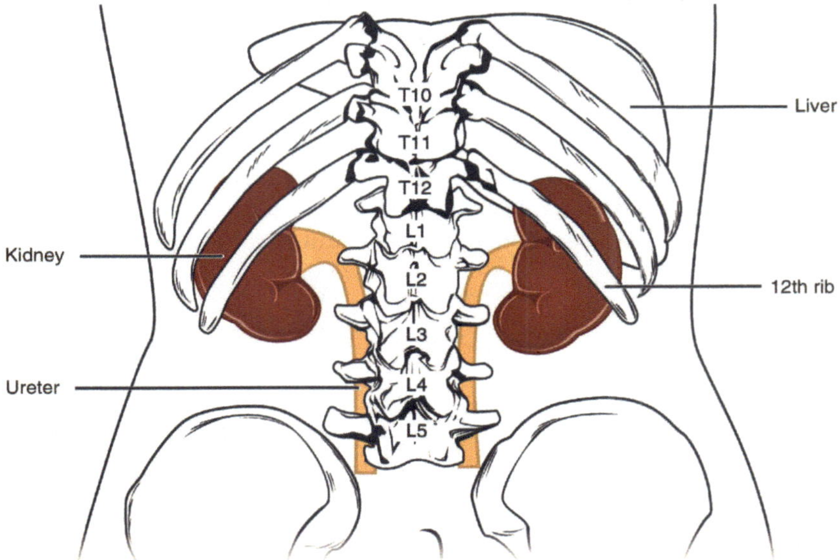

FIG. 7.1 The kidneys and their location in the abdominopelvic cavity [1].

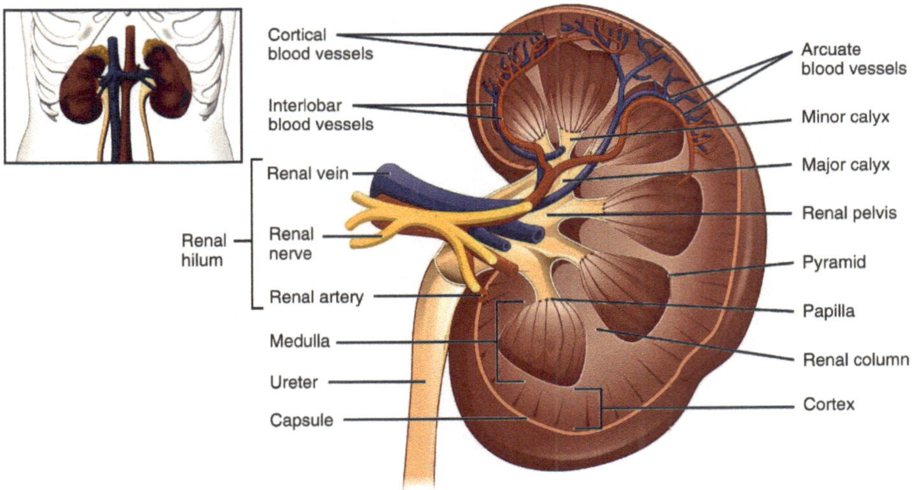

FIG. 7.2 Internal anatomy of the left kidney depicting the parts such as the pyramids, papillae, minor and major calyses, renal pelvis and the ureter as well as regions such as the cortex and the medulla [2].

separate the **renal pyramids**. The pyramids and renal columns together constitute the renal lobes. Renal pyramids are triangular wedges with their broader base facing the cortex and the narrower apex called the **papilla**. The papillae are bundles of collecting ducts that transport urine made by nephrons to the **minor calyces**. Several minor calyces, in turn, fuse to form **major calyces**, which join to become the **renal pelvis** which continues as the ureter.

7.4 Nephrons

Nephrons are both the structural and functional units of the kidneys. Each human kidney contains about 1,200,000 nephrons, a number that does not increase after birth. In situations of compensatory hypertrophy, such as after nephrectomy, the increase in kidney size is due to an increase in the size of the nephron, but not the number. The main components of the nephron are the **glomerulus**, the **Bowman's capsule** or **glomerular capsule**, the **proximal convoluted tubule**, the **loop of Henle**, and the **distal convoluted tubule** (Fig. 7.3). The glomerulus

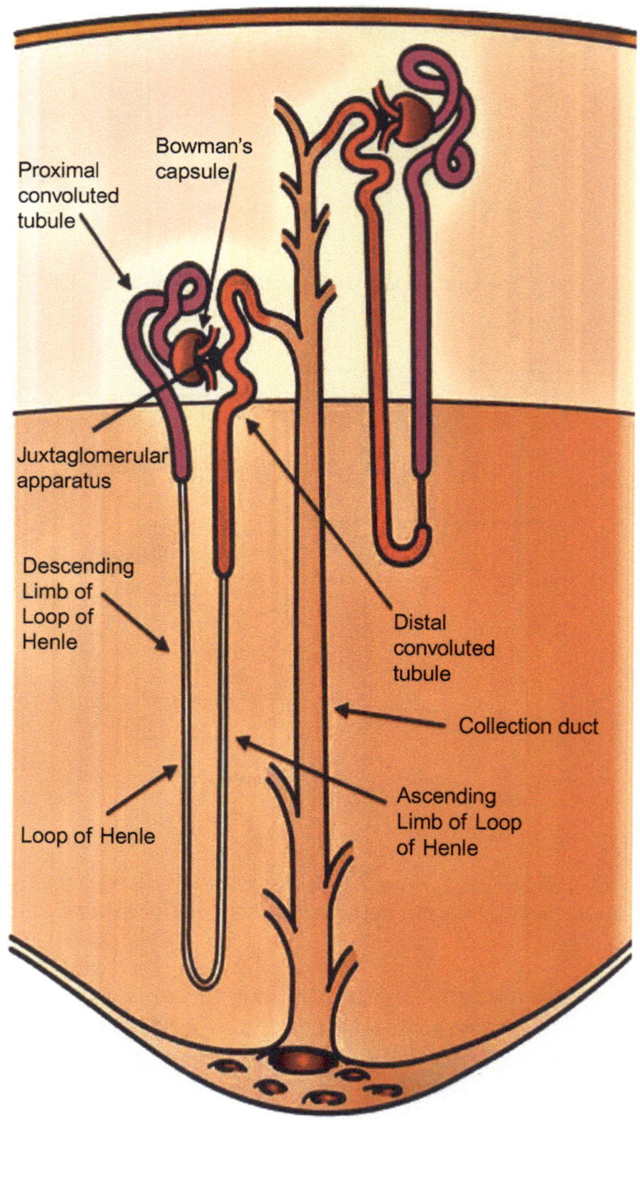

FIG. 7.3 Two types of nephrons—juxtamedullary nephron (left) and cortical nephron (right) [3].

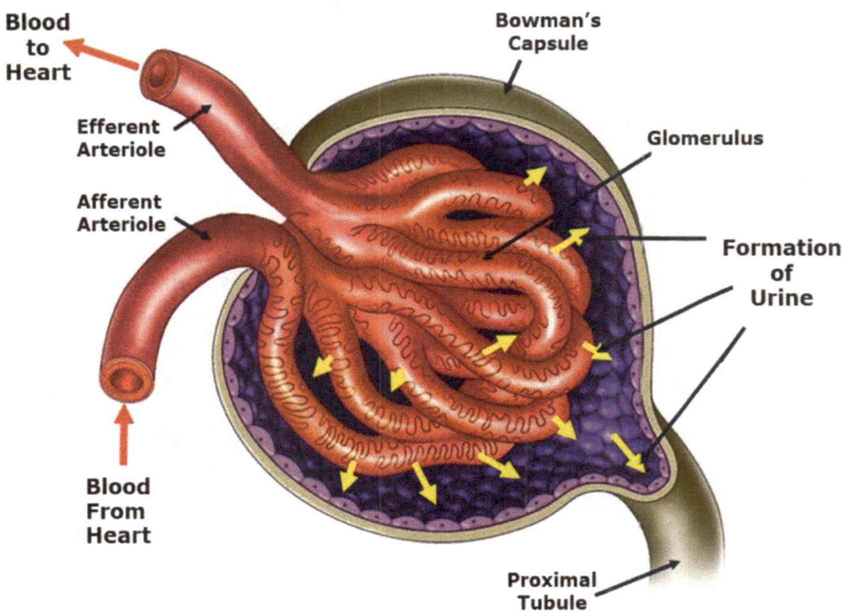

FIG. 7.4 Renal corpuscle consists of the Bowman's capsule and the glomerulus [4].

and the Bowman's capsule together are called **the renal corpuscle** (Fig. 7.4), whereas the rest of the nephron is referred to as the **renal tubule**.

The renal corpuscles, the proximal tubules, and distal tubules are situated in the cortex. Approximately 85% of the nephrons have a short loop of Henle that does not dip beyond the cortex. These nephrons are called **cortical nephrons**. The remaining 15% of nephrons have long loops of Henle that extend deep into the medulla and are called **juxtamedullary nephrons** (Fig. 7.3).

7.5 The glomerulus

The tuft of glomerular capillaries arises from an **afferent arteriole** and is drained by an **efferent arteriole**. The glomerulus filters the blood based chiefly on particle size and electrical charge. This filtration property of the glomerulus allows almost all of the solutes from the blood to be filtered except for the blood cells and proteins such as albumins. Capillary endothelial cells are fenestrated (porous), where the pore size has a diameter of about 1000 Å. The filtrate concentration resembles that of plasma due to solutes being filtered, including ions, amino acids, glucose, vitamins, and wastes. The glomeruli form approximately 200 L (189 quarts) of filtrate per day, yet less than two liters of waste are excreted in the form of urine. Because the glomerular capillaries have a unique anatomic arrangement between two sets of arteries, the hydrostatic pressure inside these capillaries can be varied by changing the tone of each of these arteriolar systems.

7.6 The Bowman's capsule

The Bowman's capsule surrounds the glomerulus (Fig. 7.4). It consists of two layers of epithelial cells—the outer **parietal** and the inner **visceral layer**. The epithelial cells that form the visceral layer of Bowman's capsule are provided with foot-like processes known as **podocytes** (Fig. 7.8). From these podocytes, a multitude of small processes arise and surround the basement membrane of the capillary. Thus, the ultrafiltrate present in the Bowman's capsule is separated from the blood in the glomerulus by three layers: the glomerular endothelium, the visceral epithelium formed by the cells of Bowman's space, and the basement membrane interposed between these two layers. The lumen of the Bowman's space is in direct continuity with that of the proximal tubule.

The hydrostatic pressure in the glomerular capillaries of about 50 mmHg pushes water and solutes into Bowman's capsule space. Proteins with molecular weights less than 40,000 can easily pass through the three layers of the glomerulus, whereas proteins with large molecular weights, after passing through the endothelium and basement membrane, are impeded by the slits between the feet of the podocytes that rest on the basement membrane.

7.7 Proximal tubule

The proximal tubule is a continuation of the Bowman's capsule and is highly convoluted. The last part of the proximal tubule that is not located next to its glomerulus extends toward the medulla and is relatively straight. The proximal convoluted tubular lining is composed of a single layer of epithelial cells. The sides of these cells facing the lumen of the tubule have a prominent brush border made up of many microvilli. These microvilli increase the surface area of the membrane in contact with the tubular fluid.

Since the proximal tubule is the first part of the tubular system of the nephron, several solutes are reabsorbed and secreted in this portion. Approximately 60%–80% of the Na^+ and Cl^- is reabsorbed in the proximal tubule. Na^+ is reabsorbed using an active transport mechanism, whereas Cl^- passively follows Na^+ to maintain electroneutrality. Glucose is also actively reabsorbed in the first part of the proximal tubule. However, there is a maximum threshold rate of glucose transfer across the tubule referred to as transport maximum. Above this threshold level, the system is saturated, and glucose appears in the urine. Other substances are also actively transported across the tubular lumen. These include K^+, PO_4^-, amino acids, creatinine, uric acid, and ketone bodies, each of which has a different threshold level. Reabsorption of approximately 65% of water results in an increased concentration of urea within the proximal tubule, which allows the diffusion of urea into the peritubular capillaries. The proximal tubules are actively involved in the formation of H^+ and ammonia. They also secrete metabolic products of endogenous hormones and exogenous drugs into the filtrate. The filtrate remains isotonic by the movement of both solutes and water at a steady concentration.

7.8 Loop of Henle

The descending portion of the loop of Henle is the continuation of the proximal tubule. It makes a hairpin turn at the deepest point of its descent. Then, it begins to ascend toward the cortex and, thus, is referred to as the ascending limb of the loop of Henle (Fig. 7.3). The descending limb of the loop of Henle is relatively impermeable to Na^+ and Cl^-. The depth that the loop dips into the medulla varies according to the location of the glomerulus. Cortical nephrons, whose glomeruli are located in the outer two-thirds of the cortex, have short loops and dip only for a short distance into the medulla before turning back to the cortex. The remaining 15% of the nephrons have long loops that penetrate deeply to near the tip of the papilla before turning back. The position of the descending limb in the medulla exposes it to a hypertonic interstitial fluid, making the water escape from the lumen, causing the filtrate to become hypertonic. On the other hand, the active transport of $Na^+/K^+/Cl^-$ from the ascending limb results in the hypotonic situation as filtrate enters the distal convoluted tubule.

7.9 Distal tubule

Like the proximal tubule, the distal tubule is convoluted and formed by simple cuboidal epithelium, but it is shorter than the proximal tubule. The positions of the afferent arteriole and the distal tubule are such that they meet at one specific part of the nephron and this specific location is where the cells are specialized and the entire arrangement is referred to as the **juxtaglomerular apparatus** (Fig. 7.3).

The cells lining the distal tubule are not as active as those in the proximal tubule. Thus there are fewer microvilli on the apical surface. The distal tubule can reabsorb Na^+ and water at different rates, exchange Na^+ for K^+ and H^+, and reabsorb Ca^{2+}, Cl^-, and water.

The osmolality and the volume of the filtrate in the distal tubule and the collecting ducts are under the control of ADH, which increases the permeability of the epithelium of the distal tubules and the collecting ducts to water, allowing a small volume of concentrated urine to be excreted. In the absence of ADH, a large amount of fluid remains in the collecting ducts, and the urinary output increases while fluid osmolality drops.

7.10 Collecting ducts

The distal ends of several convoluted tubules join together to form the collecting ducts, which, in turn, merge as they descend deeper in the medulla to empty at a papilla. Collecting ducts respond to ADH and insert **aquaporin channel proteins** into their membranes to allow water to be reabsorbed.

7.11 Renin-angiotensin-aldosterone system

The cells in the afferent arterioles in the juxtaglomerular apparatus contain renin granules and are released in response to low blood pressure or renal ischemia in an effort by the kidney to restore its blood flow. Renin is an enzyme, also referred to as a hormone, that cleaves angiotensinogen to angiotensin I. Angiotensin I is not biologically active until it is converted to angiotensin II by **angiotensin-converting enzyme** from the lungs (Fig. 7.5). Angiotensin II is a potent vasoconstrictor that helps increase blood pressure to normal by increasing

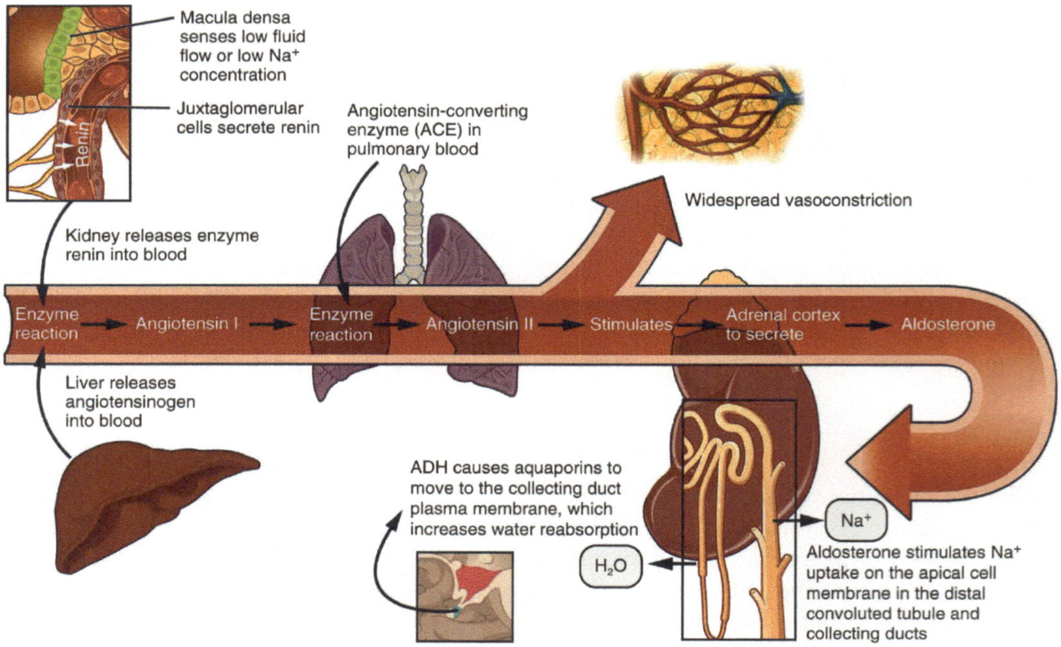

FIG. 7.5 Renin-angiotensin-aldosterone system. Renin converts angiotensinogen to angiotensin I. Angiotensin-converting enzyme (ACE), the lung-derived enzyme, converts angiotensin I into active angiotensin II [5].

peripheral resistance. Angiotensin II also triggers the release of aldosterone from the adrenal cortex. Aldosterone stimulates the reabsorption of Na^+ while causing the secretion of K^+ and the reabsorption of water by osmosis.

7.12 Filtration, reabsorption, and secretion

Nephrons first filter the blood and then modify the resulting filtrate into the urine. Many changes take place in the different parts of the nephron before the urine is created for disposal. The principal task of the nephrons is to balance the plasma to homeostatic set points and excrete potential toxins in the urine. They do this by accomplishing three key functions—**filtration**, **reabsorption**, and **secretion**. Filtration of blood is the movement of solutes and water from the glomerulus into the Bowman's capsule. The filtrate contains waste molecules as well as essential nutrients, ions, and water. These molecules must be claimed before the filtrate passes through the rest of the nephron. Those substances that are conserved from the tubular lumen back into the blood are referred to as reabsorption. Some substances such as water and salts are reabsorbed throughout the length of the tubule, whereas the reabsorption of others occurs specifically in one location. For example, 100% of the glucose is reabsorbed in the proximal tubule alone. Secretion is the release of substances into the filtrate from the blood in the peritubular capillaries or by the tubular epithelial cells (Fig. 7.6).

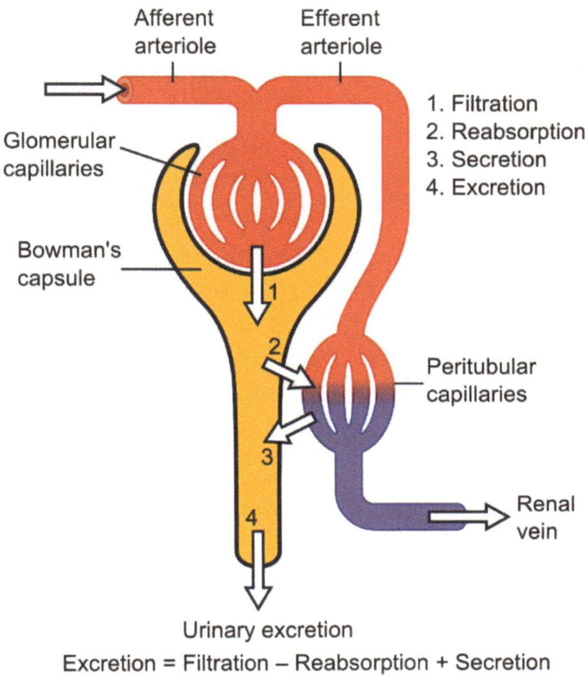

FIG. 7.6 Schematic diagram of the nephron (*yellow*), relevant circulation (*red/blue*), and the methods of altering the filtrate [6].

7.13 Glomerular filtration

The glomerulus is a high-pressure capillary bed since it is situated between afferent and efferent arterioles. It can filter substances 100–500 times greater than that of most capillaries. Bowman's capsule surrounds the glomerulus, collects, and directs this filtrate to the proximal tubule. The outermost part of Bowman's capsule, the parietal layer, is a simple squamous epithelium. The inner visceral layer wraps the glomerular capillaries. The visceral layer consists of cells that are uniquely shaped (podocytes), extending finger-like arms (**pedicels**) to cover the glomerular capillaries (Fig. 7.7). These projections interdigitate to form **filtration slits**, leaving small gaps between the digits to form a sieve. As blood passes through the glomerulus, approximately 20% of the plasma gets filtered between these sieve-like fingers to be captured by Bowman's capsule and funneled to the proximal tubule. When the fenestrae in the glomerular capillaries match the spaces between the podocyte "fingers," the only structure separating the capillary lumen and the lumen of Bowman's capsule is their shared basement membrane (Fig. 7.8). Thus, the glomerular endothelium, the visceral layer of the Bowman's capsule, and the basement membrane comprise the **filtration membrane**. This membrane permits very rapid movement of filtrate from the capillary to capsule through pores about 70 nm in diameter.

The fenestrations prevent the filtration of blood cells and large proteins but allow most other constituents to pass through. These substances cross through the capillary wall if they are less than 8 nm in size. The electrical charge of a substance is yet another factor that

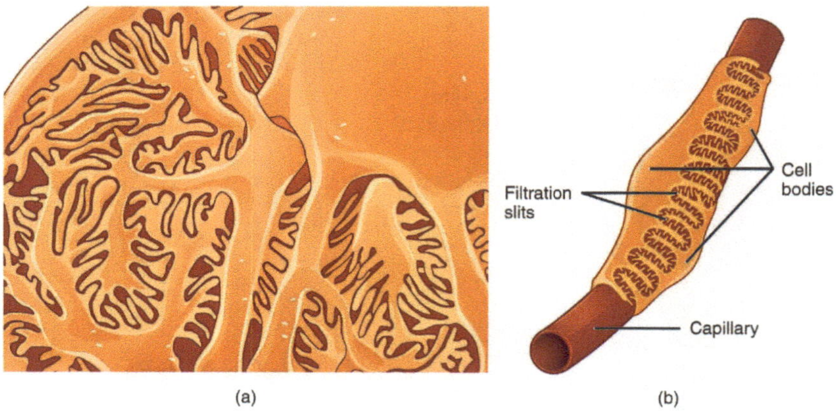

FIG. 7.7 Podocytes, lining the visceral layer of the Bowman's capsule, interdigitate with structures called pedicels and filter substances in a way similar to fenestrations. In (A), the large cell body can be seen at the top right corner, with branches extending from the cell body. The smallest finger-like extensions are the pedicels. Pedicels on one podocyte always interdigitate with the pedicels of another podocyte. (B) This capillary has three podocytes wrapped around it [7].

FIG. 7.8 Fenestrated glomerular capillary. Fenestrations allow many substances to diffuse from the blood based primarily on size [8].

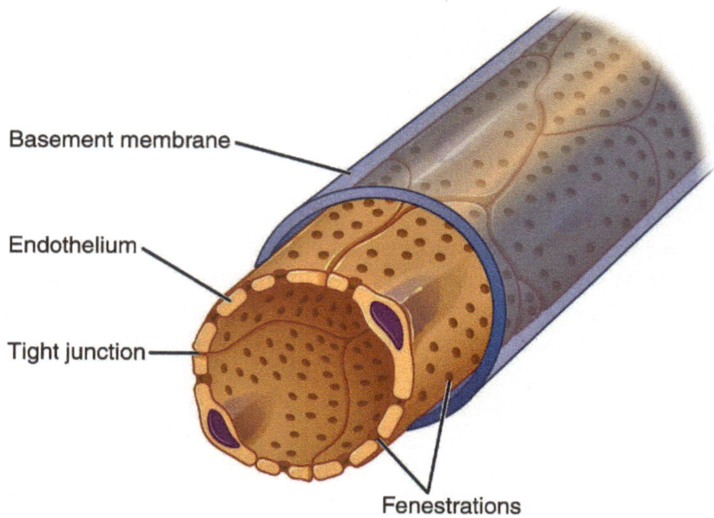

determines the passage of molecules across the glomerular wall. The proteins associated with the fenestrae are negatively charged, so they tend to repel negatively charged substances and allow positively charged substances to pass more readily. On the other hand, the basement membrane prevents the filtration of medium to large proteins. Overall, filtration is regulated by fenestrations in capillary endothelial cells, podocytes with filtration slits, membrane charge, and the basement membrane of the capillary. The result is creating a filtrate that does not contain cells or proteins and has a slight predominance of positively charged substances.

7.14 Glomerular filtration rate (GFR)

The total amount of filtrate formed by the kidneys per minute is referred to as the **glomerular filtration rate (GFR)**. The heart pumps about 5 L of blood per minute under resting conditions out of which, approximately 20% or 1 L of blood enters the kidneys to be filtered. On average, about 125 mL/min filtrate is produced in men and 105 mL/min filtrate in women. This amount equates to a volume of approximately 180 L/day in men and 150 L/day in women. Ninety-nine percent of this filtrate is reabsorbed so that only about 1–2 L of urine is produced per day.

GFR is determined by the hydrostatic and colloid osmotic pressures between the glomerulus and the Bowman's capsule. Glomerular filtration occurs when glomerular hydrostatic pressure exceeds the hydrostatic pressure of Bowman's capsule. There is also an opposing force, the colloid osmotic pressure, which is typically higher in the glomerular capillary than in other places due to more water filtering out of these capillaries than in the plasma elsewhere.

Since the fenestrae limit the size of particles crossing the membrane, the blood cells and most proteins remain in the capillary, and therefore osmotic pressure inside the glomerular capillary is higher than that of the Bowman's capsule. An average colloid osmotic pressure of 30 mmHg is observed within the capillary. The absence of proteins in Bowman's capsule results in an osmotic pressure near zero. Thus, the only pressure moving fluid across the capillary wall into the lumen of Bowman's capsule is glomerular hydrostatic pressure which is sufficient to push water through the membrane despite the colloid osmotic pressure working against it. The sum of all of the influences, both osmotic and hydrostatic pressures, results in a **net filtration pressure (NFP)** of about 10 mmHg (Fig. 7.9).

$$\begin{aligned} \text{NFP} &= \text{Glomerular blood hydrostatic pressure (GBHP)} \\ &\quad - [\text{capsular hydrostatic pressure (CHP)} + \text{blood colloid osmotic pressure (BCOP)}] \\ &= 10\,\text{mmHg}. \end{aligned}$$

$$\text{NFP} = 55 - [15 + 30] = 10\,\text{mmHg}.$$

7.15 Regulation of glomerular filtration rate

A normal concentration of solutes in the blood is vital in maintaining osmotic pressure both in the glomerulus and systemically. Any shift in osmolarity of the blood or capillary hydrostatic pressure will result in significant changes in the amount of filtrate formed. However, the kidneys can withstand a wide range of blood pressures due to the **autoregulatory** nature of smooth muscle lining the afferent and efferent arterioles. When blood pressure increases, smooth muscle fibers in the afferent arterioles contract and lead to vasoconstriction to minimize any increase in blood flow to the glomerulus and filtration rate. When blood pressure decreases, the same arterioles relax and bring about vasodilation to increase blood flow to maintain normal GFR. The net result is a relatively steady flow of blood into the glomerulus and GFR despite significant systemic blood pressure changes. There are disorders such as **glomerulonephritis**, where protein may escape from the glomerulus into the filtrate. Excess protein in the filtrate not only leads to a deficiency of plasma proteins but also increases its

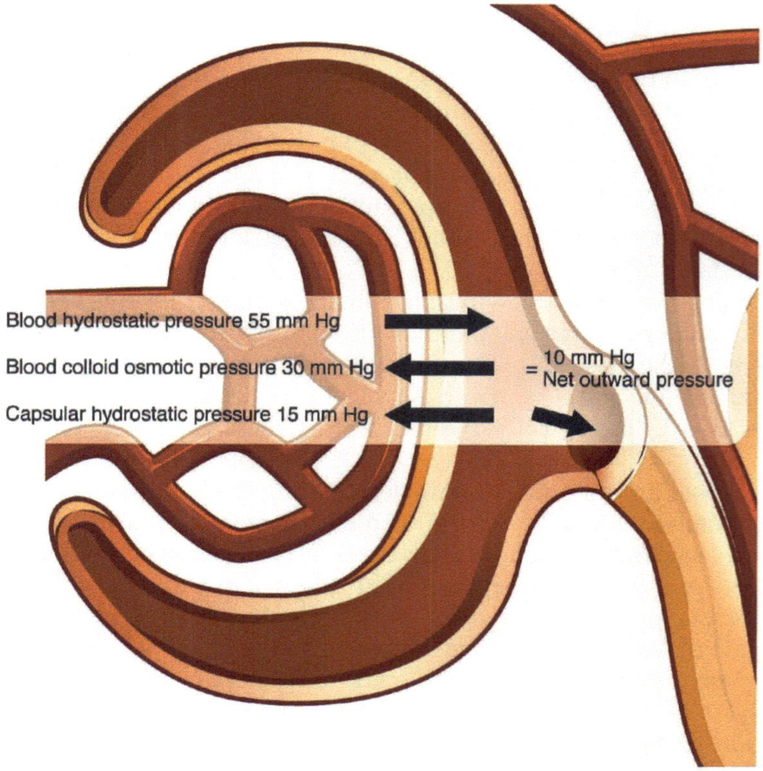

FIG. 7.9 Filtration pressures. Net filtration pressure (NFP) is the sum of blood colloid osmotic and hydrostatic pressures [9].

osmolarity of the filtrate, attracting more water and increasing urine volume. The osmotic pressure of the blood falls due to loss of protein. Less osmotic pressure pulling water back into the capillaries leaves the hydrostatic pressure to push more plasma out of the capillaries. The net effect would be water loss from the blood to interstitial tissues and cells, a condition termed **systemic edema**.

7.16 Tubular reabsorption and secretion

The function of the renal tubules can be classified into two main categories: reabsorption and secretion. Reabsorption of essential substances occurs in the direction of movements of filtrate from the tubule back into the peritubular capillaries. The mean arterial pressure drops to approximately 13 mmHg, whereas the blood osmotic pressure remains at about 30 mmHg. This ideal pressure gradient, along with the porous nature of the peritubular capillaries, promotes rapid reabsorption.

The movement of reabsorbed substances could occur in the **transcellular** or **paracellular** forms of transport. Transcellular transport is where substances travel through the cell, whereas paracellular transport occurs between the cells lining the tubule. Epithelial cells lining the tubule (tubular cells), like all other cells elsewhere, have two surfaces: **apical** or **luminal** surface facing the lumen and **basal** or **basolateral** surface facing the opposite side of the tubular cell. The numbers and types of ion pumps and channels vary in different parts of the tubule and between the apical and basilar surfaces. Na^+ is actively exchanged for K^+ using ATP on the basal surface of the tubular membrane. Most of the substances transported by a symport mechanism on the apical surface are transported by facilitated diffusion on the basal surface. Water is reabsorbed by osmosis.

Out of 180 L of filtrate passing through the nephrons every day, about 99% of the filtrate is reabsorbed between the proximal tubule, the loop of Henle, distal tubule, and the collecting ducts (Fig. 7.10). Various portions of the nephron differ in their capacity to reabsorb water and specific solutes. The amount of water and ions reabsorbed or lost is tightly regulated directly by ADH and aldosterone and indirectly by renin. Most water is recovered in the proximal tubule, loop of Henle, and distal tubule. About 10% of the remaining filtrate reaches the collecting ducts, where almost all of the water passing through them is recovered in a situation such as dehydration with the help of ADH, whereas less water is claimed when a person is well hydrated.

7.17 Reabsorption and secretion in the proximal tubule

The blood gets filtered at the glomerulus, where the filtered plasma is now referred to as filtrate which differs from blood mainly by the absence of cells and proteins. It undergoes modification throughout the tubule by means of secretion and reabsorption before leaving the tubule, at which time it is referred to as urine. The first part of the tubule where any modification occurs is at the proximal tubule. Here, some substances are reabsorbed, whereas others are secreted. Water and substances that are reabsorbed are returned to peritubular capillaries. The water movement into the peritubular capillaries and vasa recta, peritubular capillaries around the tubular portion of the juxtaglomerular nephrons,is influenced primarily by osmolarity and concentration gradient.

More substances are reabsorbed at the proximal tubule than any other portion of the nephron since it is the first part of the tubule the filtrate passes through, and it is the longest of all the other tubular segments. Many of the substances, such as amino acids and glucose, use the **symport** (a type of cotransport where two substances move in the same direction) mechanism of transport, whereas **antiport** (a type of cotransport where substances move in the opposite direction), active transport, diffusion, and facilitated diffusion are some other mechanisms by which substances are reabsorbed.

About 65% of the Na^+, Cl^-, and water are reabsorbed at the proximal tubule. Na^+ is actively transported in the proximal tubule through a pump mechanism (Na^+/K^+exchange pump), requiring ATP. On the other hand, the proximal tubule is freely permeable to water. The passive movement of water out of the tubule is generated by the osmotic gradient created by the reabsorption of Na^+ and its accompanying anions and is referred to as **obligatory water reabsorption** because water follows Na^+ (Fig. 7.11). The tubular fluid in this segment of the nephron remains isotonic to the body fluids as Na^+ and water are reabsorbed proportionally.

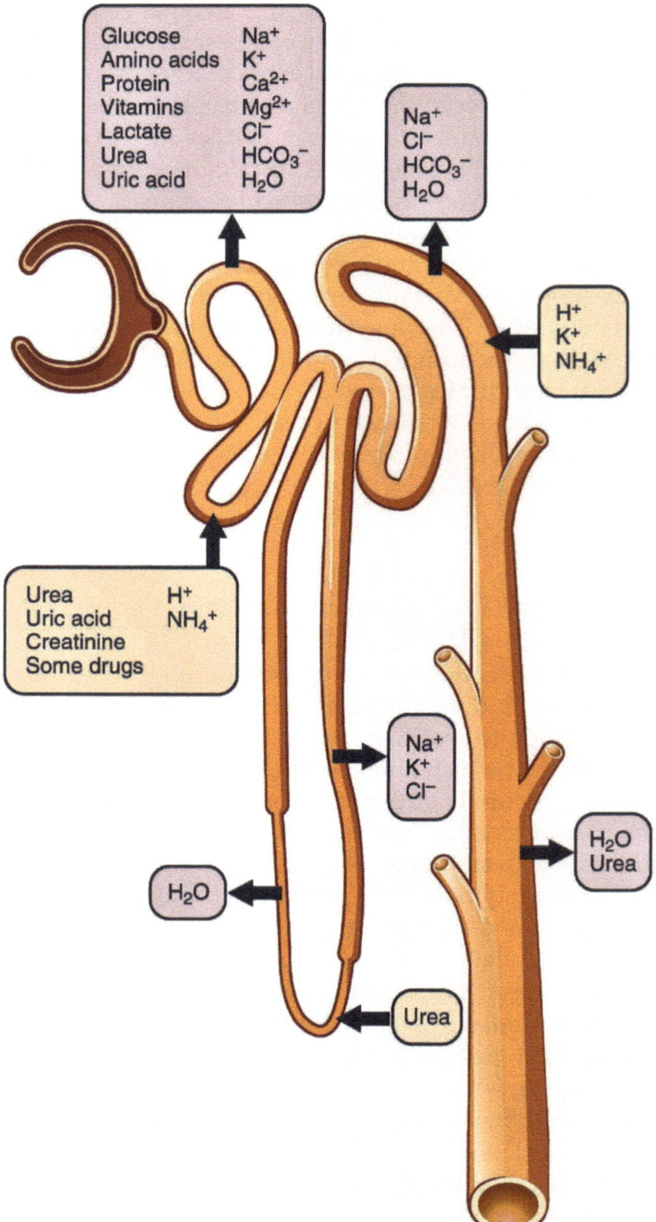

FIG. 7.10 Locations of secretion and reabsorption in the tubular parts of a nephron [10].

Approximately 50% of urea, the most abundant nitrogenous waste, is passively reabsorbed at the proximal tubule. Almost all glucose, amino acids, and vitamins are generally recovered in this portion of the nephron. Glucose is reabsorbed from the lumen into the tubular cell along with Na^+ by symport mechanism. Na^+/K^+ ATPase pumps Na^+ out of the cell on the basal surface, thus creating a gradient for Na^+ within the cell, which helps move Na^+ down

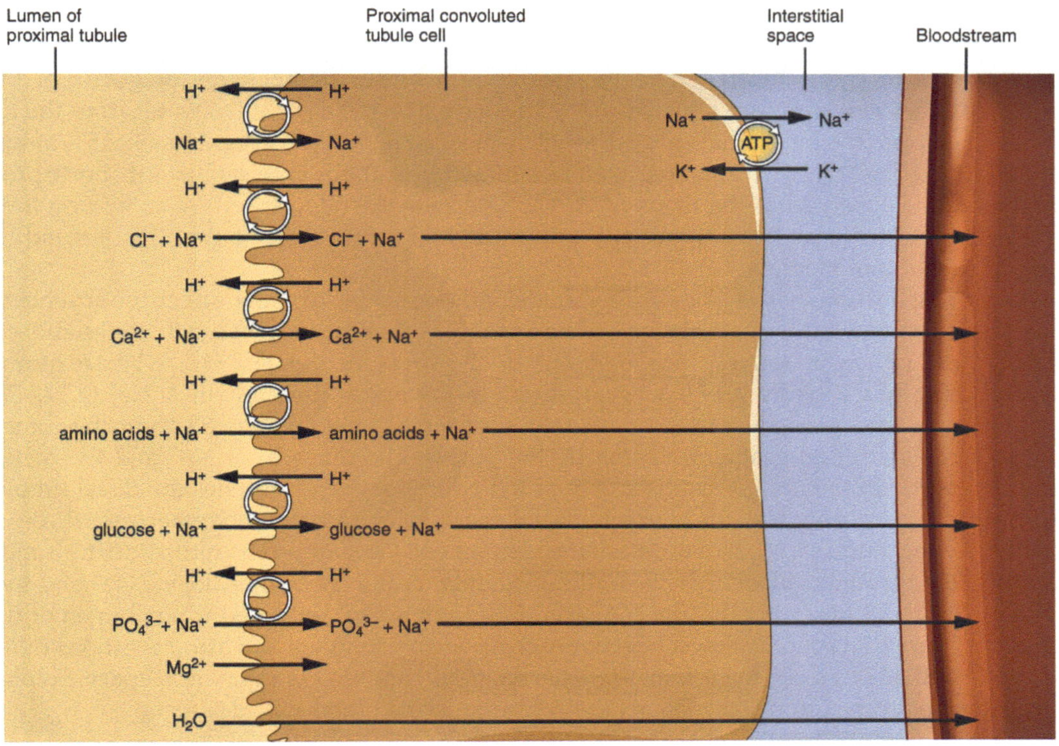

FIG. 7.11 Substances reabsorbed and secreted by the proximal convoluted tube (PCT) [11].

its electrochemical and concentration gradient into the cell. As Na^+ diffuses into the cell, it takes glucose with it (symport). At the basal surface, glucose is transported by facilitated diffusion. This glucose transport mechanism has a maximum capacity, known as **transport maximum or Tm**, at which it becomes saturated. The maximum amount of glucose that can be reabsorbed in men is about 350 mg/min, whereas it is about 300 mg/min for women, which translates to a blood glucose concentration of about 200 mg/dL. Glucose may appear in the urine if blood glucose levels are high enough that all the glucose transporters in the proximal tubule are saturated. Exceptionally high glucose intake may cause glucose to appear briefly in the urine (glycosuria).

Variable quantities of Ca^{2+}, Mg^{2+}, $HPO_4{}^{2-}$, Na_2HCO_3, H^+, $SO_4{}^-$, ascorbic acid, acetoacetic acid, and beta-hydroxybutyric acid are also recovered in the proximal tubule. The reabsorption of PO_4^{3-} varies inversely with its excretion and is regulated by the parathyroid hormone.

7.18 Reabsorption and secretion in the loop of Henle

The loops of cortical nephrons are short and are mainly seen in the cortex compared to the loops of juxtamedullary nephrons, which are long and extend deep into the medulla (Fig. 7.3).

The medullary interstitial fluid concentration is greater than in the cortex, and the concentration of the medullary tissue fluid increases steadily (from 300 to 1200 mOsm) as it reaches the deeper parts of the medulla. Since the descending limbs of the loops of Henle of the juxtamedullary nephrons are situated in the deeper parts of the medulla, water from the filtrate escapes into the interstitial space through osmosis (Fig. 7.10). Approximately 15% of water is reabsorbed in this manner. A phenomenon called **countercurrent multiplier mechanism** is observed at the loops of Henle. The salts extruded from the ascending limb of the loop of Henle creates a hypertonic condition that draws water out of the descending limb of the loop of Henle.

Ascending portions of the loops are completely impermeable to water due to the absence of aquaporin proteins but are highly specialized in recovering much of the salts in the filtrate, mainly Na^+ which is actively pumped out of the loop by the Na^+/K^+ ATPase pump. Reabsorption of Cl^- is by Na^+/Cl^- symporters in the apical membrane. A $Na^+/K^+/2Cl^-$ symporter in the apical membrane passively allows Cl^- to enter the cell from the lumen due to a concentration gradient created by the pump. The removal of Na^+ and Cl^- while retaining water in the lumen results in a hypotonic filtrate when it reaches the distal tubule. Additionally, pumping Na^+ into the interstitial space contributes to the hyperosmotic environment in the medulla. Most of the K^+ that enters the cell via symporters returns to the lumen via diffusion through leaky channels in the apical membrane. Since there is one Na^+ and two Cl^- in the interstitium surrounding the ascending limb when compared to the lumen of the loop, the interstitial space is now a negatively charged environment and attracts cations (Na^+, K^+, Ca^{2+}, and Mg^{2+}) from the lumen via a paracellular route to the interstitial space. Solutes and water recovered from these loops are returned to the blood via vasa recta.

7.19 Reabsorption and secretion in the distal tubule

Approximately 80% of filtered water is recovered by the time the filtrate enters the distal tubule. The distal tubule will recover an additional 10%–15% before the filtrate enters the collecting ducts (Fig. 7.10). Distal tubules and the collecting ducts are the targets for many hormones in regulating the blood composition of water and salts. Aldosterone increases the number of Na^+/K^+ ATPase in the basal membrane of the distal tubule and collecting duct. The pumping of Na^+ out of the lumen creates a negative charge that promotes Cl^- to diffuse out of the lumen into the interstitial space by a paracellular route. Peritubular capillaries receive the solutes and water, returning them to the circulation.

Cells of the distal tubule also recover Ca^{2+} from the filtrate. Receptors for parathyroid hormone are found in distal tubule cells and, when bound to parathyroid hormone, induce the insertion of Ca^{2+} channels on their luminal surface. The channels enhance Ca^{2+} recovery from the filtrate. Also, as Na^+ is pumped out of the cell, the resulting electrochemical gradient attracts Ca^{2+} into the cell. Finally, calcitriol (1,25 dihydroxy vitamin D, the active form of vitamin D) is very important for Ca^{2+} recovery. It induces the production of calcium-binding proteins to transport Ca^{2+} into the cell. These binding proteins are also crucial for the movement of Ca^{2+} inside the cell and aid in the exocytosis of Ca^{2+} across the basolateral membrane. Any Ca^{2+} not reabsorbed at this point is lost in the urine.

7.20 Reabsorption and secretion in the collecting duct

The collecting ducts are important in regulating the normal osmolarity of blood by varying the amount of water reabsorption. If there is an increase in plasma osmolarity, more water is reabsorbed, and the opposite is true when plasma osmolality decreases. ADH is a key hormone in the regulation of plasma osmolarity. For example, when a person is dehydrated, the osmotic concentration of the blood rises. Osmoreceptors in the hypothalamus detect this change and stimulate ADH release from the posterior pituitary. In the presence of ADH, aquaporin channels are inserted into the apical surface of the cells lining the collecting ducts, thus allowing more reabsorption of water. As the filtrate moves through the medullary portion of the collecting ducts, it is exposed to the higher osmotic concentration of the tissue fluid, allowing more water to escape from the tubule into the interstitial space.

The cells in the collecting duct also express receptors for aldosterone. Aldosterone is released in response to the renin–angiotensin mechanism, low blood Na^+ or increased plasma K^+ levels. When aldosterone is released, it promotes the addition of Na^+/K^+ ATPase pumps on the basal membrane of the cells, allowing more Na^+ recovery from the filtrate. Water follows Na^+ through osmosis. As the Na^+/K^+ ATPase pump recovers Na^+, it also pumps K^+ into the filtrate.

Urea escapes partially from the collecting duct and contributes to the higher osmotic concentration of the medullary tissue fluid (Fig. 7.10). Urea reenters the filtrate through the descending limb of the loop of Henle.

7.21 Natriuretic hormones

Natriuretic hormones are peptides released by the atrial and ventricular muscles in the heart in response to stretch experienced due to high pressure within these chambers. They stimulate the kidneys to excrete Na^+ by inhibiting aldosterone release and therefore inhibiting Na^+ reabsorption in the collecting ducts. If Na^+ remains in the filtrate, its osmotic force will cause a concurrent loss of water. Natriuretic hormones also inhibit ADH release, which will result in less water recovery. Thus, natriuretic peptides inhibit the reabsorption of both Na^+ and water. One example from this family of hormones is **atrial natriuretic hormone (ANH)**, a 28-amino acid peptide produced by heart atria in response to overstretching of the atrial wall.

7.22 Diuretics and fluid volume

A diuretic is a compound that increases loss of water by interfering with the recapture of solutes and water from the filtrate, increasing urine volume. Coffee, tea, and alcohol are well-known examples of diuretics. The caffeine in coffee and tea promotes vasodilation in the nephron, increasing GFR. Alcohol brings about its effect by inhibiting ADH release from the posterior pituitary. Diuretics, often in the form of salts such as thiazides, are prescribed to patients with hypertension to reduce blood volume and, thereby, blood pressure.

References

[1] 2608 Kidney Position in Abdomen.jpg. Wikimedia Commons, the free media repository. Updated October 3, 2020. Accessed March 16, 2021. https://commons.wikimedia.org/w/index.php?title=File:2608_Kidney_Position_in_Abdomen.jpg&oldid=479892236.

[2] 2610 The Kidney.jpg. Wikimedia Commons, the free media repository. Updated October 3, 2020. Accessed March 16, 2021. https://en.wikipedia.org/wiki/File:2610_The_Kidney.jpg.

[3] Kidney Nephron.png. Wikimedia Commons, the free media repository. Updated September 20, 2020. Accessed March 16, 2021. https://commons.wikimedia.org/w/index.php?title=File:Kidney_Nephron.png&oldid=465532166.

[4] de Bowman.jpg. Wikimedia Commons, the free media repository. Updated October 28, 2020. Accessed March 16, 2021. https://commons.wikimedia.org/w/index.php?title=File:C%C3%A1psula_de_Bowman.jpg&oldid=504817690.

[5] 2626 Renin Aldosterone Angiotensin.jpg. Wikimedia Commons, the free media repository. Updated October 18, 2020. Accessed March 16, 2021. https://commons.wikimedia.org/w/index.php?title=File:2626_Renin_Aldosterone_Angiotensin.jpg&oldid=493486992.

[6] Physiology of Nephron.png. Wikimedia Commons, the free media repository. Updated September 9, 2020. Accessed March 16, 2021. https://commons.wikimedia.org/w/index.php?title=File:Physiology_of_Nephron.png&oldid=452930156.

[7] 2613 Podocytes.jpg. Wikimedia Commons, the free media repository. Updated October 17, 2020. Accessed March 16, 2021. https://commons.wikimedia.org/w/index.php?title=File:2613_Podocytes.jpg&oldid=492228684.

[8] 2614 Fenestrated Capillary.jpg. Wikimedia Commons, the free media repository. Updated October 12, 2020. Accessed March 16, 2021. https://commons.wikimedia.org/w/index.php?title=File:2614_Fenestrated_Capillary.jpg&oldid=487760549.

[9] 2617 Net Filtration PressureN.jpg. Wikimedia Commons, the free media repository. Updated October 15, 2020. Accessed March 16, 2021. https://commons.wikimedia.org/w/index.php?title=File:2617_Net_Filtration_PressureN.jpg&oldid=490561000.

[10] 2618 Nephron Secretion Reabsorption.jpg. Wikimedia Commons, the free media repository. Updated October 18, 2020. Accessed March 16, 2021. https://commons.wikimedia.org/w/index.php?title=File:2618_Nephron_Secretion_Reabsorption.jpg&oldid=493604836.

[11] 2619 Substances Reabsorbed and Secreted by the PCT.jpg. Wikimedia Commons, the free media repository. Updated September 17, 2020. Accessed March 16, 2021. https://commons.wikimedia.org/w/index.php?title=File:2619_Substances_Reabsorbed_And_Secreted_By_The_PCT.jpg&oldid=473292386.

CHAPTER

8

Blood pressure, hypertension, and exercise

OUTLINE

Biology of Cardiovascular and Metabolic Diseases
https://doi.org/10.1016/B978-0-12-823421-1.00007-X

OBJECTIVES

- Define arterial blood pressure and how it is generated.
- Explain how the values of the blood pressure measurements can be used to calculate mean arterial blood pressure.
- Describe how arterial systolic, diastolic, mean, and pulse pressure are affected by changes in the cardiac output and total peripheral resistance.
- Explain the causes of essential (primary) hypertension.
- Explain the mechanisms involved in the development of hypertension.
- Identify the factors that regulate arterial blood pressure.
- Describe how the baroreceptor reflex helps compensate for a fall in blood pressure. Map the aortic and carotid artery baroreceptor reflexes. Explain its significance in the regulation of blood pressure.
- Differentiate how sympathetic and parasympathetic activation affects blood pressure.
- Explain the chemoreceptor reflex in the regulation of blood pressure.
- Name the hormones that help regulate blood pressure and the mechanisms used to accomplish the same.
- Identify the mechanisms associated with exercise in lowering blood pressure.
- List the recommended types of exercise for hypertensive patients.

8.1 Introduction

Left ventricular systole and the ejection of blood into the aorta increases pressure and distends the blood vessel wall, which, in turn, creates a tension wave that is transmitted along the wall of the artery creating a pressure pulse. Thus arterial blood pressure is the force exerted by the blood on the walls of the arteries. With each stroke, the left ventricle ejects approximately 80 mL of blood during systole. As the aortic valve closes, blood flow continues toward the peripheral arteries because of its initial momentum and the elastic recoil of the arteries.

8.2 Components of arterial blood pressure

8.2.1 Systolic and diastolic pressure

Arterial blood pressure consists of several distinct components—**systolic** and **diastolic pressures, pulse pressure**, and **mean arterial pressure** (Fig. 8.1). The systemic arterial blood pressure in the normal healthy young adult is 110–120 mmHg systolic and 70–80 mmHg diastolic. The systolic pressure indicates the arterial pressure resulting from the ejection of blood during ventricular contraction, whereas the diastolic pressure represents the arterial pressure of blood during ventricular relaxation. These values change with age, being lower in infants and children than in adults. Moreover, the blood pressure values of women are

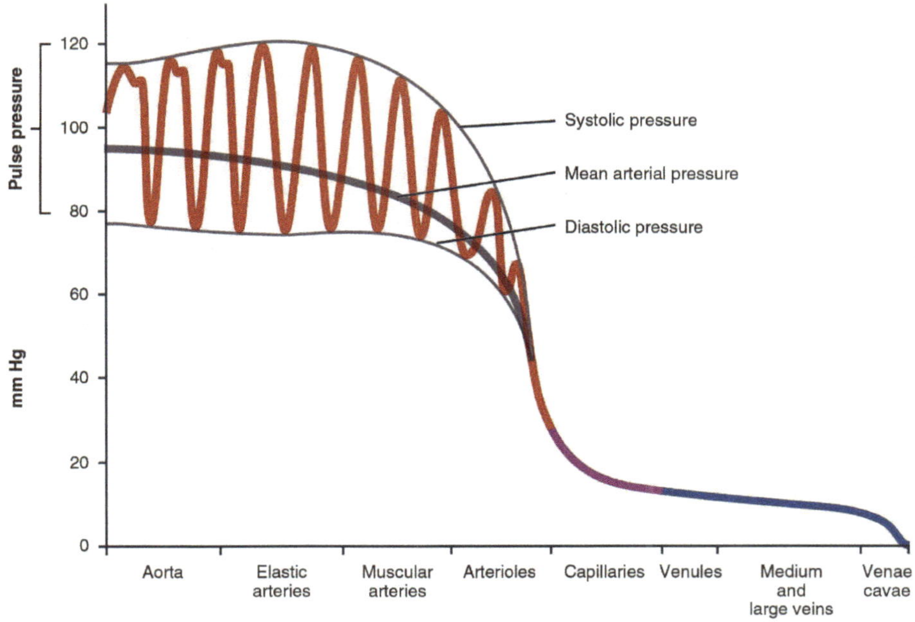

FIG. 8.1 Systemic blood pressure [1]. The graph shows the systolic, diastolic, mean arterial, and pulse pressures in the different types of blood vessels.

5–10 mmHg lower than those of males until the age of 50, after which there is no appreciable difference. In the elderly, an increase in the systolic and diastolic pressures is partially due to loss of elasticity of the arteries.

8.2.2 Pulse pressure

As shown in Fig. 8.1, the difference between the peak systolic pressure and diastolic pressure in the aorta is the **pulse pressure**. For example, a person with a systolic pressure of 120 mmHg and a diastolic pressure of 80 mmHg would have a pulse pressure of 40 mmHg. Pulse pressure should be at least 25% of the systolic pressure. The higher pulse pressure is observed in healthy people following strenuous exercise; if their resting pulse pressure is 30–40 mmHg, it may increase temporarily to 100 mmHg as stroke volume increases.

8.2.3 Mean arterial pressure

Mean arterial pressure (MAP) is the average blood pressure throughout the cardiac cycle and is the adequate pressure that drives blood throughout the circulatory system. MAP is nearer to the diastolic pressure than to the systolic pressure. Although complicated to measure directly, MAP can be approximated by adding the diastolic pressure to one-third of the pulse pressure (systolic pressure minus the diastolic pressure):

$$MAP = \text{diastolic BP} + (\text{systolic} - \text{diastolic BP})/3$$

In Fig. 8.1, the MAP is approximately $80+(120-80)/3$, which is 93.33 mmHg.

8.2.4 Pulse

When blood is ejected into the aorta from the left ventricle during systole, elastic fibers in the arteries allow expansion of the vessel size to accommodate the stroke volume, then recoil to maintain a high-pressure gradient. This expansion and recoiling effect, known as the **pulse**, can be palpated manually and measured electronically. Although the effect diminishes over distance from the heart, elements of the systolic and diastolic components of the pulse are still evident at the level of the arterioles.

Since each pulse represents one heartbeat and can be detected, it is clinically used to measure heart rate. The strength of the pulse can also be detected and is of significance as it can reflect the strength of ventricular contraction and cardiac output. If the pulse is strong, then systolic pressure is high, and if the pulse is weak, systolic pressure has fallen, and medical intervention may be warranted.

Pulse can be detected manually by placing the tips of the fingers across an artery that runs close to the body surface and gently pressing. Although radial artery in the wrist or the common carotid artery in the neck is often accessed, any superficial artery that can be palpated may be used (Fig. 8.2). A variety of commercial electronic devices are available to measure the pulse.

8.3 Regulation of blood pressure

The arterial blood pressure depends mainly on cardiac output and peripheral vascular resistance. Cardiac output is, in turn, dependent on stroke volume and heart rate. Therefore, any changes in these components will directly affect arterial blood pressure.

8.3.1 Autonomic regulation of blood pressure

Neural regulation of blood pressure and blood flow depends on the **cardiovascular centers** located in the medulla oblongata. A cluster of neurons in the medulla oblongata respond to changes in the blood pressure and blood concentrations of O_2, CO_2, and H^+. The cardiovascular center contains three distinct paired components:

- The **cardiac accelerator centers** stimulate cardiac function by regulating heart rate and stroke volume via sympathetic stimulation from the cardiac accelerator nerve.
- The **cardiac inhibitory centers** slow cardiac function by decreasing heart rate and stroke volume via parasympathetic stimulation from the vagus nerve.
- The **vasomotor centers** control vessel tone or contraction of the smooth muscle in the tunica media. Changes in the blood vessel diameter affect peripheral resistance, pressure, and flow, which, in turn, affect cardiac output.

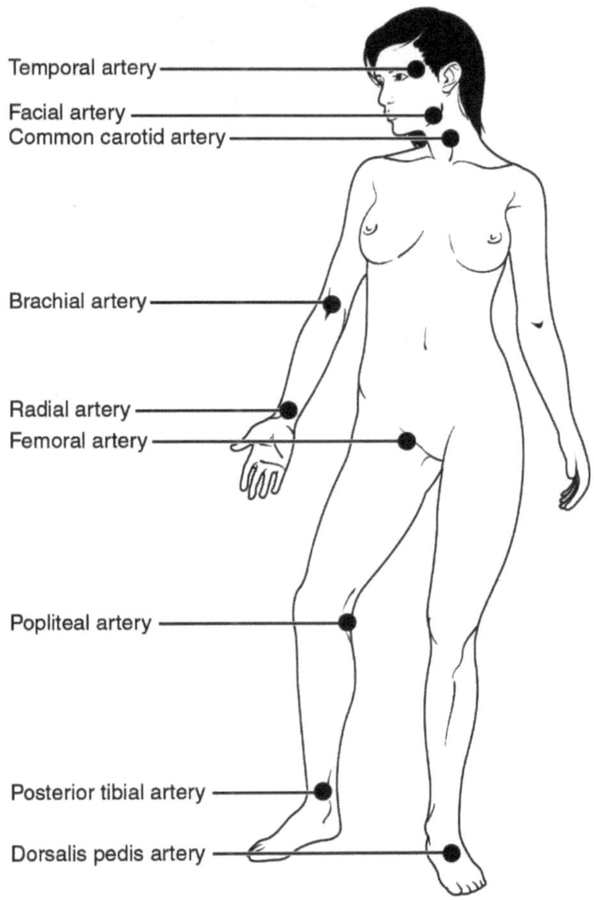

FIG. 8.2 Pulse sites [2]. The pulse is most readily measured at the radial artery but can also be detected at any of the pulse points shown.

8.3.2 Baroreceptor reflex

Vascular baroreceptors are mechanoreceptors that detect the change in tension in the arterial wall and respond to changes in blood pressure. They are found in the aortic arch and carotid sinus and form the afferent part of the baroreceptor reflex. In response to the change in pressure, baroreceptors send sensory signals to the cardiovascular center. For example, when blood pressure increases, the baroreceptors are stretched more tightly and initiate action potentials at a higher rate and trigger the parasympathetic stimulation of the heart which leads to a slower heart rate. As a result, cardiac output falls. In addition, the baroreceptors can inhibit the vasomotor center and, therefore, the inhibition of the action of the sympathetic system on peripheral vascular resistance resulting in vasodilation. Together, these activities cause blood pressure to fall. At lower blood pressures, the degree of stretch is less and the rate of firing is slower. When the cardiovascular center receives this input, it will trigger the cardioaccelerator center, increasing the sympathetic stimulation of the heart, causing

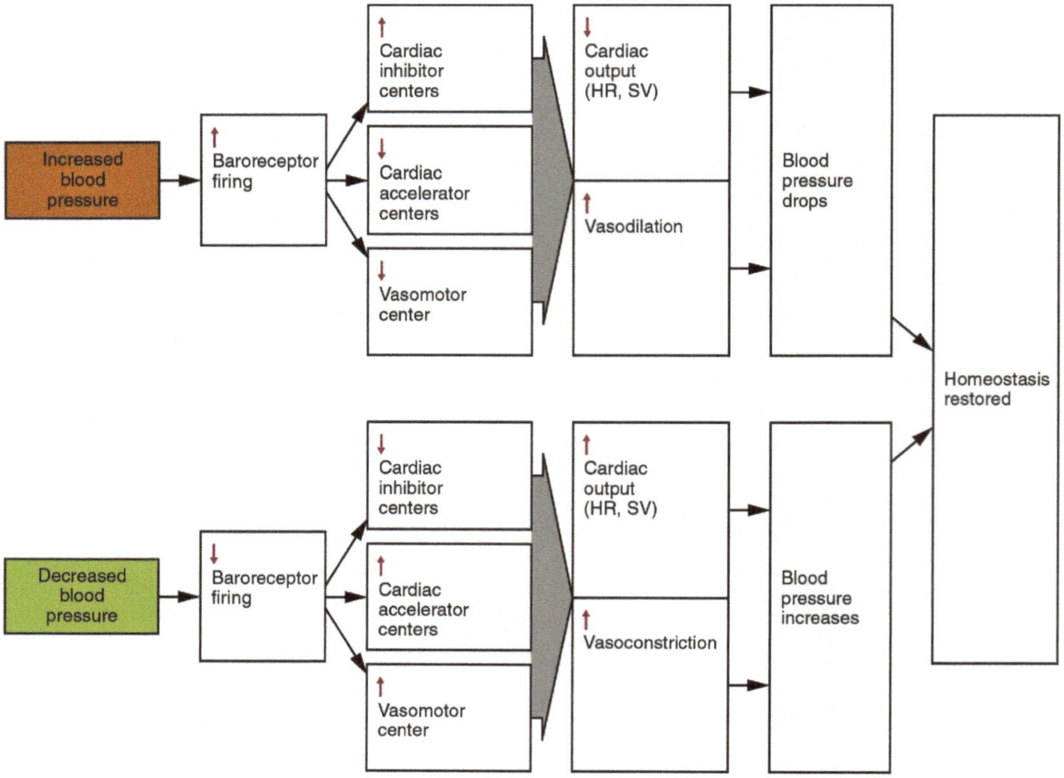

FIG. 8.3 Baroreceptor reflex for maintaining vascular homeostasis [3]. Increased blood pressure results in increased rate of baroreceptor firing, whereas decreased blood pressure results in slower rate , both initiating the homeostatic mechanism to restore blood pressure.

the cardiac output to increase. It will also trigger the vasomotor center, which activates the sympathetic tone and leads to increased vascular resistance. The combination of the activation of the sympathetic system and the inactivation of the parasympathetic system results in increased blood pressure (Fig. 8.3).

The baroreceptors in the venae cavae and the right atrium monitor blood pressure as the blood returns to the heart from the systemic circulation. Typically, blood flow into the aorta is the same as blood flow back into the right atrium. If blood is returning to the right atrium more rapidly than it is being ejected from the left ventricle, the atrial receptors will stimulate the cardiovascular centers to increase sympathetic firing and increase cardiac output until homeostasis is achieved. The opposite is also true. If a person rises quickly from a supine position, the venous return to the heart tends to be diminished which would lead to hypotension and fainting if rapid reflex mechanisms did not compensate by increasing the heart rate, the peripheral resistance, and venous tone.

8.3.3 Chemoreceptor reflex

Chemoreceptors monitor O_2, CO_2, and H^+ (pH) levels and, thereby, contribute to blood pressure regulation. The peripheral arterial chemoreceptors are located near the baroreceptors in the aortic and carotid sinuses and signal the cardiovascular and respiratory centers in the medulla oblongata when they detect any changes in the blood levels of O_2, CO_2, and H^+.

Tissues consume O_2 and produce CO_2 and acids as waste products. When the body is more active, O_2 levels fall, and CO_2 levels rise because the cells undergo cellular respiration to meet the increased energy needs, which also causes more H^+ to be produced and the blood pH to drop. Chemoreceptors respond to greater than normal levels of CO_2 and H^+ (falling pH) by stimulating the cardiac accelerator and vasomotor centers, increasing cardiac output, and constricting peripheral vessels. The cardiac inhibitory centers, on the other hand, are suppressed. Compared to the exercise period, during the resting period, O_2 levels are higher, CO_2 levels are lower, and less H^+, thus maintaining normal pH. With falling CO_2 and H^+ levels (increasing pH), the cardiac inhibitory centers are stimulated, and the cardiac accelerator and vasomotor centers are suppressed, decreasing cardiac output and causing peripheral vasodilation.

8.3.4 Endocrine regulation of blood pressure

As shown in Fig. 8.4, the neural regulation of blood pressure is possible within seconds, whereas the endocrine regulation offers long-term blood pressure regulation. Endocrine control of blood pressure involves several hormones, among which the major ones are epinephrine, norepinephrine, renin, angiotensin I and II, aldosterone, antidiuretic hormone, and erythropoietin.

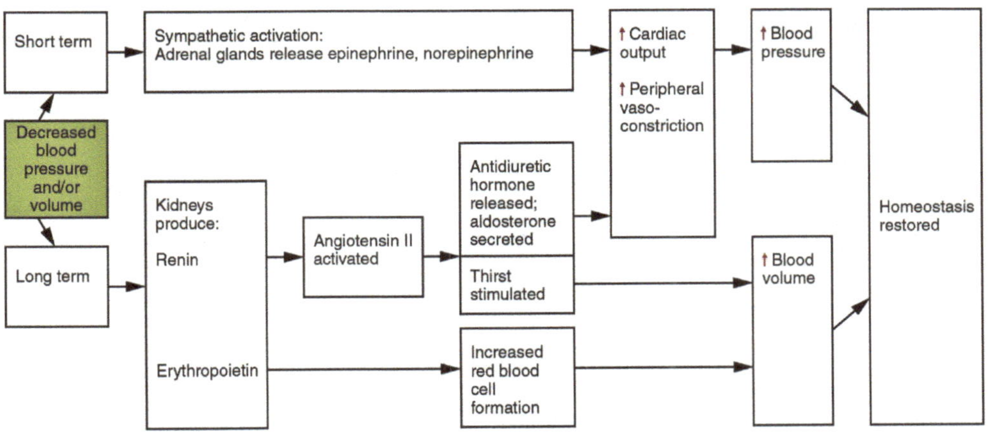

FIG. 8.4 Hormones involved in renal control of blood pressure [4]. While neural mechanisms alter blood pressure over the short term, the endocrine mechanism regulates blood pressure over the long term.

8.3.5 Epinephrine and norepinephrine

Epinephrine and norepinephrine are considered neurotransmitters when released by the neurons within a neural circuit and as hormones when released by the adrenal medulla directly into the blood supply in response to sympathetic stimulation. These two hormones bring about an increase in heart rate and an increase in the force of contraction (Fig. 8.4) while constricting blood vessels to organs not essential for flight-or-fight responses such as the gastrointestinal and renal structures.

8.3.6 Renin-angiotensin-aldosterone system

The renin-angiotensin-aldosterone system plays a vital role in the cardiovascular system (Fig. 8.4). Angiotensin II, a potent vasoconstrictor, contributes significantly to an increase in blood pressure. In addition, angiotensin II stimulates the release of **aldosterone** by the adrenal cortex, which, in turn, increases the reabsorption of Na^+ by the renal tubules drawing water back into the vascular system and increasing blood volume and blood pressure (see Chapter 7).

8.3.7 Antidiuretic hormone (ADH)

ADH, also called **vasopressin**, is synthesized by the neurons in the hypothalamus and stored in the posterior pituitary until released upon neural stimulation. An increase in osmolarity of the blood is a trigger prompting the hypothalamus to release ADH, which signals cells in the kidney tubules to reabsorb more water by inserting **aquaporins**, thus preventing additional fluid loss in the urine. This action of ADH will increase overall fluid levels within the body and help restore blood volume and blood pressure. In addition, ADH constricts peripheral vessels and therefore gets its other name, vasopressin.

8.3.8 Erythropoietin (EPO)

EPO is a peptide hormone made by the kidneys in response to less than the normal amount of O_2 available in its blood supply. EPO targets the red bone marrow to trigger erythrocyte production, which is expected to increase the O_2-carrying capacity within the body. In addition, EPO is a vasoconstrictor. Overproduction or the intake of synthetic EPO to enhance athletic performance (which is illegal) causes not only vasoconstriction but also increases viscosity which decreases blood flow and increases resistance and blood pressure.

8.4 Hypertension

The blood pressure in the range of 140/90 mmHg is considered hypertension. However, with recent epidemiological evidence, the definition of what is normal and hypertension has been lowered. The American College of Cardiology/American Heart Association guidelines replace "prehypertension" with "elevated blood pressure" for the blood pressure ranges of 120–129/80–89 mmHg [5,6]. In addition, the cutoff for the definition of hypertension has been lowered to ≥130/80 mmHg with a blood pressure range of 130–139/80–89 mmHg, now classified as stage 1 hypertension. Research has shown that for each 20/10-mmHg increase in blood pressure, the risk of mortality from stroke, ischemic heart disease, and other vascular causes doubles (Fig. 8.5) [7]. This finding appears to be the same for both men and women.

Hypertension can be **primary** (without a known cause) or **secondary** (caused by disease or drugs). Approximately 29% of the adult population is hypertensive, out of which 90%–95% of the patients suffer from primary or **essential** hypertension, and 5% are said to have **secondary** hypertension. Systolic hypertension is the most significant cause of organ damage. Thus, blood pressure must be maintained within a narrow range to prevent undesirable consequences such as end-organ destruction.

During the early developmental phase, cardiac output and blood volume may be elevated, but these variables usually become normal after hypertension is established. On the other hand, during an established phase of hypertension, an increase in total peripheral resistance (TPR) is a common condition. The increased TPR associated with established hypertension may be due to the pronounced structural adaptations that occur in the peripheral vascular bed, increased activity of the vascular smooth muscle fibers, increased sensitivity and reactivity of the vascular smooth muscle cells to external vasoconstrictor stimuli, and diminished production or effect of endogenous vasodilator substances.

Although both neural and hormonal influences may help initiate primary hypertension, the chronic elevation of blood pressure does not appear to be due to a sustained elevation in sympathetic vasoconstrictor discharge, nor is it due to a sustained elevation of any

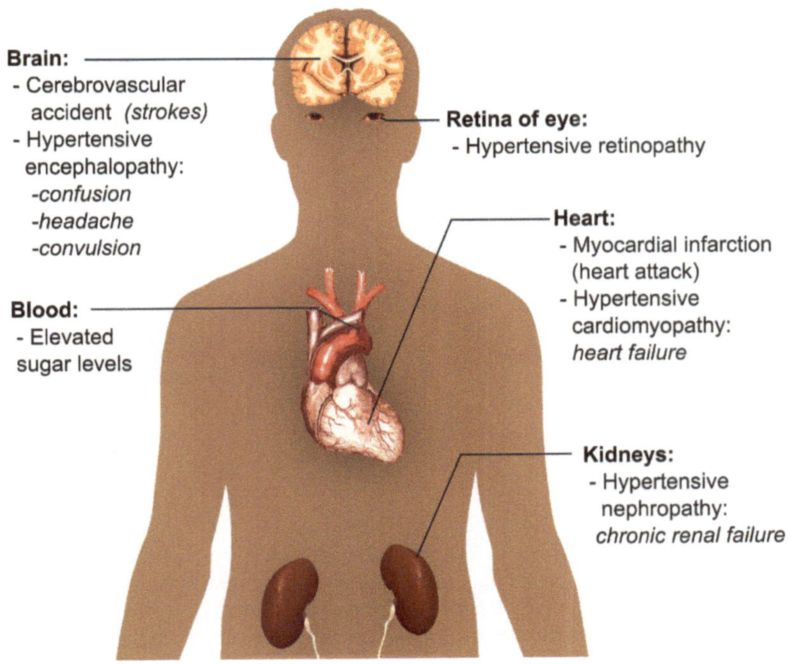

FIG. 8.5 Overview of major complications of persistent high blood pressure [8].

blood-borne vasoconstrictive factor. Instead, both baroreceptor reflexes and renal-dependent reflexes that are involved in regulating blood pressure become adapted or "reset" to regulate blood pressure at a higher than normal level.

Signs and symptoms of hypertension vary from having no symptoms in the early stages to nonspecific symptoms such as headache, palpitations, and exertional dyspnea. Long-term implications are target organ damage.

8.5 Pathogenesis of essential hypertension

Hypertension is a multifactorial disorder resulting from the cumulative effects of multiple genetic conditions as well as interacting environmental factors (Fig. 8.6). The major contributors to the pathogenesis of hypertension are discussed as follows.

8.5.1 Genetic factors

Genetic links for high blood pressure are well documented. Several single-gene disorders cause relatively rare forms of hypertension by altering net sodium reabsorption in the kidney. It is also suspected that variations in blood pressure may result from the cumulative effects of polymorphisms of several genes that affect blood pressure. For example, sequence variants in both angiotensinogen and angiotensin receptor genes have been associated with hypertension in some studies.

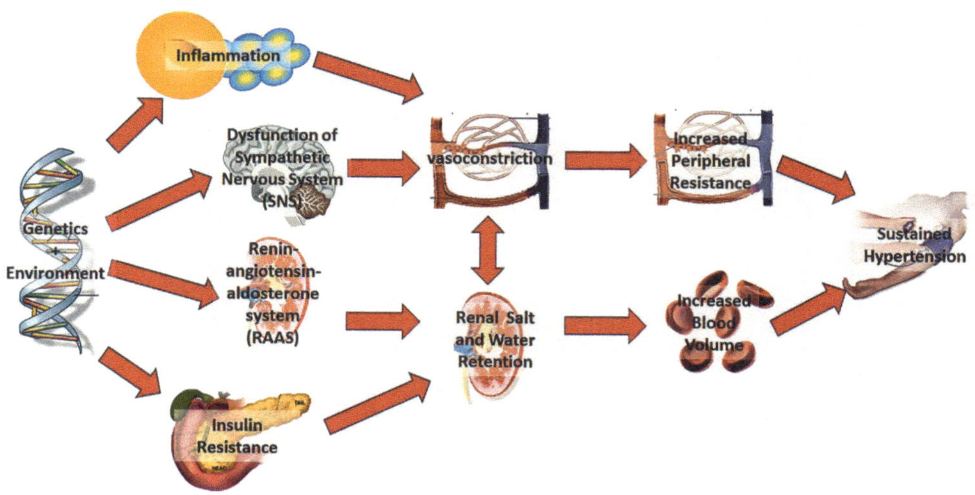

FIG. 8.6 Pathophysiology of hypertension [9–11]. Several genetic factors have been linked to hypertension, and these, in combination with environmental risks, could cause dysfunction of the sympathetic nervous system (SNS), renin-angiotensin-aldosterone system (RAAS), and trigger inflammation and insulin resistance. Such neurohumoral dysfunctions could contribute to sustained systemic vasoconstriction and, therefore, increased peripheral resistance. Inflammation combined with neurohumoral alterations may result in renal salt and water retention and, thus, increased blood volume. The two primary causes of sustained hypertension are increased peripheral resistance and increased blood volume.

8.5.2 Increased sympathetic activity

Factors that lead to vasoconstriction or stimuli that cause structural changes in the vessel wall can increase peripheral resistance, which may play a role in essential hypertension (Fig. 8.7). Elevated sympathetic activity increases peripheral resistance and causes vascular remodeling, which leads to narrowing of the vessels and triggers vasospasm of arteries. In addition, sympathetic activation contributes to insulin resistance which is associated with endothelial dysfunction and decreased production of vasodilators, such as nitric oxide. The sympathetic nervous system also has procoagulant properties, often resulting in thrombosis. All of these factors contribute to sustained increases in blood pressure (Fig. 8.7).

8.5.3 Activation of the renin-angiotensin-aldosterone system

Abnormal chronic activation of the renin-angiotensin-aldosterone system where elevated angiotensin II causes blood vessels to constrict, resulting in increased blood pressure. Aldosterone, another hormone released as part of the renin-angiotensin-aldosterone system activation, contributes to hypertension by promoting retention of Na^+ and water while secreting increased amounts of K^+ [12].

8.5.4 Inflammation and endothelial dysfunction

High blood pressure continuously stresses the blood vessel wall. Local inflammation resulting from the stress induced by high blood pressure causes the release of several chemical agents by the endothelial cells. Abnormally high levels of inflammatory agents such as

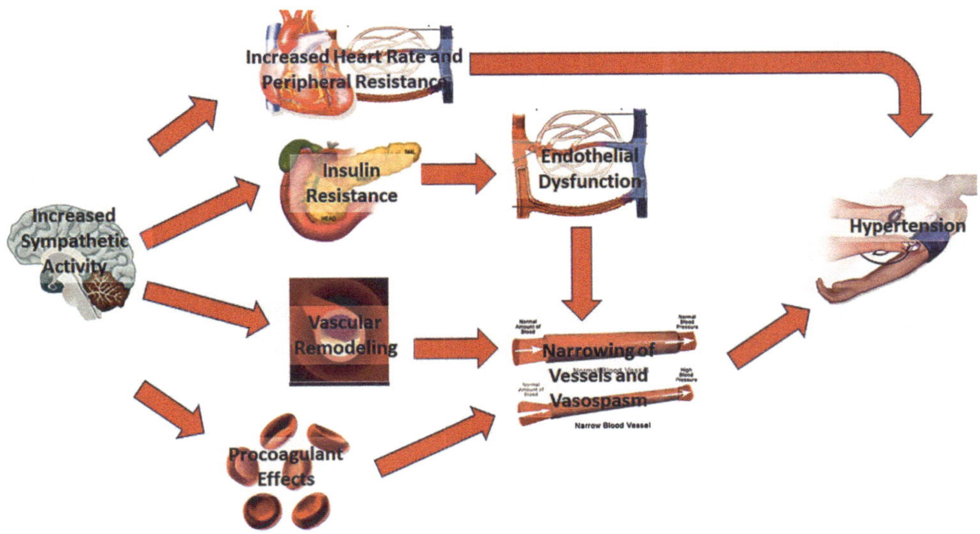

FIG. 8.7 The role of the sympathetic nervous system in the pathogenesis of hypertension [9–11]. Increased activity of the sympathetic nervous system (SNS) increases heart rate and peripheral resistance and causes vascular remodeling. In addition, increased activation of the SNS contributes to insulin resistance, which causes endothelial dysfunction, decreased production of vasodilators, such as nitric oxide, and procoagulant properties, which could cause vascular spasms and thrombosis. All of these factors contribute to sustained increases in blood pressure.

C-reactive protein, TNF-α, and IL-6 may also cause hypertension. Some of these inflammatory agents increase the growth of the smooth muscle lining of the blood vessel wall, thus causing changes in its connective tissue matrix. Additional changes include the proliferation of the endothelial cells, which leads to vascular luminal narrowing and leakage of plasma, causing local edema and activation of platelets resulting in the local formation of blood clots (thrombosis). The repair process followed by injury results in a thicker vascular lining and thus a narrower lumen. These changes manifest themselves in the form of ischemia of the brain, heart, or kidneys, pulmonary hypertension, and the subsequent right ventricular wall stress contributing to myocardial remodeling.

8.5.5 Environmental factors

Many environmental factors, including obesity, smoking, lack of physical activity, and heavy salt consumption, are implicated in the development of hypertension. The vast majority of adults eat more Na^+ than they should—an average of more than 3400 mg each day. According to the most recent Dietary Guidelines for Americans, it is recommended that Americans consume less than 2300 mg of Na^+ per day as part of a healthy eating pattern. Adults with high blood pressure, African Americans, the middle aged, and the elderly should consume no more than 1500 mg a day, according to the dietary guidelines from the U.S. Department of Agriculture. Diets that include the recommended five servings of fruits and vegetables per day combined with weight loss are beneficial for treating hypertension.

8.6 Hypertension and exercise

Approximately one in three American adults suffer from hypertension. One approach to minimize the development of hypertension is by reducing and eliminating risk factors such as restricting Na^+ intake, increasing K^+ intake, restricting saturated fat, adjusting caloric intake to maintain optimal weight, cessation of smoking, and an exercise program that promotes aerobic or cardiovascular exercise. In fit individuals and those undergoing training, blood pressure rises more slowly compared to those in poor cardiorespiratory condition, in whom blood pressure rises rapidly, steep, and large immediately after beginning or even before exercise.

The physiological effects of exercise vary in patients with hypertension and are classified as **acute** and **postexercise** or **chronic**. Genetic background, the cause of hypertension, pharmacodynamics, and pharmacokinetics may explain the different blood pressure responses to exercise among hypertensive patients. In addition, there is an inverse dose-response relationship between total quantity and intensity of exercise and cardiovascular mortality in both healthy and those with cardiovascular disease.

8.6.1 Acute physiological response to exercise

Endurance exercise typically causes an increase and redistribution of cardiac output to maintain perfusion of active muscles. Neurohormonal and hydrostatic mechanisms mainly trigger this response. While systolic blood pressure increases, diastolic pressure either stays the same or decreases due to decreased peripheral resistance, facilitating perfusion of large muscle groups. In resistance exercise, both systolic and diastolic blood pressure increases due to baroreceptor reflex where intramuscular pressure during contraction exceeds arterial blood pressure interrupting muscle blood flow. As a result, arterial pressure increases to overcome this resistance to muscle perfusion. If inappropriately used, resistance exercise can cause systolic blood pressure to rise to 320 mmHg and diastolic blood pressure up to 250 mmHg during a single repetition at the maximum workload.

8.6.2 Chronic or postexercise physiological response

Systolic and diastolic blood pressure can be reduced by approximately 6–10 mmHg with regular aerobic exercise for many previously sedentary men and women regardless of age. Blood pressure falls within minutes or hours after exercise in hypertensive and prehypertensive patients but to a lesser degree in normotensive subjects. This decrease in pressure could last up to 22 h after exercise. Thus regular aerobic exercise as preventative medicine contributes to controlling blood pressure to increase over time in individuals at risk for hypertension. The effects of exercise training on blood pressure are impressive in most patients with moderate or "borderline" hypertension.

Several factors influence the hypotensive response, including the duration and type of exercise and the individual's clinical status, age, ethnicity, and physical fitness. The duration appears to influence the magnitude and the length of the fall in blood pressure. More extended but less intense periods of exercise seem to produce the same results as shorter and more intense sessions. Intermittent exercise protocols and those that use more muscle mass may also enhance the blood pressure-lowering effect. Postexercise blood pressure response appears to differ with ethnicity. For example, blood pressure in African American hypertensive women does not necessarily fall after exercise.

The aerobic response has been shown to reduce hypertension incidence and lower blood pressure in hypertensive subjects, with systolic blood pressure falling by 5–15 mmHg (less in the elderly). The overall reduction is directly proportional to the duration and frequency of exercise. It appears to be independent of body mass index and is more significant in middle-aged hypertensive patients and uncontrolled hypertensive individuals. During rhythmic muscular activity such as jogging, swimming, and bicycling, dilation of the blood vessels in the active muscles decreases total peripheral resistance, thus enhancing blood flow through most of the peripheral vasculature. The alternate contraction and relaxation of the muscles also provide an effective force to move blood through the vascular circuit and return it to the heart. The increased blood flow during rhythmic steady-rate exercise causes systolic pressure to rise quickly during the first few minutes of exercise. The blood pressure then levels off at approximately 140–160 mmHg, with probably no difference between sexes. As exercise continues, systolic pressure may gradually decline as the arterioles in the muscles continue to dilate, and there is a reduction in peripheral resistance to blood flow. The diastolic blood pressure remains relatively unchanged during this exercise.

Resistance exercise has also been shown to reduce both systolic and diastolic blood pressure by 3–5 mmHg by increasing parasympathetic tone, reducing peripheral resistance when the exercise regime concentrates on muscle resistance. However, given that the fall in blood pressure following both endurance and resistance exercise is relatively small, it is clear that exercise does not cure hypertension, but it serves as an excellent adjunctive therapy.

Exercise requiring a large muscle mass and greater relative strain elicits the most significant blood pressure response. This response is due to greater stimulation of the cardiovascular center, by the active areas of the motor cortex and extensive peripheral feedback to this center from the contracting muscle mass. The acute cardiovascular strain with heavy resistance exercise could be harmful to individuals who have heart and vascular disease, particularly for those untrained in this form of exercise. For these people, more rhythmic forms of moderate exercise are beneficial.

Although resistance exercise causes a significant rise in blood pressure than lower-intensity dynamic movement, it does not bring about any long-term increase in the resting blood pressure. Furthermore, a regular program of resistance training blunts the blood pressure response to this form of exercise. Trained bodybuilders, for example, show smaller increases in systolic and diastolic blood pressure with resistance exercise than both novice bodybuilders and untrained individuals. Most research indicates that standard resistance training is less effective in lowering resting blood pressure than programs of regular aerobic exercise, although a few positive effects of resistance training have been reported. Heavy resistance exercise is not recommended as the only form of training to lower blood pressure in hypertensive patients.

8.6.3 Graded exercise and blood pressure

After the initial rapid rise from the resting level, systolic blood pressure increases linearly with exercise intensity, whereas diastolic pressure remains stable or increases slightly at the higher levels of exercise. This response is similar to both trained and sedentary subjects. During maximum exercise performed by healthy fit men and women, the systolic blood pressure may increase to 200 mmHg or higher, despite significant reductions in total peripheral resistance. This response is most likely due to the large output of blood from the heart during maximal exercise.

8.6.4 Upper body exercise and blood pressure

Both systolic and diastolic blood pressures are considerably higher when work is performed with the arms than with the legs. This is because the smaller muscle mass and vasculature of the arms offer greater resistance to blood flow than the larger mass and vasculature of the legs. This form of exercise clearly represents greater cardiovascular strain because the work requirement of the myocardium is increased considerably. These observations support the use of exercise for individuals with cardiovascular dysfunction, requiring large muscle groups such as walking, bicycling, and running, in contrast to unregulated exercises that engage a somewhat limited muscle mass such as shoveling, overhead hammering, or even arm-crank exercises. After a sustained submaximal exercise, systolic blood pressure is temporarily reduced below preexercise levels for both normotensive and hypertensive individuals. This hypotensive response to the previous exercise can last for up to 12 h into recovery and occurs in response to low- and moderate-intensity aerobic exercise.

8.6.5 Mechanisms involved in the blood pressure-lowering and cardioprotective effects of regular exercise

The precise mechanism of lowering blood pressure by exercise is not well understood. However, significant contributing factors may include (A) decreased activity of the sympathetic nervous system would decrease peripheral resistance to blood flow and a subsequent reduction in blood pressure; (B) altered renal function facilitates the elimination of Na^+ by the kidneys to reduce fluid volume and blood pressure; (C) decreased levels of angiotensin II, endothelin, and receptors in the central nervous system reduce peripheral resistance and increase baroreceptor reflex sensitivity; and (D) vasodilator effects of prostaglandins and nitric oxide released during exercise may contribute to a decrease in blood pressure.

Structural adaptations include vascular remodeling and angiogenesis. One mechanism proposed for postexercise hypotension is that when aerobic exercise ceases, there is a prolonged period during which a significant quantity of blood remains pooled in the visceral organs and lower limbs. Excessive blood in the visceral organs and lower limbs reduces the central blood volume, which causes a lowering of systemic arterial blood pressure. These findings further support the cause of exercise as a crucial nonpharmacologic line of defense in treating hypertension and would justify participating in several periods of moderate physical activity interspersed throughout the day.

Appropriate medical screening is essential before participating in exercises. Caution should be taken to identify contraindications to particular types of exercise, limitations in the practice of particular types of exercise, and to tailor the exercise regime to each clinical condition to minimize complications and promote rehabilitation. For hypertensive patients, the specific concerns to diagnose silent cardiovascular disease, to identify individuals at high cardiovascular risk, such as those with uncontrolled hypertension, unstable angina, or decompensated diabetes, who should not exercise until their clinical condition has been stabilized and to prepare a personalized plan designed to avoid activities that will tend to raise blood pressure. For patients with systolic pressure of 160 or greater and a diastolic pressure of 100 or higher, blood pressure must be controlled first before exercise.

8.6.6 Prescription of exercise

Exercise type: Endurance exercise is most preferred. Rhythmic aerobic exercise engaging the main muscle groups, such as walking, running, cycling, and swimming, reduces BP by 5–15 mmHg. The response to resistance training is marked (5 mmHg), but it can be more significant if it involves more repetitions of lighter loads. Diastolic blood pressure rises with greater static work, which should be kept to a minimum. A resistance training program should be reviewed if the program increases DBP more than 20 mmHg over a more than 120 mmHg baseline.

Exercise frequency: Exercise for 3–5 days a week reduces BP. However, there is evidence that 7 days a week may be more effective; three sessions a week have 75% of the antihypertensive effect of seven sessions.

Exercise intensity: Exercise intensity of less than 70% of VO_2 reserve appears to have a more marked BP-lowering effect that does not depend on improved maximal aerobic capacity.

Exercise duration: Continuous exercise from 30 to 60 min per session.

In summary, exercise is recommended as a treatment for hypertension, particularly cardiovascular exercise for 20–60 min, 3–5 days a week, at an intensity of 40%–70% VO_2 max. Unfortunately, not all hypertensive patients respond in the same way to exercise; around 25% show no blood pressure fall, although they still derive other benefits. Nevertheless, regular exercise helps reduce body weight, visceral adiposity, and inflammation and prevents or reduces age-related arterial stiffness, and improves endothelial function, thus explaining its overall benefit on cardiovascular risk profile independently of reductions in blood pressure.

References

[1] 2109 Systemic Blood Pressure.jpg. Wikimedia Commons, the free media repository. Updated October 15, 2020. Accessed March 17, 2021. https://commons.wikimedia.org/w/index.php?title=File:2109_Systemic_Blood_Pressure.jpg&oldid=490560768.

[2] 2110 Pulse Sites.jpg. Wikimedia Commons, the free media repository. Updated October 6, 2020. Accessed March 17, 2021. https://commons.wikimedia.org/w/index.php?title=File:2110_Pulse_Sites.jpg&oldid=482449511.

[3] 2116 Baroreceptor Reflex Flow Art.jpg. Wikimedia Commons, the free media repository. Updated September 26, 2020. Accessed March 17, 2021. https://commons.wikimedia.org/w/index.php?title=File:2116_Baroreceptor_Reflex_Flow_Art.jpg&oldid=471706548.

[4] 2117 Renin Angiotensin Aldosterone Pathway.jpg. Wikimedia Commons, the free media repository. Updated September 11, 2020. Accessed March 17, 2021. https://commons.wikimedia.org/w/index.php?title=File:2117_Renin_Angiotensin_Aldosterone_Pathway.jpg&oldid=454831451.

[5] P.K. Whelton, R.M. Carey, W.S. Aronow, et al., 2017 ACC/AHA/AAPA/ABC/ACPM/AGS/APhA/ASH/ASPC/NMA/PCNA Guideline for the prevention, detection, evaluation, and management of high blood pressure in adults: executive summary: a report of the American College of Cardiology/American Heart Association task force on clinical practice guidelines. Practice guideline systematic review, Circulation 138 (17) (2018) e426–e483, https://doi.org/10.1161/CIR.0000000000000597.

[6] Correction to: 2017 ACC/AHA/AAPA/ABC/ACPM/AGS/APhA/ASH/ASPC/NMA/PCNA guideline for the prevention, detection, evaluation, and management of high blood pressure in adults: executive summary: a report of the American college of cardiology/American heart association task force on clinical practice guidelines. Published erratum, Hypertension 72 (3) (2018) e33, https://doi.org/10.1161/HYP.0000000000000080.

[7] S. Lewington, R. Clarke, N. Qizilbash, R. Peto, R. Collins, Age-specific relevance of usual blood pressure to vascular mortality: a meta-analysis of individual data for one million adults in 61 prospective studies. Meta-analysis research support, Non-U.S. Gov't research support, U.S. Gov't, P.H.S, Lancet 360 (9349) (2002) 1903–1913, https://doi.org/10.1016/s0140-6736(02)11911-8.

[8] Main complications of persistent high blood pressure.png. Wikimedia Commons, the free media repository. Updated September 7, 2020. Accessed March 17, 2021. https://commons.wikimedia.org/w/index.php?title=File:Main_complications_of_persistent_high_blood_pressure.png&oldid=451160424.

[9] Blausen 0761 RedBloodCells.png. Wikimedia Commons, the free media repository. Updated October 25, 2020. Accessed March 19, 2021. https://commons.wikimedia.org/w/index.php?title=File:Blausen_0761_RedBloodCells.png&oldid=501278904.

[10] Blausen 0322 DNA 2.png. Wikimedia Commons, the free media repository. Updated October 25, 2020. Accessed March 19, 2021. https://commons.wikimedia.org/w/index.php?title=File:Blausen_0322_DNA_2.png&oldid=501290261.

[11] Blausen 0092 BloodPressureFlow.png. Wikimedia Commons, the free media repository. Updated October 25, 2020. Accessed March 19, 2021. https://commons.wikimedia.org/w/index.php?title=File:Blausen_0092_BloodPressureFlow.png&oldid=501295838.

[12] V.G. DeMarco, A.R. Aroor, J.R. Sowers, The pathophysiology of hypertension in patients with obesity, Nat. Rev. Endocrinol. 10 (6) (2014) 364–376, https://doi.org/10.1038/nrendo.2014.44.

CHAPTER

9

Cellular metabolism

OUTLINE

OBJECTIVES

- Define metabolism and its two aspects.
- Discuss the role of ATP in metabolism.
- Compare oxidation and reduction reactions.
- Explain the role of NAD and FAD in oxidation-reduction reactions.

Biology of Cardiovascular and Metabolic Diseases
https://doi.org/10.1016/B978-0-12-823421-1.00003-2

- Identify the hormones that are important in altering metabolism and how they bring about any changes.
- Describe how dietary carbohydrates are digested and absorbed.
- Define glycolysis and locate where in the cell it occurs and the net yield of this reaction.
- Explain the cause for the conversion of glucose to lactic acid.
- Summarize the metabolism of glucose within the mitochondrion, including the net yield of products.
- Identify the part of the aerobic pathway where oxygen is utilized and how.
- Compare aerobic and anaerobic reactions.
- Define gluconeogenesis, glycogenesis, and glycogenolysis.
- Describe how proteins can be used as energy sources.
- Define essential and nonessential amino acids.
- Identify the significance of deamination and transamination.
- Spot the location and the significance of the urea cycle.

9.1 Introduction

The body is made up of approximately two-thirds of water, whereas the rest consists of proteins, lipids (fats), carbohydrates, nucleic acids, minerals, and vitamins. Food is the source of all of these components. Several chemical reactions occur to obtain nutrients from the food, convert them into different forms, and obtain energy to allow such reactions to be carried out.

Metabolism refers to the sum of all chemical reactions within the body and is subdivided into **catabolism** (*cata* =downward) and **anabolism** (*ana* =upward). For example, the reactions governing the breakdown of food to obtain energy are **catabolic** reactions. Conversely, **anabolic** reactions are reactions that synthesize large molecules from smaller ones using energy that is obtained by catabolic reactions. Thus the two types of chemical reactions—catabolic and anabolic—are coupled to one another. Metabolic processes occur constantly and are essential in maintaining normal functions within the body.

Catabolic reactions produce energy and anabolic reactions consume energy. If the net energy change is positive, i.e., catabolic reactions releasing more energy than the anabolic reactions use, then the body stores excess energy by building glycogen and lipid molecules. On the other hand, if the net energy change is negative where catabolic reactions release less energy than anabolic reactions use, the body targets stored energy where lipid and glycogen molecules are broken down to their simpler forms to compensate for the deficiency of energy released by catabolism.

The rate of metabolism varies depending on age, gender, activity level, food intake, and lean body mass. Aging tends to decrease the metabolic rate. Men have a lean body mass and a higher basal metabolic rate (metabolic rate at rest) compared to women. Additionally, an individual's inherent metabolic rate is determined by their genetic makeup, which permits specific proteins and lipids to be derived from their genetic background. For example, specific lipoproteins are expressed in higher or lower concentrations based on the genetic condition. It also depends on the different types of food consumption. Thus one may modify metabolic events by changing diet and exercise regimen.

9.2 Catabolic reactions

Catabolic reactions break down a large molecule such as glucose into smaller molecules, CO_2, and water, releasing energy. The energy released is not 100% efficient since the amount of energy released is less than the actual amount contained in the molecule. Approximately 40% of the energy released from catabolic reactions is directly transferred to a high-energy molecule, **adenosine triphosphate** (**ATP**), a major energy currency in the body that can be used immediately to support functions within the cell. Excess ATP can also be stored to some extent to fulfill future energy demands. The remaining 60% of the energy released from catabolic reactions is given off as heat, which tissues and body fluids absorb and help maintain normal body temperature.

Structurally, an ATP molecule consists of a nitrogenous base called **adenine**, a five-carbon sugar called **ribose** (a monosaccharide), and three phosphate groups (Fig. 9.1). The chemical bond that links the second and third phosphate groups is called a high-energy bond and contains the most significant energy source compared to other bonds within the ATP molecule. Catabolic enzymes often break the first bond when cells require energy to do work, and this process is referred to as **dephosphorylation**. The products of dephosphorylation are a molecule of **adenosine diphosphate** (**ADP**) and a phosphate group (P). ATP, ADP, and P are constantly being cycled between reactions. The building of ATP through a reaction is called **phosphorylation** and is used in the storage of energy, and reactions that break down ATP release energy (Fig. 9.1). The energy from ATP drives all bodily functions, such as muscle contraction, breathing, and absorption of food in the gastrointestinal tract.

There are four major macromolecules. Those are **carbohydrates**, **lipids**, **proteins**, and **nucleic acids**. Carbohydrates are the most common source of energy to fuel the body and are divided into **monosaccharides**, **disaccharides**, and **polysaccharides**. Monosaccharides are simple sugars and include **glucose** (Fig. 9.2), **galactose**, and **fructose**, where glucose is the most common fuel for ATP production in cells. Both galactose and fructose are converted to glucose within the liver. Disaccharides include **sucrose** (table sugar that is made up of glucose and fructose), **lactose** (milk sugar consisting of glucose and galactose), and **maltose** (malt sugar composed of two glucose molecules). Polysaccharides consist of **glycogen** (Fig. 9.3) and **starch** (long chains of glucose). Another polysaccharide is **cellulose**, but humans do not carry enzymes to break down this fibrous molecule. Although indigestible polysaccharides do not provide any nutritional value, they provide dietary fiber, which helps propel food through the alimentary canal. There are several hormones involved in the regulation of glucose concentration in the bloodstream. Excess glucose is stored as glycogen, an energy reserve in the liver and muscles, or converted into **triglyceride** (**fat**) in the liver or fat tissue.

Dietary lipids, commonly known as fat, appear in two forms, **saturated** and **unsaturated**. Saturated fat is different from unsaturated fat by having only single covalent bonds within the hydrocarbon chains of their fatty acids, unlike in unsaturated fat, where there may be one or more double covalent bonds. Moreover, saturated fat remains solid at room temperature, whereas unsaturated fat holds its liquid form. A common source of saturated fat is red meat, egg yolk, butter, and some plant products such as coconut and palm oil. On the other hand, unsaturated fat is rich in sunflower, corn, canola, soybean, flaxseed oils, walnuts, and fish.

Based on their structures, lipids can be classified into three groups. Triglycerides are the most abundant types of lipids. Structurally, as the name suggests, they are made of three **fatty acids** and one **glycerol** molecule. Fatty acids vary in size, but they typically contain an even

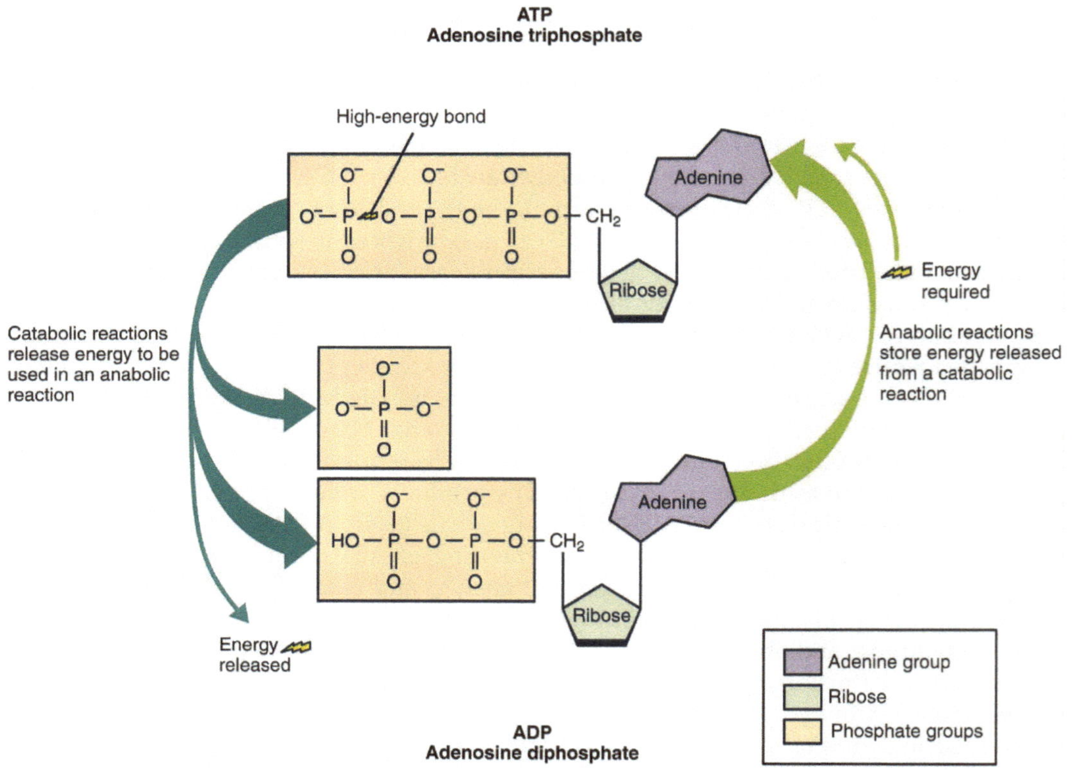

FIG. 9.1 Catabolic reactions break down a large molecule such as glucose into smaller molecules, CO_2, and water, releasing energy [1]. However, the energy released is not 100% efficient since the amount of energy released is less than the actual amount contained in the molecule.

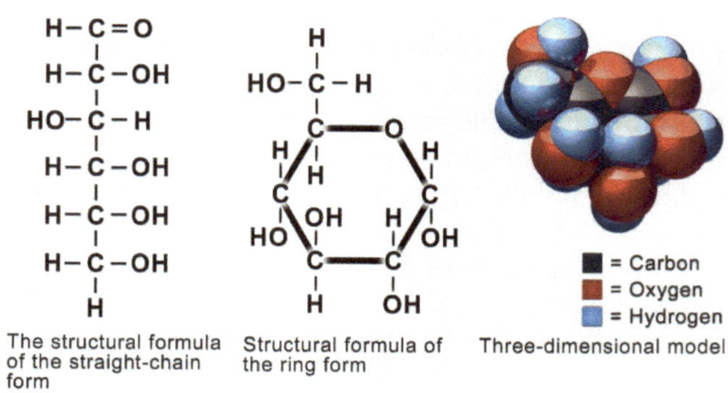

FIG. 9.2 The structure of glucose. The diagram shows the structural arrangement of atoms in glucose [2]. Glucose consists of 6 carbons, 12 hydrogens, and 6 oxygen atoms with the chemical formula of $C_6H_{12}O_6$.

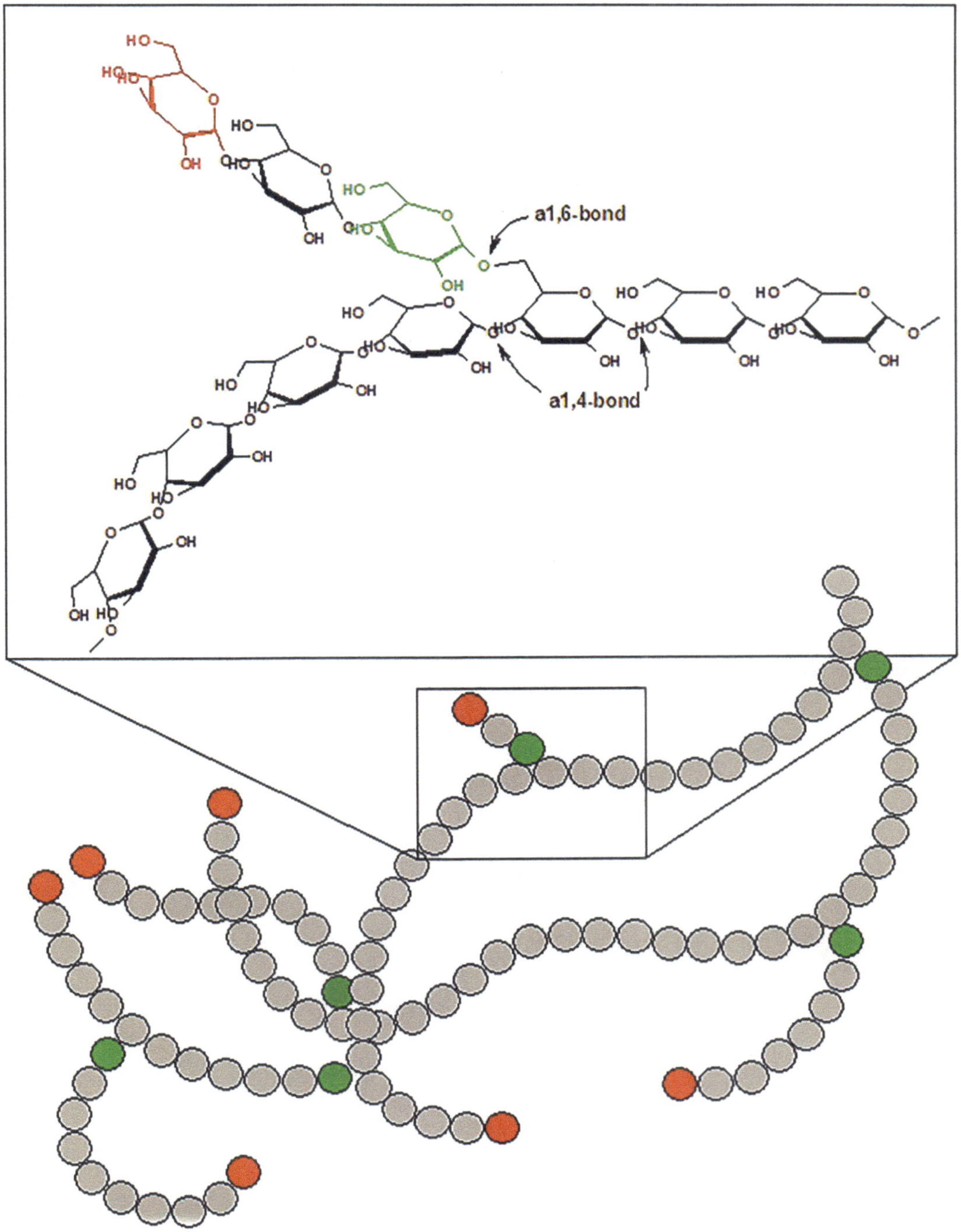

FIG. 9.3 The straight chain of glucose residues (lower) and a branched chain indicate a partial structure of glycogen (upper) figure [3].

number of carbon atoms in a straight chain. For example, **stearate** contains 18 carbon atoms, and **palmitate** contains 16 carbon atoms. Both stearate and palmitate are **saturated** fatty acids containing single covalent bonds between carbon and hydrogen, allowing a maximum number of hydrogen atoms. Fatty acids such as **linoleate** and **oleate** have a single **unsaturation** between C9 and C10 and are referred to as **essential fatty acids** since they must be obtained through diet. Both linoleate and oleate are found in abundance in nuts and seeds. Triglycerides are stored in the adipose tissue and serve as a major energy reserve. About one-half of excess fat is stored in adipocytes that accumulate in the subcutaneous tissue (under the skin), whereas the rest is found lining internal organs. **Phospholipids** are a type of lipids that are important in the structure of membranes where they form a lipid bilayer and regulate the movements of substances across the membrane. An example of a phospholipid bilayer is a cell membrane. All other membranes within the cell, including nuclear membrane and endoplasmic reticulum, consist of similar arrangements. **Steroids** are another class of lipids and are important in the synthesis of hormones and bile salts. **Cholesterol** belongs to the steroid family (Fig. 9.4).

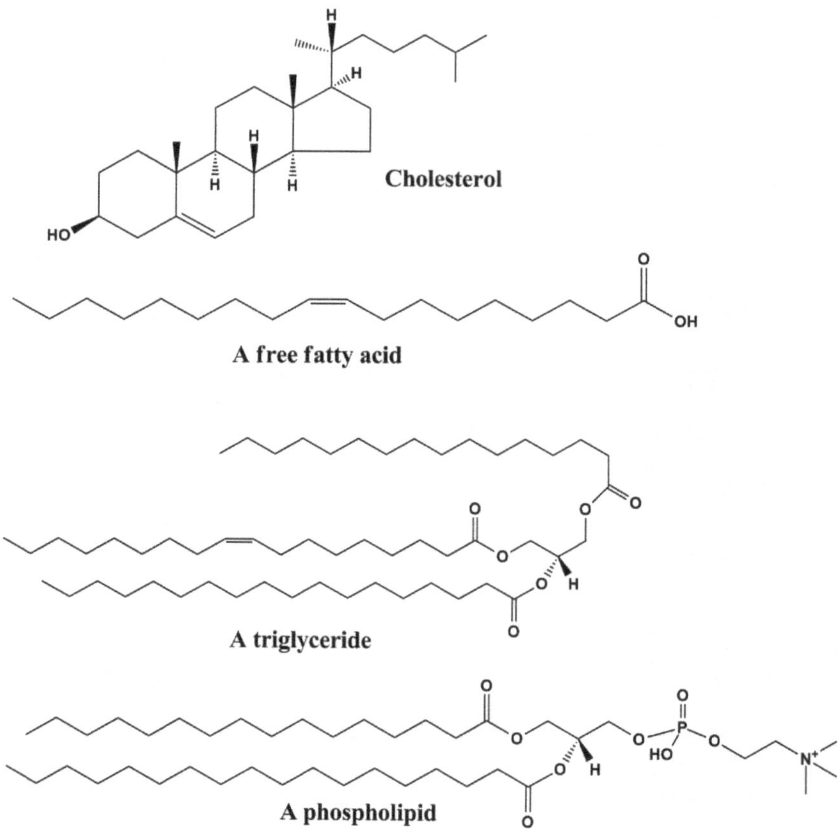

FIG. 9.4 Structure of different types of lipids [4]. Cholesterol is a ring-like molecule that represents the steroid family of lipids. Free fatty acid represents three fatty acids used in the construction of a triglyceride, a common type of lipid. A phospholipid consists of a phosphate group as part of its lipid structure. https://upload.wikimedia.org/wikipedia/commons/2/20/Common_lipids_lmaps.png.

Proteins are made of **amino acids** that are linked together by peptide bonds to form long chains. Digestion of proteins releases amino acids which can be used as building blocks of new proteins or broken down further to produce ATP (Fig. 9.3). Approximately 14% of the body consists of proteins that can be found in many forms, such as the structure of collagen fibers, contractile proteins, enzymes, and receptors. If an individual is starving for weeks, the use of amino acids for energy production occurs by the breakdown of proteins which can cause wasting away of the body structures (Fig. 9.5).

Nucleic acids are found in most of the foods we consume. During digestion, nucleic acids such as **deoxyribonucleic acid** (**DNA**) and **ribonucleic acid** (**RNA**) are broken down into their simpler forms called **nucleotides**. Each nucleotide consists of a nitrogenous base, a monosaccharide, and a phosphate group. These nucleotides are readily absorbed and transported throughout the body to be used by individual cells.

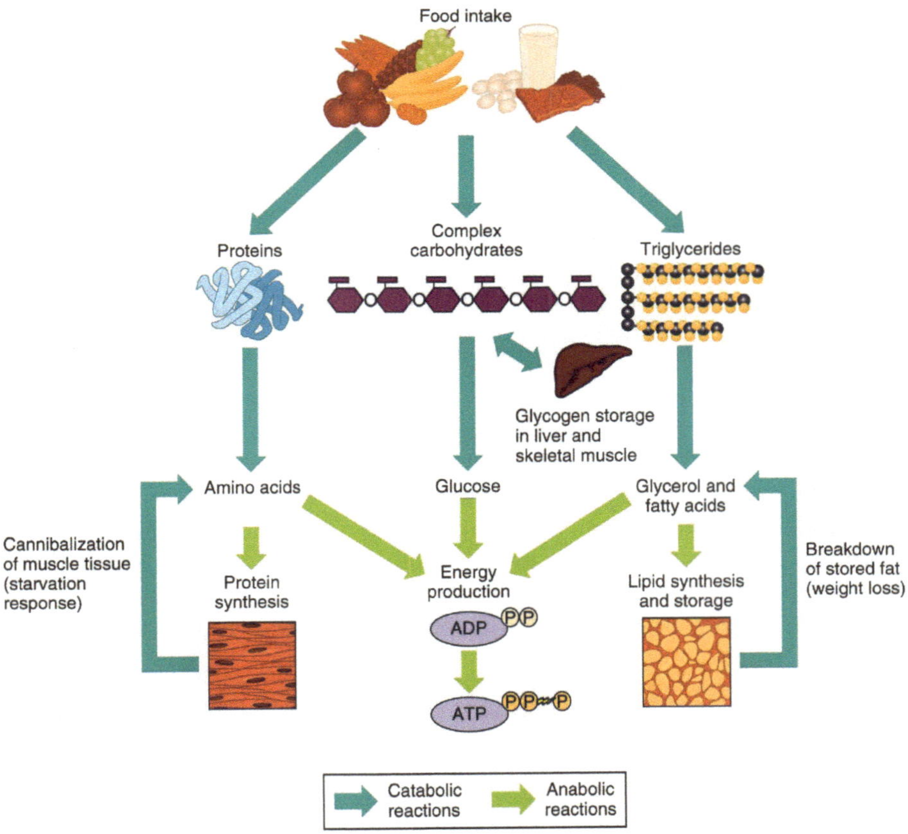

FIG. 9.5 Sources of ATP during catabolic reactions [5]. Proteins are broken down into amino acids, polysaccharides are broken down into glucose, and triglycerides are broken down into fatty acids and glycerol. Amino acids, fatty acids, and glucose are broken down further to release ATP. During anabolic reactions, amino acids, glucose, and fatty acids are used to build larger molecules such as proteins, glycogen, and triglycerides.

9.3 Anabolic reactions

Anabolic reactions, in contrast to catabolic reactions, involve the building of larger molecules from smaller ones. For example, anabolic reactions combine glucose to form polysaccharides, fatty acids to triglycerides, amino acids to proteins, and nucleotides to nucleic acids. In addition, anabolic reactions require energy in the form of ATP generated by catabolic reactions. Thus anabolic reactions create new molecules essential in constructing cell structures and, ultimately, new cells and tissues.

9.4 Oxidation and reduction reactions

Metabolic events often involve **oxidation** and **reduction** reactions. The oxidation reaction is associated with the loss of an electron, whereas the reduction reaction gains an electron. When a molecule gives up a hydrogen atom (H), it is released in a hydrogen ion (H^+) and an electron. The loss of H in the case of carbon would allow carbon (C) to bind to O_2, and it is oxidized. On the other hand, when H is bound to C, there is less binding of O_2 to C, and therefore it is reduced. Oxidation-reduction reactions are catalyzed by enzymes whose actions may be enhanced by coenzymes. The two most common coenzymes of oxidation-reduction reactions are **nicotinamide adenine dinucleotide** (**NAD**) and **flavin adenine dinucleotide** (**FAD**). Their respective reduced coenzymes are **NADH** and **$FADH_2$**, which are energy-containing molecules used to transfer energy during the creation of ATP.

9.5 Hormonal regulation of metabolism

Several hormones are important in the regulation of metabolism, and those are **cortisol** (**hydrocortisone**), **glucagon**, **epinephrine** (**adrenaline**), **growth hormone**, **insulin**, **testosterone**, and **estrogen**. Some hormones are catabolic, whereas others are anabolic.

Cortisol, glucagon, and epinephrine are the major catabolic hormones. They stimulate the breakdown of molecules and the production of energy (Table 9.1). Cortisol is secreted by the adrenal cortex in response to stress. It increases blood glucose levels through **gluconeogenesis** (breaking down fats and proteins). Glucagon is released by the pancreas in response to decreased blood glucose levels. Glucagon increases blood glucose levels by breaking down

TABLE 9.1 Catabolic hormones.

Hormone	Function
Cortisol	Released from the adrenal cortex in response to stress; increases blood glucose levels through gluconeogenesis
Glucagon	Released from the pancreas either when starving or when the blood glucose levels are low; promotes the breakdown of glycogen to increase blood glucose levels. Its effect is the opposite of insulin
Epinephrine	Released in response to stress; stimulates gluconeogenesis

TABLE 9.2 Anabolic hormones.

Hormone	Function
Growth hormone	Synthesized in the anterior pituitary; stimulates growth
Insulin	Produced by the pancreas; plays a vital role in carbohydrate and fat metabolism, regulates blood glucose levels, stimulates the uptake of glucose into cells, and promotes glucose storage in the liver and muscle as glycogen. Its effect is the opposite of glucagon
Testosterone	Made by the testes in males and the ovaries in females; promotes an increase in muscle mass and strength, promotes the growth and strengthening of bone
Estrogen	Produced primarily by the ovaries and some in the adrenal cortex. Its anabolic functions include increasing metabolism and fat deposition

glycogen (glycogenolysis) and gluconeogenesis. The adrenal medulla releases epinephrine in response to stress, and it also promotes gluconeogenesis.

Anabolic hormones are required in the synthesis of large molecules such as glycogen, proteins, and triglycerides. These hormones include growth hormone, insulin, testosterone, and estrogen (Table 9.2). Growth hormone is made in the anterior pituitary. Its actions are anabolic in terms of growth, but the same hormone also brings about catabolic effects to raise blood glucose levels. It promotes the breakdown of fat, for example. On the other hand, insulin, made in the pancreas, promotes the production of glycogen, lipids, and proteins and prevents the breakdown of fat and glycogen. Similarly, estrogen and testosterone, produced by gonadal tissues, enhance the functions of both growth hormone and insulin in bringing growth.

9.5.1 Carbohydrate metabolism

Carbohydrates are the main energy source of the nervous system and are used by the skeletal muscle when energy needs are acute. The breakdown and storage of carbohydrates, in particular glucose and glycogen, are important to our understanding of diabetes and metabolic diseases are most commonly seen by health professionals.

Glucose (Fig. 9.2), also called blood sugar is a six-carbon molecule found in foods or through the digestion of more complex carbohydrates. Glycogen and starch are common polysaccharides. During resting conditions, the body stores glucose in the muscle and liver as glycogen (Fig. 9.3) to be used when the blood sugar becomes too low. Approximately 400 g (1600 kcals) of glycogen is stored in the skeletal muscle and 100 g (400 kcals) in the liver.

9.6 Carbohydrate digestion and absorption

Carbohydrates ingested in the form of food are first digested into their simplest forms, monosaccharides. Their digestion begins in the mouth with the action of the enzyme called

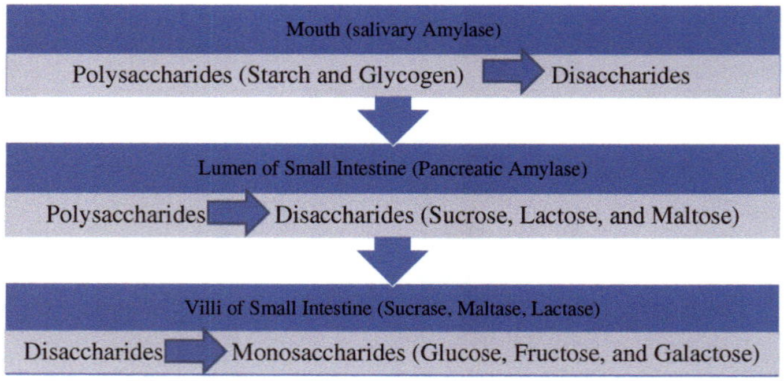

FIG. 9.6 Carbohydrate digestion. Carbohydrates are broken down into their monomers in a series of steps. First, salivary amylase breaks complex carbohydrates such as starch and glycogen into less complex disaccharides in the mouth. Further digestion occurs in the small intestine, where the remaining polysaccharides and disaccharides are digested into monosaccharides.

salivary amylase. Only a small amount of carbohydrates is digested in the mouth since food stays in the mouth for a short period. As the partially digested food reaches the stomach, salivary amylase gets inactivated due to the acidic environment within this region. Digestion of carbohydrates, however, continues in the small intestine with the help of several enzymes. **Pancreatic amylase,** an enzyme secreted by the exocrine portion of the pancreas, is brought to the small intestine via the pancreatic duct. It is similar to salivary amylase in digesting larger carbohydrates such as glycogen and starch to smaller ones such as disaccharides and monosaccharides. **Sucrase**, **maltase**, and **lactase** are enzymes found in the villi of the small intestinal wall. These enzymes digest sucrose, maltose, and lactose, respectively, into their individual monosaccharides (Fig. 9.6). For example, sucrose is digested to release one glucose and one fructose. As monosaccharides are released, they are absorbed first into the epithelial cell lining the small intestine (enterocyte) and ultimately into the blood. A particular circulation called the **hepatic portal system** carries glucose-rich blood directly into the liver. The handling of glucose by the liver depends on the blood levels of glucose and the hormones influencing the metabolic events within the liver.

Glucose is the most readily available nutrient to be used for energy purposes. Cells are equipped with enzymes and substrates to break down glucose to release energy in the form of ATP. There are four key sets of reactions that exist within the cell in the presence of O_2. These are **glycolysis**, production of **acetyl coenzyme A** (**acetyl CoA**), **Krebs cycle**, and **electron transport system**. In the absence of O_2, on the other hand, glucose is converted to **lactic acid**.

9.7 Glycolysis

Glycolysis, the breakdown of glucose, is a central pathway involving a series of steps in almost all prokaryotic and eukaryotic cells. Cells take up circulating glucose (which in some cases depends on the presence of insulin) and, as it is broken down, transfer some of the energy to ADP to form ATP. Two ATP molecules are required (endergonic reaction) during the first few steps of glycolysis. However, as the reaction continues, there are four ATPs generated. Therefore the net yield of ATP is two ATPs per glucose molecule through glycolysis (Fig. 9.7).

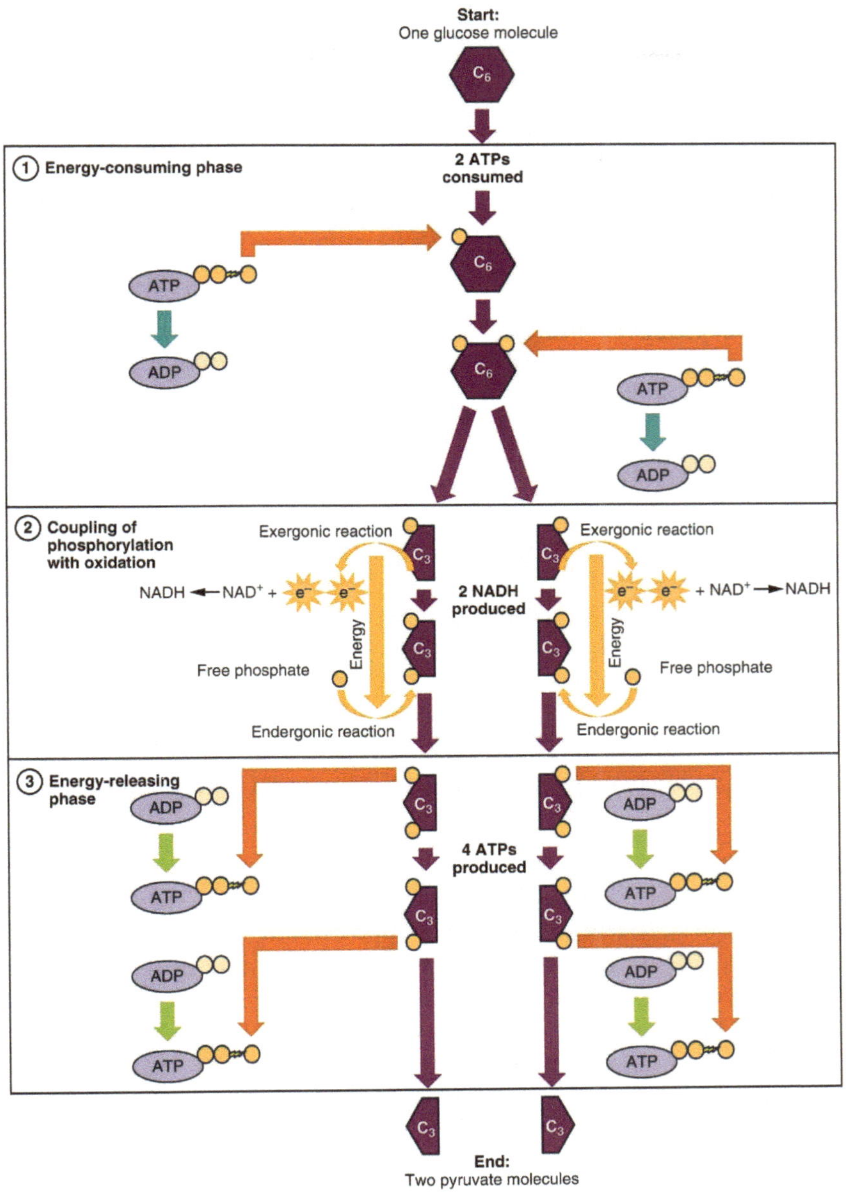

FIG. 9.7 Glycolysis overview [6]. Two ATPs are consumed during the energy-requiring phase of glycolysis, transferring two phosphates to the glucose molecule. Subsequently, the glucose molecule splits into two three-carbon compounds, each containing a phosphate. In a later step, an additional phosphate is added to each of the three-carbon compounds. The energy for this reaction requiring energy is provided by the removal (oxidation) of two electrons from each three-carbon compound. During the energy-releasing phase, the phosphates are removed from both three-carbon compounds and used to produce four ATP molecules.

Glycolysis begins with the phosphorylation of glucose in the cytoplasm to form **glucose-6-phosphate** (G6P) and uses ATP. This conversion process to G6P has a vital purpose, i.e., it prevents glucose from escaping through the plasma membrane, thus allowing glycolysis to proceed. It also maintains a concentration gradient with higher glucose levels in the blood than in the tissues. By establishing this concentration gradient, glucose in the blood will flow from an area of higher concentration (the blood) into an area of lower concentration (the tissues) to be either used or stored. **Phosphatase**, an enzyme that removes phosphate from G6P, is limited to hepatocytes and thus prevents the trapping of glucose in these cells (Fig. 9.7).

In the next step, G6P is converted into **fructose-6-phosphate** (F6P). A second ATP donates its phosphate group to F6P, forming **fructose-1,6-bisphosphate** (F16BP). Fructose in the diet can enter glycolysis at this step, where it is converted into F16BP. F16BP splits to form two phosphorylated three-carbon molecules, **glyceraldehyde-3-phosphate** (G3P), which is further phosphorylated to form the three-carbon molecule **1,3-bisphosphoglycerate**. This reaction releases an electron that NAD (a coenzyme that serves as an electron acceptor) then picks up to create an NADH molecule. NADH helps produce three ATPs per molecule through the electron transport system. Because there are two G3P molecules, two NADH molecules are synthesized during this step. When lipids are digested, one of their metabolites, glycerol, could enter the glycolytic pathway as G3P. Amino acids may also join the glycolytic pathway once their amino group is removed. Ultimately, each 6-carbon glucose molecule during glycolysis is broken down into two molecules of 3-carbon structures, **pyruvate** or **pyruvic acid**.

Glycolysis takes place without depending on the availability of O_2 and is thus referred to as an **anaerobic** reaction. It can be expressed as the following equation:

$$\text{Glucose} + 2\,\text{ATP} + 2\,\text{NAD} + 2\,\text{ADP} + 2\,\text{P} \rightarrow 2\,\text{Pyruvate} + 4\,\text{ATP} + 2\,\text{NADH} + 2\text{H}^+ + 2\text{H}_2\text{O}$$

As per this equation, glucose, combined with ATP, NAD, and inorganic phosphate, splits into two pyruvate molecules, generating four ATP molecules—for a net yield of two ATPs—and two energy-containing NADH. The NADH produced in this process will be used to produce ATPs in the mitochondria by a separate process called **oxidative phosphorylation**. Excess glucose may be stored as **glycogen** in the muscle cells and the hepatocytes, and this process is called **glycogenesis**. Glycogen is converted to glucose by a process called **glycogenolysis** when glucose levels in the blood are less than normal (Fig. 9.8).

9.8 Fate of pyruvate

In the presence of O_2, pyruvate is processed further within the mitochondria where it gets converted to acetyl CoA. In the absence of O_2, pyruvate is converted to **lactic acid**. Pyruvate may also be used in the synthesis of new glucose through a process called **gluconeogenesis** (Fig. 9.8).

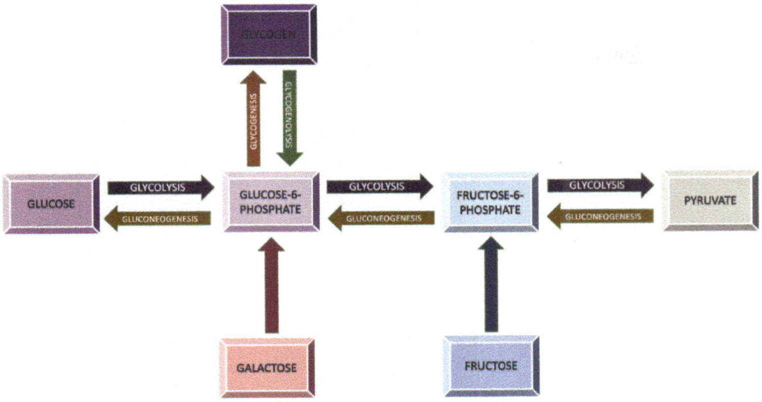

FIG. 9.8 Diagram of the relationship between the process of carbohydrate metabolism, including glycolysis, gluconeogenesis, glycogenesis, glycogenolysis, fructose metabolism, and galactose metabolism [7].

9.9 Anaerobic respiration

When there is a limited amount of O_2, such as during exercise, muscles use ATP in greater amounts, but the delivery of O_2 could be limited. Thus they depend on anaerobic respiration for rapid ATP production. NAD is not released in sufficient quantities, and its low levels threaten to stop glycolysis and ATP production. During such periods, pyruvate enters an alternate pathway where it is converted into lactic acid (Fig. 9.9). This pathway helps keep the pyruvate concentration low so that glycolysis can continue. It also allows the oxidation of NADH to NAD, which is required in the continuation of glycolysis. In this reaction, lactic acid serves as the final electron acceptor, unlike aerobic respiration, where O_2 serves that role. Thus anaerobic respiration occurs in most cells when O_2 is limited, or mitochondria are absent. For example, erythrocytes lack mitochondria and therefore generate their ATP from anaerobic respiration alone. The lactic acid produced from anaerobic respiration is eventually converted back to pyruvate in the liver and is called the **Cori cycle**.

9.10 Aerobic respiration

In the presence of O_2, the three-carbon pyruvate molecule moves from the cytoplasm into the mitochondria, where it is converted into a two-carbon **acetyl coenzyme A (acetyl CoA)** molecule releasing CO_2 and transferring two electrons to NAD, which now becomes NADH $+H^+$ (Fig. 9.10).

9.10.1 Krebs cycle/citric acid cycle/tricarboxylic acid cycle

The Krebs cycle is considered one of the main pathways in cellular metabolism. Acetyl CoA enters the Krebs cycle by combining with a four-carbon molecule, **oxaloacetate**, and

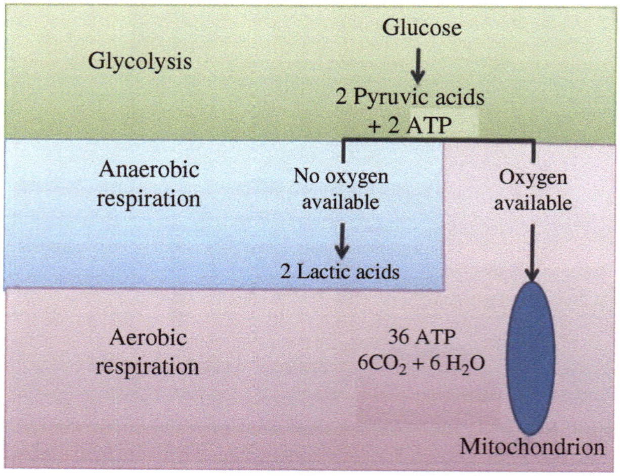

FIG. 9.9 Aerobic versus anaerobic respiration. The process of anaerobic respiration converts glucose into two lactate molecules in the absence of oxygen (or within erythrocytes that lack mitochondria). During aerobic respiration, glucose is oxidized into two pyruvic acid molecules.

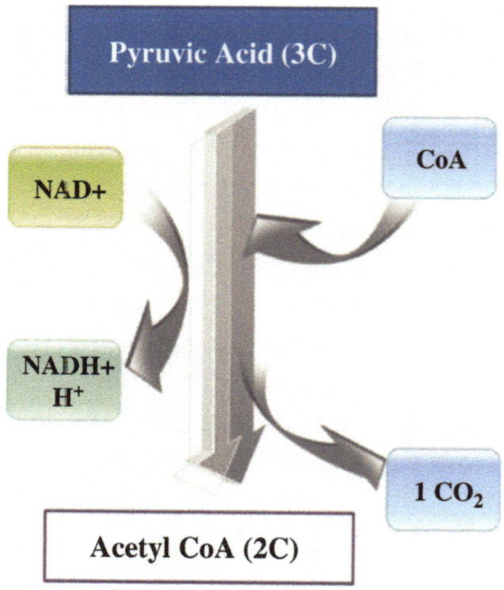

FIG. 9.10 Formation of acetyl CoA. A 3-carbon structure, pyruvic acid, is converted to a 2-carbon structure called acetic acid, which combines with coenzyme A to become acetyl coenzyme A (acetyl CoA). During this process, one molecule of CO_2 is released. NAD becomes $NADH+H^+$.

becomes six-carbon-molecule **citrate**, or **citric acid**. It is the formation of citric acid that gives its name, the citric acid cycle. The citrate molecule is systematically converted to a five-carbon molecule and a four-carbon molecule, ending with oxaloacetate (Fig. 9.11). The two carbons from citrate to oxaloacetate are released in the form of two molecules of CO_2. Each citrate molecule causes the release of one ATP, one $FADH_2$, and three NADH coenzymes along the way. In this reaction, coenzyme A molecule is released, combining with another pyruvate molecule to start a new cycle as needed. The $FADH_2$ and NADH will enter the oxidative phosphorylation system located in the inner mitochondrial membrane.

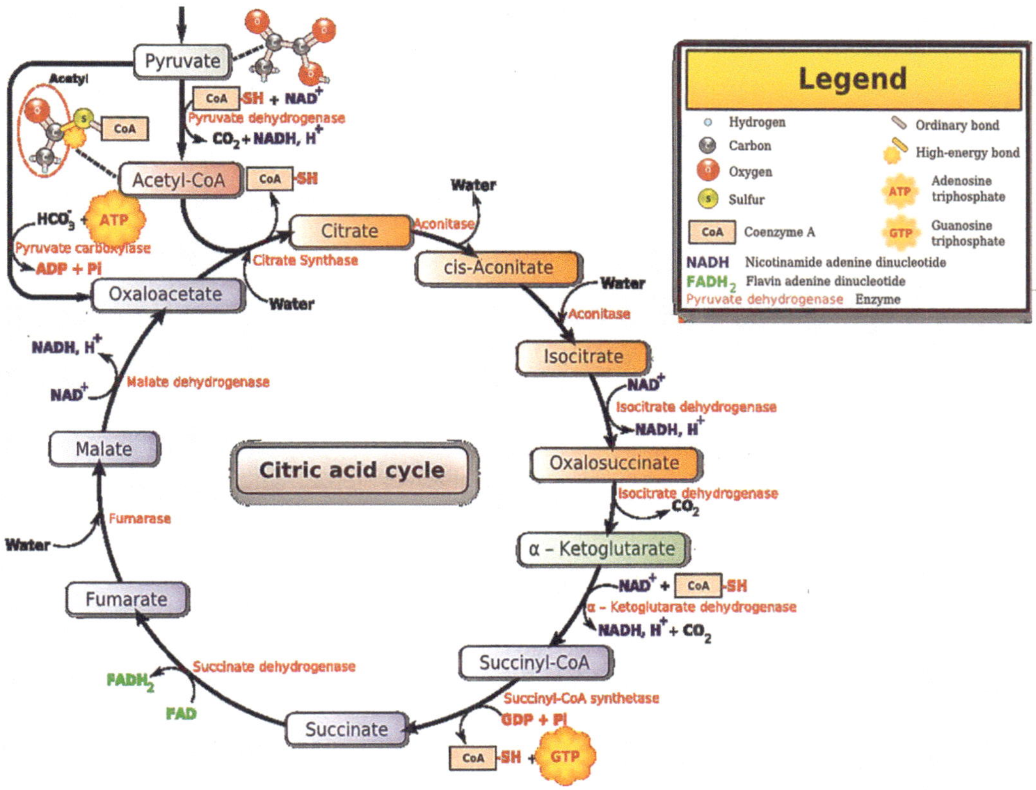

FIG. 9.11 Krebs cycle(citric acid cycle or tricarboxylic acid cycle) [8]. One acetyl CoA is processed in each cycle to yield high-energy NADH, $FADH_2$, and ATP molecules. Two CO_2 molecules are released as waste products.

9.10.2 Electron transport system and oxidative phosphorylation

The **electron transport system** consists of an arrangement described as the **electron transport chain** (ETC). ETC is a series of four enzyme complexes (**Complex I–Complex IV**) and two coenzymes (**coenzyme Q** and **cytochrome** *c*) arranged in a specific order in the inner mitochondrial membrane and serve as electron carriers and proton pumps used in the transfer of H^+ into the space between the inner and outer mitochondrial membranes (Fig. 9.12). Their arrangement is such that high-energy electrons assume successively lower energy levels as they are transferred from one carrier to another. When the electrons have passed through the entire ETC, they must be removed from the ETC. The role of O_2 that is brought to the cell through the respiratory mechanisms is to serve as the terminal electron acceptor. Each oxygen atom traps two electrons and two H^+ from the matrix to form a new H_2O molecule. Thus O_2 serves as the final electron acceptor and is the basis for breathing in O_2. Without O_2, electron flow through the ETC ceases, and so does energy production. In summary, the ETC couples the transfer of electrons between a donor (like NADH) and an electron acceptor (like O_2) with

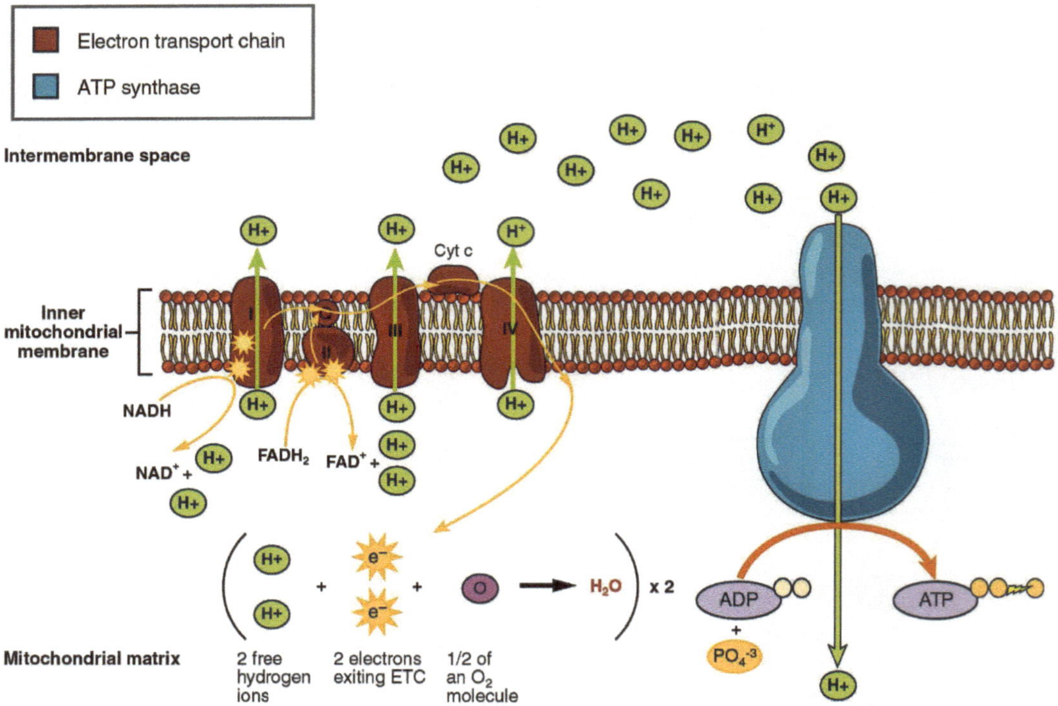

FIG. 9.12 Electron transport chain [9]. The electron transport chain (ETC) consists of a series of electron carriers (Complex I–Complex IV) and two coenzymes (coenzyme Q and cytochrome c). As an electron moves through the chain, the proton gets accumulated in the intermembrane space. At the end of the ETC, an electron is trapped by oxygen. The diffusion of protons into the matrix through ATP synthase (shown in *blue*) allows the production of ATPs.

the transfer of protons (H^+) across the inner mitochondrial membrane, enabling the process of **oxidative phosphorylation** and the production of ATPs.

The ETC uses the NADH and $FADH_2$ produced by the Krebs cycle to generate ATPs. H^+ released from NADH and $FADH_2$ have considerable energy stored and contain electrons at high energy levels. These electrons are transferred through protein complexes embedded in the inner mitochondrial membrane by a series of enzymatic reactions. The electrons released from NADH and $FADH_2$ are passed along the chain by each of the carriers, which are reduced when they receive the electron and oxidize when passing it onto the next carrier. Each of these reactions releases a small amount of energy. At three sites in the electron transport chain, complexes I, III, and IV, the energy released during the transport of electrons is used to transport protons across the inner membrane. The accumulation of these protons in the space between the membranes creates a proton gradient with respect to the mitochondrial matrix. This H^+ gradient generated by the electron transport system supplies the energy that drives ATP synthesis by the membrane-bound mitochondrial enzyme **ATP synthase**. It is the last protein embedded in the inner mitochondrial membrane (shown in *blue*; Fig. 9.12). As a result of releasing H^+ and electrons, NADH and $FADH_2$ are converted back to NAD and FAD.

Each NADH allows the production of 3 ATP molecules. The shorter pathway followed by an electron pair released from $FADH_2$ synthesizes 2 ATP molecules. Thus a total of 30 ATP molecules from 10 NADH and 4 ATP molecules are obtained from 2 $FADH_2$ adding up to a total of 34 ATP molecules that are produced between NADH and $FADH_2$ using ETC.

The total number of ATP molecules produced per glucose molecule from the entire aerobic respiration is as follows:

- A net gain of two ATP molecules is produced through glycolysis.
- Each Krebs cycle results in one ATP. With two pyruvates, each glucose can turn on two Krebs cycles.
- In the ETC, ATP synthase provides a rich yield of 32 to 34 more ATP molecules for each glucose molecule processed (10 NADH and 2 $FADH_2$). Approximately 3 ATPs are synthesized as a pair of electrons released by NADH travel through the ETC or a total of 30 ATPs from NADH.
- The electrons from $FADH_2$ produce fewer ATPs (2 ATPs per $FADH_2$ for a total of 4 ATPs) because they start at a lower point in the ETC (Complex II) compared to the electrons from NADH (Complex I).
- In some cells, two electrons from the NADs enter the electron transport chain at complex II instead of complex I, thus generating 28 ATPs in total instead of 30.
- For every glucose molecule that enters aerobic respiration, a net total of 36–38 ATP molecules are produced (Fig. 9.13).

9.11 Gluconeogenesis, glycogenolysis, and glycogenesis

Gluconeogenesis refers to synthesizing glucose molecules from noncarbohydrates such as lactate, pyruvate, fatty acids, and amino acids (Fig. 9.14). Many vital organs, including the brain, solely depend on glucose as an energy source. Therefore the body must maintain a minimum blood glucose concentration. Gluconeogenesis occurs primarily in the liver during low blood glucose concentrations, such as fasting, starvation, and low carbohydrate diets.

Glycogenolysis is the breakdown of glycogen into glucose, whereas **glycogenesis** is the formation of glycogen from glucose. Glycogen can be stored in the hepatocytes and muscle cells. Both of these processes are regulated by hormones and neural signals. For example, glycogenolysis is triggered by hormones glucagon and epinephrine, whereas insulin stimulates glycogen production (Table 9.2). Glycogenesis depends on the demand for blood glucose concentrations and ATPs. If both are present in sufficient amounts, excess glucose is then converted into glycogen.

9.11.1 Protein metabolism

Much of the body is made of proteins, and these proteins take on a myriad of forms such as receptors, signaling molecules, structural members, enzymes, intracellular trafficking components, extracellular matrix scaffolds, ion pumps, ion channels, and O_2 and CO_2 transporters

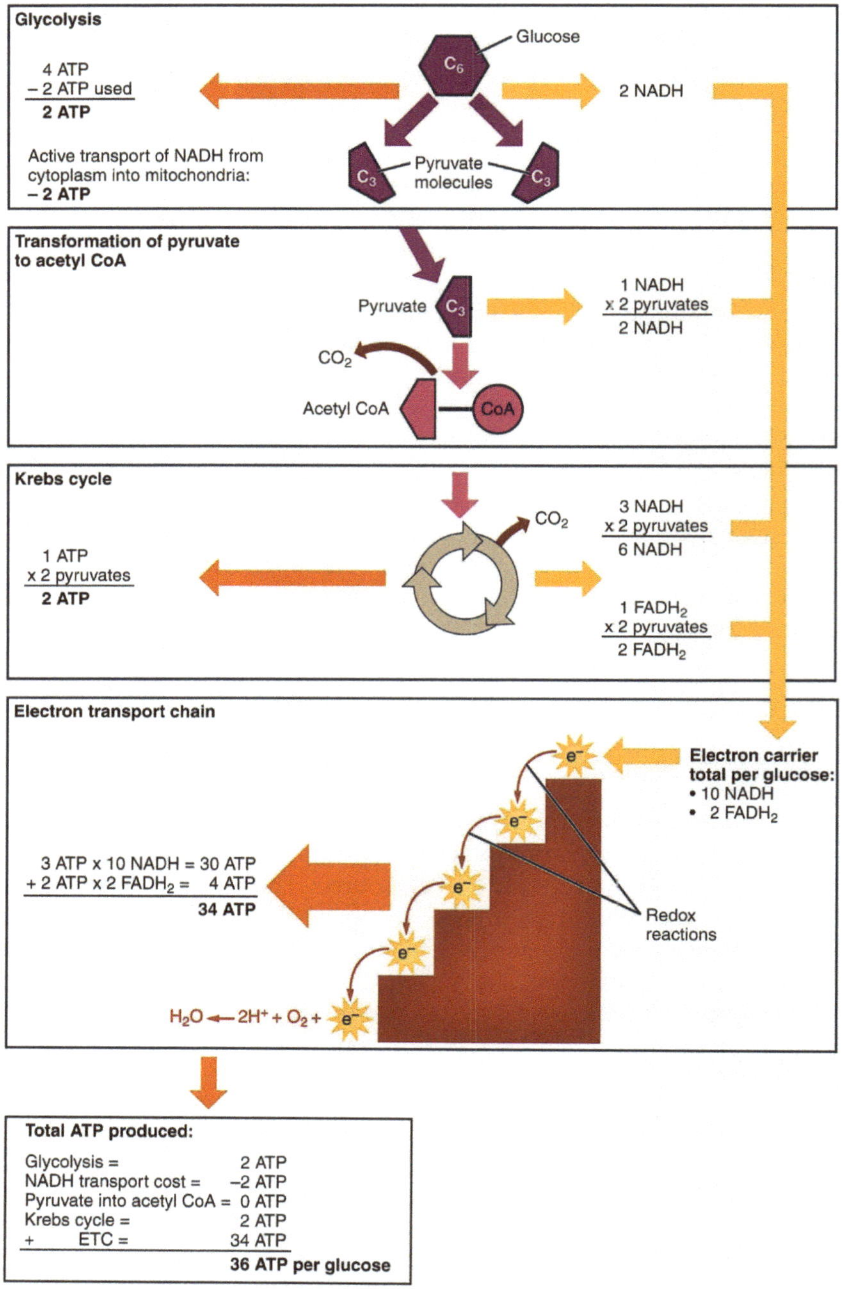

FIG. 9.13 ATP molecules produced during aerobic respiration of glucose [10].

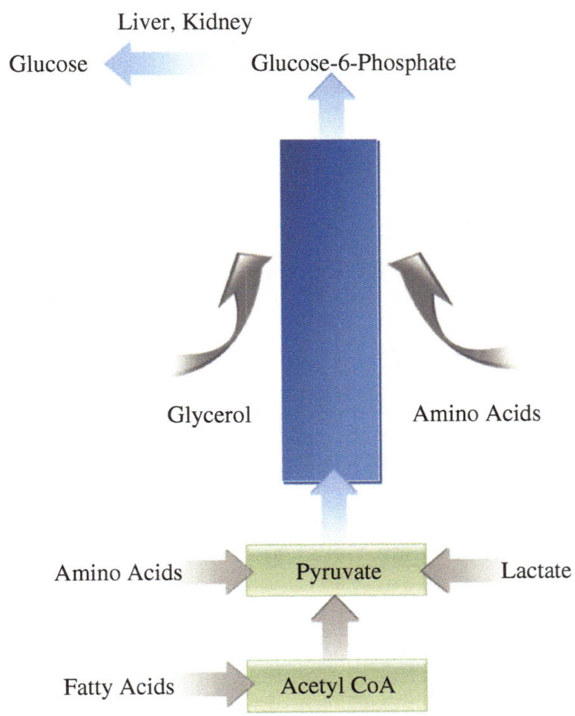

FIG. 9.14 Gluconeogenesis is the production of glucose from noncarbohydrate sources such as lactate, pyruvate, fatty acids, and amino acids.

(hemoglobin). In addition, collagen fibers within the bones, tendons, and cartilages, and muscle proteins within the muscle fibers are also protein structures. Amid all these crucial functions, proteins also hold the potential to serve as a metabolic fuel source. Approximately 10%–15% of total energy comes from the catabolism of amino acids, far less than lipids (20%–35%) and carbohydrates (45%–65%).

Proteins are made of amino acids which are linked by **peptide bonds**. The name varies from a **peptide** for a short chain of amino acids to a **polypeptide**, which is approximately 50-amino acid long, and a **protein**, a longer chain than that of a polypeptide. While a protein is being constructed, its shape changes sequentially (Fig. 9.15). First, it assumes a long chain of amino acids and is referred to as its **primary structure**. Then, interactions among adjacent amino acids fold the straight chain in different forms, and this intermediate form is the **secondary structure or alpha-helix**. Additional folding due to the possibility of more interactions among the amino acids in their new positions results in the complex three-dimensional structure referred to as **tertiary structure**. For most of the proteins, tertiary structures are their final forms. However, in some cases, more than one polypeptide chain is held

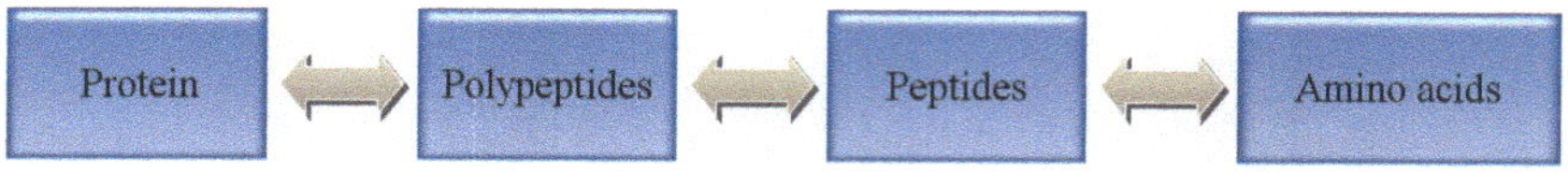

FIG. 9.15 Digestion of a protein may result in polypeptides, peptides, and amino acids, whereas amino acids can be joined together to form peptides, polypeptides, and proteins.

together to form a protein, and such a structure is called a **quaternary structure**. For example, **insulin** is made of two polypeptide chains, whereas **hemoglobin** is made of four polypeptide chains.

9.12 Digestion and absorption of proteins

The primary sources of dietary proteins are meat, eggs, dairy products, nuts, and beans. Digestion of protein reduces them to their constituent amino acids. The breakdown of protein starts in the stomach, where hydrochloric acid and **pepsin**, an enzyme, break proteins into polypeptides or peptides, which then travel to the small intestine. Chemical digestion in the small intestine is carried out by various pancreatic enzymes such as **trypsin** and **chymotrypsin**. Furthermore, brush border cells (enterocytes with folds in their cell membrane) secrete enzymes called **peptidases** that break small peptides to their individual amino acids. Amino acids, the products of protein digestion, are small enough to enter the bloodstream. They are transported directly to the liver through the hepatic portal circulation.

9.13 Essential and nonessential amino acids

Among the 20 different amino acids that exist in nature, some can be formed within the body and are referred to as **nonessential** amino acids. For example, **glutamine** is formed by the combination of glutamate and ammonia. However, some amino acids must be obtained from the diet since they cannot be synthesized in the body and are referred to as **essential** amino acids. **For example, histidine** is an essential amino acid that must be obtained in the diet as it cannot be synthesized within the body. Other essential amino acids are **isoleucine, leucine, lysine, methionine, phenylalanine, threonine, tryptophan,** and **valine.**

9.14 Regulation of protein metabolism

Several hormones regulate protein metabolism. Growth hormone, insulin, and testosterone promote the synthesis of protein (Table 9.2). On the other hand, cortisol causes the breakdown of proteins (Table 9.1). The breakdown of extrahepatic tissue proteins

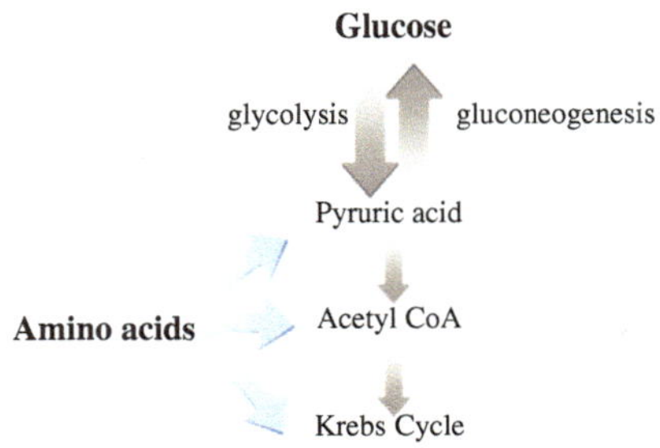

FIG. 9.16 Amino acids as energy sources. Amino acids can be broken down into precursors for glycolysis or the Krebs cycle. Amino acids can enter the metabolic pathway at multiple levels.

increases plasma amino acid levels, serving as the source of gluconeogenesis and ketogenesis. During **postabsorptive state** or fasting, decreased glucose levels in the blood trigger the release of glucagon, a hormone that promotes the conversion of amino acids to glucose by the process of gluconeogenesis. During this process, the carbon backbone of an amino acid will be converted into one of many metabolic intermediates of glycolysis or Krebs cycle, including pyruvate, acetyl CoA, acetoacetyl CoA, oxaloacetate, and α-ketoglutarate. Thus amino acids can serve as a source of energy through glycolysis or the Krebs cycle (Fig. 9.16).

9.15 Deamination and transamination

Approximately 60% of the amino acids are converted to carbohydrates and used for energy purposes. This is achieved due to two processes related to amino acids, which are **deamination** and **transamination**. As the name suggests, deamination is where the amino group of an amino acid is removed, and the remaining structure of the amino acid will now serve as an intermediate molecule in the glycolytic pathway or the Krebs cycle. For example, alanine, an amino acid, is converted to pyruvate by removing its amino group, which is changed to ammonia and eventually into a less toxic substance, **urea**.

If the amino group from alanine is removed and is transferred to **α-ketoglutarate** to obtain **glutamate**, another amino acid, this process is called **transamination**. Thus transamination is where one of the reactants is an amino acid and the other a **keto acid**. The amino group (NH_2) of one molecule is exchanged for the keto group (=O) of another molecule. Transamination is essential in the production of nonessential amino acids.

Typically, an amino acid is converted to major metabolic intermediates, which are converted to glucose or oxidized using aerobic pathways. Amino acids that are degraded to acetyl CoA are **ketogenic** since they are precursors of ketone bodies, whereas the amino acids that form glucose or progress through the Krebs cycle are termed **glucogenic**.

9.16 Urea cycle

Urea is a major nitrogenous waste accounting for 80%–90% of the nitrogen-containing substances excreted in the urine. It is synthesized in the liver, secreted into the bloodstream, and removed by the kidneys in urine. During exercise, sweat glands also excrete a small amount of urea.

Freely available amino acids are used in the production of proteins. Thus there are very few free amino acids within the cells. If amino acids exist in excess, the body has no capacity or mechanism for their storage. Thus they are converted into glucose or ketones, or they are decomposed. Amino acid degradation occurs in the liver mainly with protein excess.

Amino acid decomposition produces hydrocarbons and nitrogenous waste. The amino group is reduced to ammonia (NH_3) by the addition of a hydrogen atom. If ammonia accepts an electron (as it does during acidic conditions), it is called **an ammonium ion** (NH_4^+). The levels of ammonia must be carefully regulated in the body as it can cross the blood-brain barrier and is toxic to the central nervous system.

The **urea cycle** processes the amino group and facilitates its excretion from the body. The liver is the primary organ where a complete urea cycle is expressed, a set of biochemical reactions that produces urea from ammonium ions. The first few steps are mitochondrial, and the rest are cytosolic.

9.17 Nitrogen balance

When the dietary intake is equal to the amount that is excreted, the individual is in **nitrogen balance**. Sickness or injury could result in **negative nitrogen balance**, whereas increased protein intake or disease that interferes with the removal of nitrogenous waste may lead to **positive nitrogen balance**. Young children with an adequate protein diet are often in positive nitrogen balance.

References

[1] 2501 The Structure of ATP Molecules.jpg. Wikimedia Commons, the free media repository. Updated September 27, 2020. Accessed February 2, 2021. https://commons.wikimedia.org/w/index.php?title=File:2501_The_Structure_of_ATP_Molecules.jpg&oldid=473450980.

[2] Blausen 0434 Glucose.png. Wikimedia Commons, the free media repository. Updated October 25, 2020. Accessed March 19, 2021. https://commons.wikimedia.org/w/index.php?title=File:Blausen_0434_Glucose.png&oldid=501286945.

[3] Glycogen.png. Wikimedia Commons, the free media repository. Updated October 9, 2020. Accessed March 19, 2021. https://commons.wikimedia.org/w/index.php?title=File:Glycogen.png&oldid=484923139.

[4] Common Lipids Lmaps.png. Wikimedia Commons, the free media repository. Updated November 2, 2020. Accessed March 19, 2021. https://commons.wikimedia.org/w/index.php?title=File:Common_lipids_lmaps.png&oldid=509364616.

[5] 2502 Catabolic Reactions.jpg. Wikimedia Commons, the free media repository. Updated September 27, 2020. Accessed March 19, 2021. https://commons.wikimedia.org/w/index.php?title=File:2502_Catabolic_Reactions.jpg&oldid=473174046.

[6] 2504 Glycosis Overview.jpg. Wikimedia Commons, the free media repository. Updated September 29, 2020. Accessed March 19, 2021. https://commons.wikimedia.org/w/index.php?title=File:2504_Glycosis_Overview.jpg&oldid=475270737.
[7] Carbohydrate Metabolism.png. Wikimedia Commons, the free media repository. Updated October 29, 2020. Accessed March 19, 2021. https://commons.wikimedia.org/w/index.php?title=File:Carbohydrate_Metabolism.png&oldid=506071572.
[8] Citric Acid Cycle with Aconitate 2.svg. Wikimedia Commons, the free media repository. Updated November 17, 2020. Accessed March 19, 2021. https://commons.wikimedia.org/w/index.php?title=File:Citric_acid_cycle_with_aconitate_2.svg&oldid=513074663.
[9] 2508 The Electron Transport Chain.jpg. Wikimedia Commons, the free media repository. Updated October 20, 2020. Accessed March 19, 2021. https://commons.wikimedia.org/w/index.php?title=File:2508_The_Electron_Transport_Chain.jpg&oldid=495207373.
[10] 2509 Carbohydrate Metabolism.jpg. Wikimedia Commons, the free media repository. Updated October 3, 2020. Accessed March 19, 2021. https://commons.wikimedia.org/w/index.php?title=File:2509_Carbohydrate_Metabolism.jpg&oldid=478975680.

CHAPTER

10

Lipid metabolism

OUTLINE

OBJECTIVES

- Summarize the process of digestion and absorption of lipids.
- Define lipoproteins. List the different types of lipoproteins and their compositions.
- Identify the sites where lipoproteins are synthesized.
- Explain the metabolism of chylomicrons, low-density lipoproteins, and high-density lipoproteins.
- Define beta-oxidation and its significance in the body.
- Define ketone bodies and summarize the importance of these substances in the body.
- Distinguish between the terms lipogenesis, lipolysis, ketogenesis, ketosis, and ketoacidosis.

https://doi.org/10.1016/B978-0-12-823421-1.00013-5

10.1 Introduction

Lipids (fats) are essential organic molecules in the body. Unlike carbohydrates and proteins, lipids are not soluble in water but are soluble in organic solvents and therefore require carrier proteins for their transportation. There are many types of lipids in the body (see Chapter 9). The three major types of lipids are **triglycerides**, **phospholipids**, and **steroids**. **Cholesterol** serves as the precursor in the synthesis of steroids. Approximately 15% of cholesterol is obtained through diet, whereas the remaining 85% is manufactured within the liver.

10.2 Digestion overview

Digestion of food begins in the mouth, where the teeth break down food into smaller particles during mastication (chewing). The salivary glands are located near the oral cavity and secrete saliva, which begins chemical digestion and keeps the food moist. As food is swallowed, food particles travel as **bolus** down the pharynx and enter the esophagus, connecting the pharynx to the stomach. Cells in the stomach secrete hydrochloric acid and pepsinogen that chemically digest the bolus. Contractions initiated by the smooth muscle lining in the stomach churn its contents to break down the bolus further and mix it with stomach secretions to form a thick liquid called **chyme**. Chyme enters the small intestine, the primary site for nutrient digestion and absorption. The small intestine consists of three parts: the duodenum, jejunum, and ileum (Fig. 10.2). Bile from the liver and gallbladder and pancreatic juice from the pancreas, empty into the duodenum to assist in digestion. Most nutrients are absorbed mainly at the duodenum and jejunum, from where they are released into blood except for the lipids which are released into lymph. Chyme, not digested or absorbed in the small intestine, reaches the large intestine where water and salts are absorbed. The undigested food, referred to as the **feces**, is stored in the rectum until neural stimulation initiates the **defecation reflex**, resulting in elimination through the anal canal (Fig. 10.1).

10.3 Fate of dietary fat

Unlike carbohydrates and proteins, dietary fat is not digested much until it reaches the small intestine due to its large size. A small portion of chyme is released from the stomach into the duodenum which causes the enteroendocrine cells of the duodenum wall to release the hormones **cholecystokinin** (CCK) and **secretin** into the blood. CCK stimulates the liver to synthesize bile and causes contraction of the gallbladder, which stores bile in a concentrated form, to release through the common bile duct into the duodenum (Fig. 10.2). Fats are **emulsified** using bile salts within the duodenum. Emulsification is the breaking of large fat globules into smaller fat droplets (Fig. 10.3). The tiny fat droplets (Fig. 10.4) offer a large surface area for the action of the fat-digesting enzyme, **pancreatic lipase**, produced in the pancreas and delivered to the duodenum by the pancreatic duct. Fat molecules are broken down by pancreatic lipase into monoglycerides and free fatty acids (see Fig. 10.5). **Micelles** are a

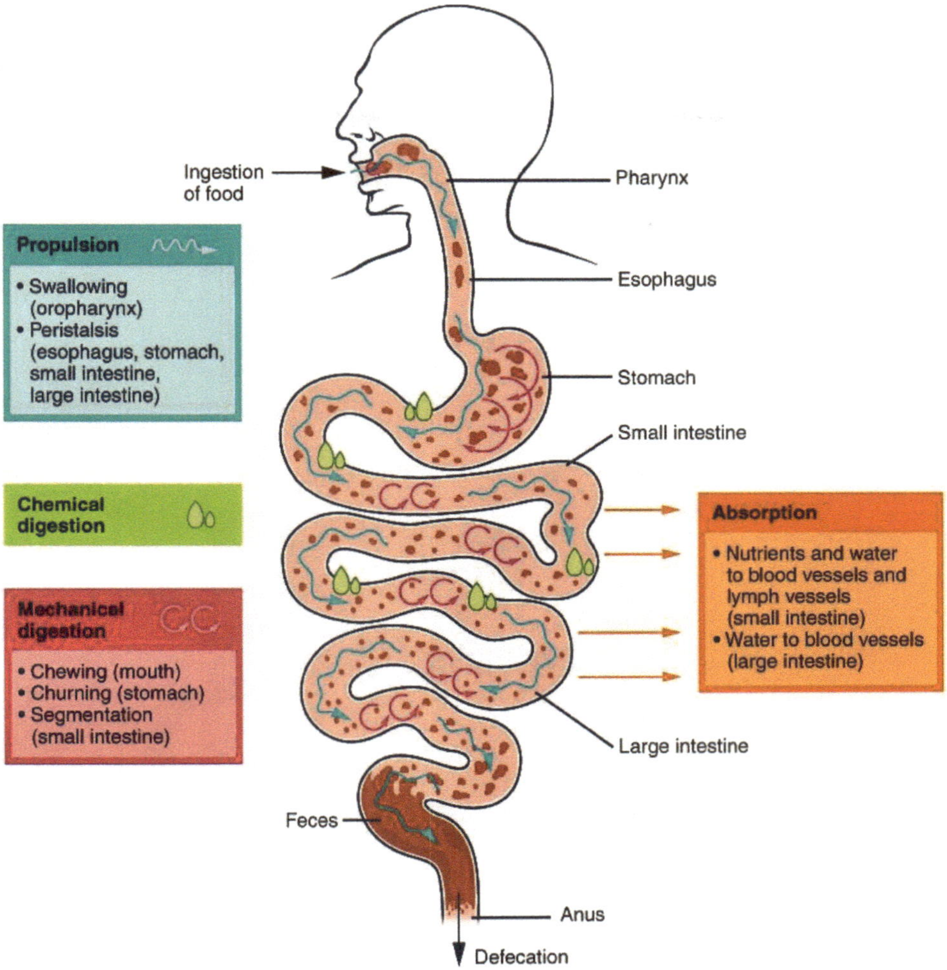

FIG. 10.1 Summary of the digestive process [1]. The digestive processes are ingestion, propulsion, mechanical digestion, chemical digestion, absorption, and defecation.

combination of proteins, monoglycerides, and fatty acids that are formed in the lumen of the small intestine to help transport lipids into the intestinal cells or **enterocytes**.

10.4 Triglycerides

Triglycerides are either ingested as food or synthesized within the body mainly by adipocytes as a form of storage and by hepatocytes. The oxidation of fatty acids is essential in obtaining energy in the form of ATP or synthesizing new lipids from smaller constituent molecules. However, despite many lipids available as fuel, the process of lipid utilization is

FIG. 10.2 Biliary lipid transport after a meal. Bile is formed in the liver from cholesterol and stored in the gallbladder. Following a meal, the contraction of the gallbladder releases bile into the duodenum to emulsify fat.

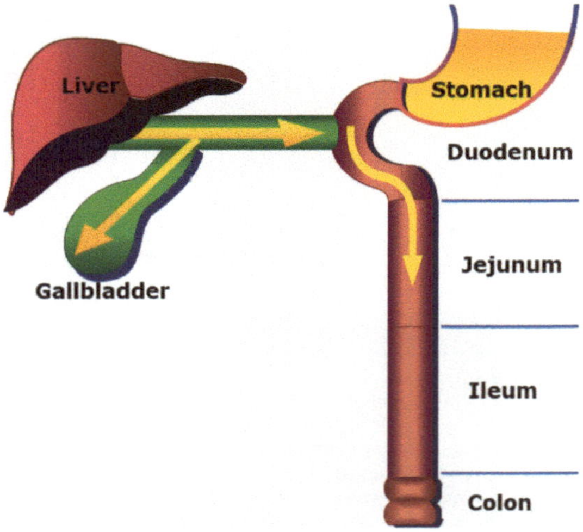

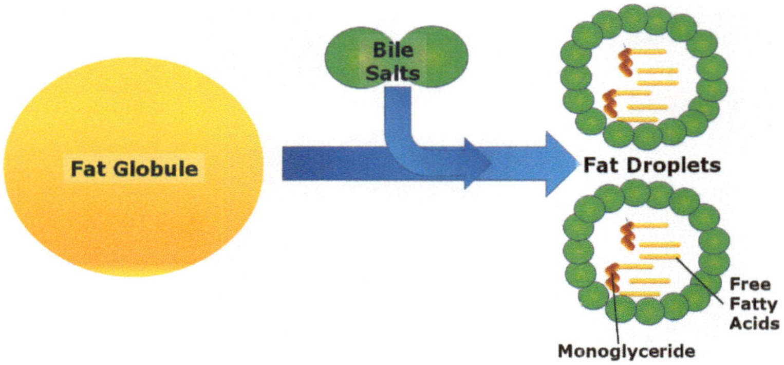

FIG. 10.3 Bile salts break apart large fat molecules into smaller globules to facilitate lipid digestion by emulsification.

FIG. 10.4 Structure of a micelle. Bile *(green)* is formed in the liver and stored in the gallbladder. Bile emulsifies fat globules which are then acted upon by pancreatic lipase *(red)* to release monoglycerides and free fatty acids that are bound to proteins resulting in a new structure called micelle.

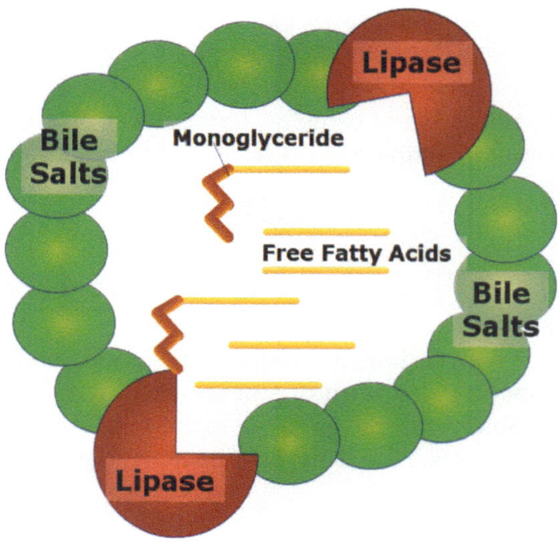

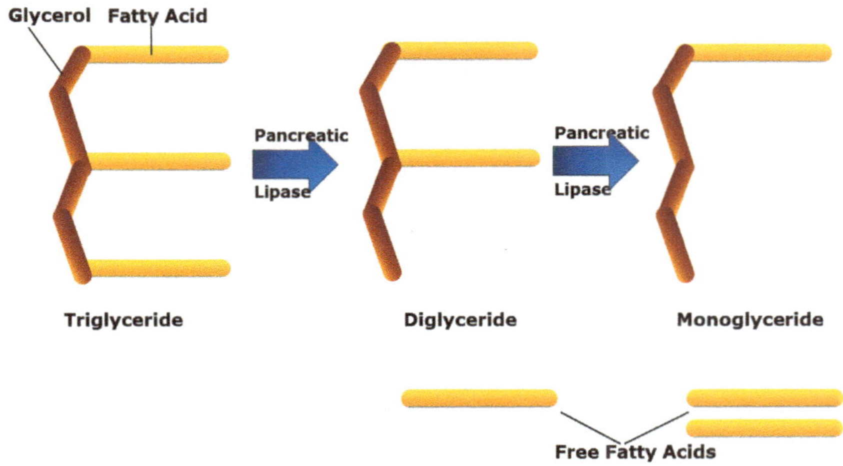

FIG. 10.5 The digestion of a triglyceride. Pancreatic lipase releases one fatty acid to form a diglyceride and a free fatty acid. Over time, the pancreatic lipase removes another fatty acid to form monoglyceride with two free fatty acids.

slow to be activated and proceeds at rates significantly slower than the process of glucose catabolism.

The metabolic pathways of lipids are tied to the metabolic pathways of glucose, such as glycolysis and the Krebs cycle. One standard process is the breakdown of long fatty acid structures into two carbon structures resulting in the formation of acetic acid molecules. An acetyl CoA is formed by combining coenzyme A with the two-carbon acetic acid, turning on the Krebs cycle. Similarly, a glycerol molecule, the backbone of triglyceride, could be converted to glyceraldehyde-3-phosphate, which can then enter the glycolytic pathway.

10.5 Digestion and absorption of lipids

Once triglycerides are broken down into fatty acids and monoglycerides, they are transported within the lumen of the small intestine by binding to a plasma protein in the form of a micelle. Cholesterol is also absorbed along with the free fatty acids and monoglycerides, but it must be de-esterified (broken down) first, which is accomplished by esterases, primarily secreted by the pancreas. Free cholesterol is then incorporated into micelles, a physical process that is required for cholesterol absorption.

Micelles that are transporting monoglycerides and free fatty acids interact with the cell membrane of the enterocyte lining the wall of the small intestine, and its contents are transferred into the enterocyte via diffusion (movement from high concentration to low concentration). Monoglycerides and fatty acids then move into the endoplasmic reticulum, where they are repackaged back into a triglyceride (Fig. 10.6). The newly formed triglycerides are brought to the Golgi where a protein coat surrounds them. Dietary cholesterol and phospholipids are also added, and the resulting structure is called a **chylomicron** (Fig. 10.7). Chylomicron reaches the basal membrane of the enterocyte where it docks, and through a process called

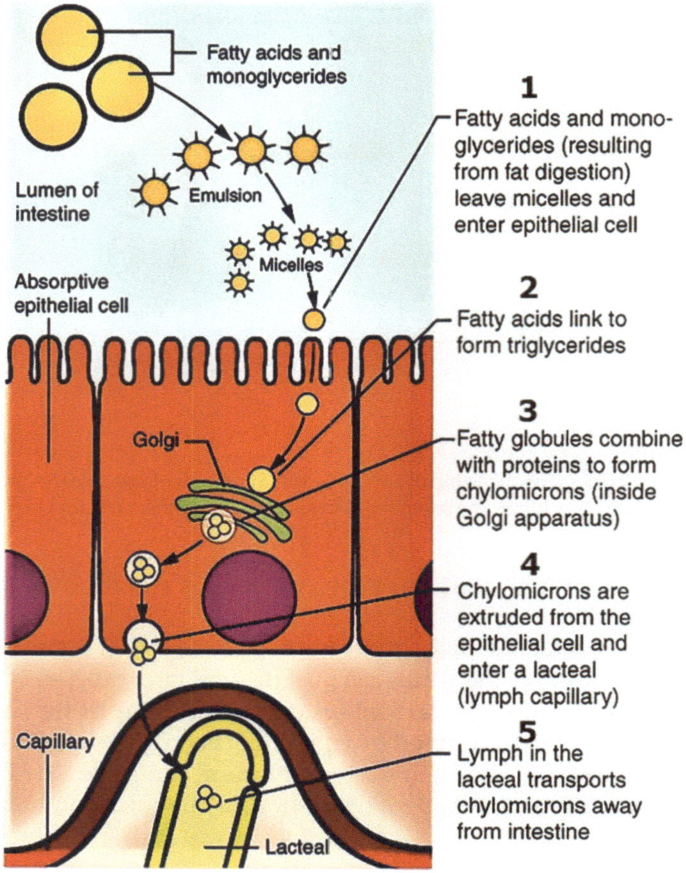

FIG. 10.6 Fat digestion and absorption in the small intestine [2].

exocytosis (exiting), it exits the small intestinal wall. Next, chylomicrons enter the lymphatic system via **lymphatic capillaries**, called **lacteals**, in the small intestinal villi, reach the thoracic duct, and ultimately join the circulatory system. Thus the lymphatic system constitutes the primary route by which dietary lipids enter the blood. However, a small number of fatty acids, particularly short-chain structures, are absorbed into the hepatic portal system and reach the liver directly.

10.6 Types of lipoproteins

Lipoproteins are classified based on their density (Fig. 10.8). Increased lipid content reduces the density of lipoproteins. Chylomicron is a newly synthesized lipoprotein consisting of dietary lipids. Each chylomicron consists of 99% of lipid and 1% of protein and has very low density among other lipoproteins. **Very low-density lipoprotein** (VLDL) consists of 92%

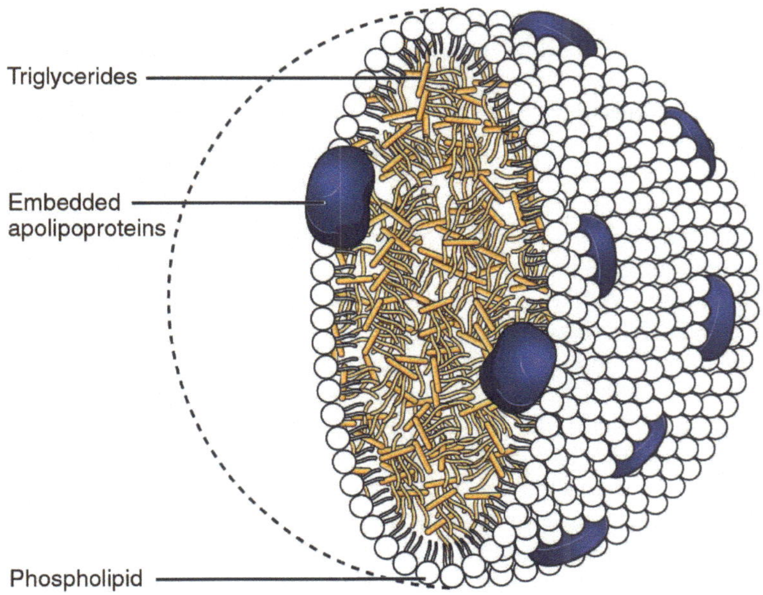

FIG. 10.7 Chylomicrons contain triglycerides, cholesterol molecules, and apolipoproteins (protein molecules) [3].

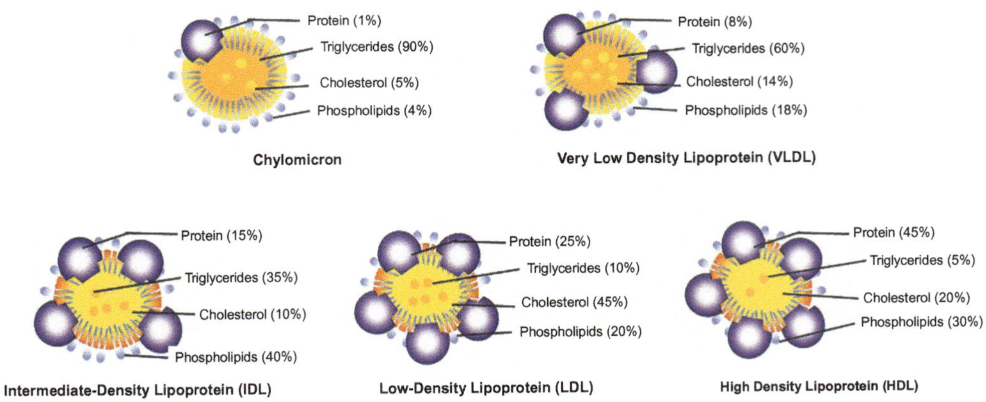

FIG. 10.8 The relationship between lipoprotein density and its diameter [4]. The density of the lipoprotein is inversely related to its size. **Chylomicrons:** 99% lipid (triglyceride, cholesterol, phospholipids) and 1% protein (extremely low density) come from dietary fat. **VLDL**: 92% lipid, 8% protein; endogenous lipids released by the liver. They are formed mostly in between meals and deliver triglycerides. VLDLs are converted into IDLs with the removal of some triglycerides. **IDL**: 85% lipid, 15% protein. IDL is then stripped up of other triglycerides to form LDL. **LDL**: 75% lipid, 25% protein. LDL transports cholesterol to cells. **HDL**: 55% lipid, 45% protein, transports excess cholesterol from cells to the liver.

lipid, mainly triglycerides, 14% cholesterol, and 8% protein. As they are transported in the blood, lipoprotein lipase (LPL) releases some fatty acids into tissues, resulting in the conversion of VLDL to **intermediate-density lipoproteins** (IDL). IDL consists of 85% lipid and 5% protein. Their fate is such that they are either brought back to the liver to be recycled or allow the release of more free fatty acids by the action of LPL and assume the composition of **low-density lipoprotein** (LDL; Fig. 10.8). LDL is high in cholesterol compared to chylomicron, VLDL, and IDL. They consist of 80% lipid and 20% protein (Fig. 10.8) and engage in cholesterol transport to peripheral cells that express specific receptors to LDL. Any cholesterol not used by the cell diffuses out of the cell into the blood, which is typically absorbed by **high-density lipoproteins** (HDL).

HDL is similar to LDL in its structure, with a layer of phospholipids, free cholesterol, and a hydrophobic core consisting mainly of cholesteryl esters and some triglycerides. HDL consists of 50% lipid and 50% protein. HDL transports cholesterol from peripheral cells back to the liver. Within the liver, HDLs are absorbed, and their cholesterol is extracted. HDL particles are smaller and contain different apolipoproteins, mainly **apolipoprotein A1** (ApoA1), which makes up 70% of the protein of HDL, and the other major apolipoprotein is **ApoA2**. Both these apolipoproteins have properties that protect the lipids against oxidative modification. In addition, some of the proteins transported by HDL, such as **paraoxonase**, have antioxidant properties. Thus LDL is very susceptible to oxidative modification, whereas HDL is relatively resistant to it and is one reason underlying the antiinflammatory properties of HDL.

10.7 Lipoprotein metabolism

Once in circulation, chylomicrons are acted upon by the enzyme LPL to release some fatty acids, and the remaining chylomicron structure is called **chylomicron remnant**, which is smaller in size and is relatively rich in cholesterol and cholesteryl esters. Chylomicron remnants reach the liver, where some of the remnant cholesterol is used in synthesizing bile acid and is released into bile. In between meals and during fasting, the liver uses chylomicron remnants to maintain lipid levels by producing other lipoproteins (Fig. 10.9). One such lipoprotein made by the liver is VLDL, which serves as a transporter of endogenous lipids (mainly triglycerides). The endothelial lining in the capillaries of various tissues (muscle, liver, adipose tissue) releases LPL, which breaks down fatty acids from the VLDLs to be stored as triglycerides in the adipose tissue and skeletal muscle. VLDL remnants that remain in circulation after their triglycerides being removed are referred to as IDL. IDL is acted upon by LPL which removes additional triglycerides. The resulting structure is an LDL. LDL binds to the LDL receptor on the surface of many cells. Once within the tissue, LDL is broken down into amino acids, free fatty acids, and cholesterol (Fig. 10.10). Any unused cholesterol is released into the blood, which is taken up by circulating HDL using the **lecithin-cholesterol acyltransferase enzyme (LCAT)** bound to HDL. This cholesterol-loaded HDL returns to the liver where it releases cholesterol in two ways: *direct release* of cholesterol and the *indirect release* where it exchanges cholesterol to triglyceride with IDL in a 1:1 ratio in the synthesis of new LDL. The rest of the cholesterol is used in the synthesis of bile salts. Bile salts help with emulsification in the small intestine and are subsequently either reabsorbed to be reused in the emulsification process or excreted in the form of feces. Upon exchanging cholesterol with IDL, HDL returns to circulation to absorb additional cholesterol.

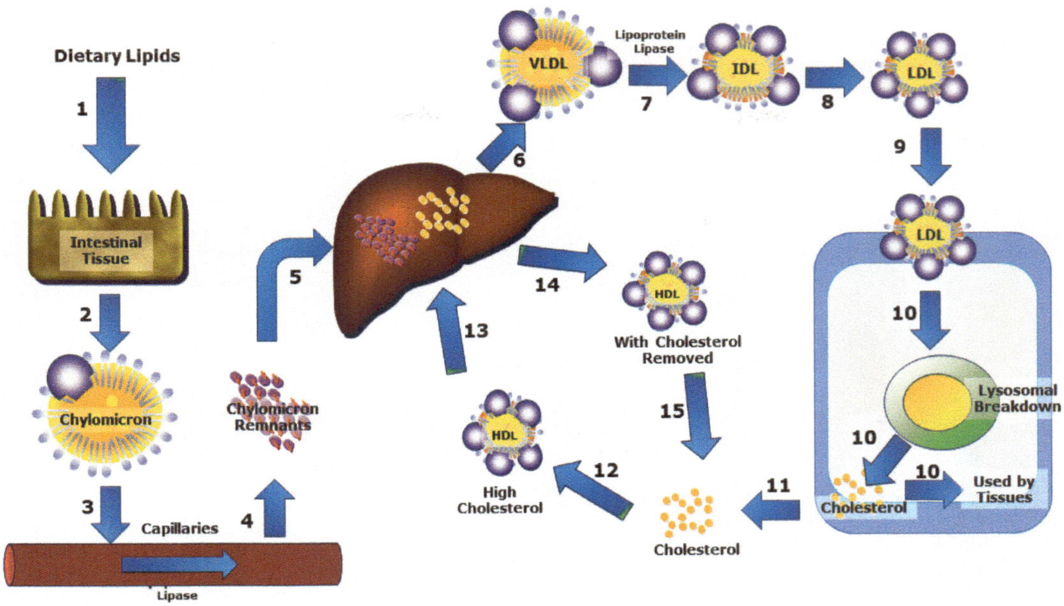

FIG. 10.9 Summary of lipoprotein metabolism [5]. (1) Dietary lipids are packaged into chylomicrons in the small intestine. (2) Chylomicrons enter circulation. (3) Chylomicrons in the blood are acted upon by lipoprotein lipase. Some fatty acids are released into blood to be used by the cells or stored in the adipose tissue. (4) As they release fatty acids, chylomicrons become smaller chylomicron remnants. (5) Chylomicron remnants reach the liver. (6) VLDL is synthesized using the chylomicron remnants and cholesterol and released back into circulation. (7) In peripheral capillaries, lipoprotein lipase removes some fatty acids and become IDLs. (8) Additional fatty acids are removed from the IDLs and are now referred to as LDLs. (9) LDLs reach peripheral tissues. (10) LDLs enter cells using receptor-mediated endocytosis where lysosomes release cholesterol to be used by the cell. (11) Any excess cholesterol, not used by the cell, diffuses out of the cell into the bloodstream, (12) Cholesterol in the blood is absorbed by HDLs. (13) HDL Returns cholesterol to the liver. (14) In the liver, HDLs are absorbed, and their cholesterol is extracted. (15) HDL travels back into peripheral tissues and absorb additional cholesterol.

Proteins such as ApoA1, produced by the liver, combine with cholesterol in the blood to form HDL (Fig. 10.11). Lipid-poor ApoA1 binds to specific receptors and becomes a mature HDL. Besides this reverse cholesterol transport, HDL competes with LDL for binding sites on the membranes and prevents the internalization of LDL cholesterol in the smooth muscle fibers of the arterial walls. HDL also helps remove cholesterol from the arterial walls by removing cholesterol from the foam cells. Thus, higher levels of HDL correlate with a reduced risk of atherosclerosis.

10.8 Lipolysis

Triglycerides must first be broken down into their two components—fatty acids and glycerol—to obtain energy. This process is called **lipolysis**, and the enzymes associated with the breakdown of triglycerides are called **lipases**. LPL and its action was discussed in the previous sections. Other examples of lipases are **hepatic lipase**, which is found in the liver, and

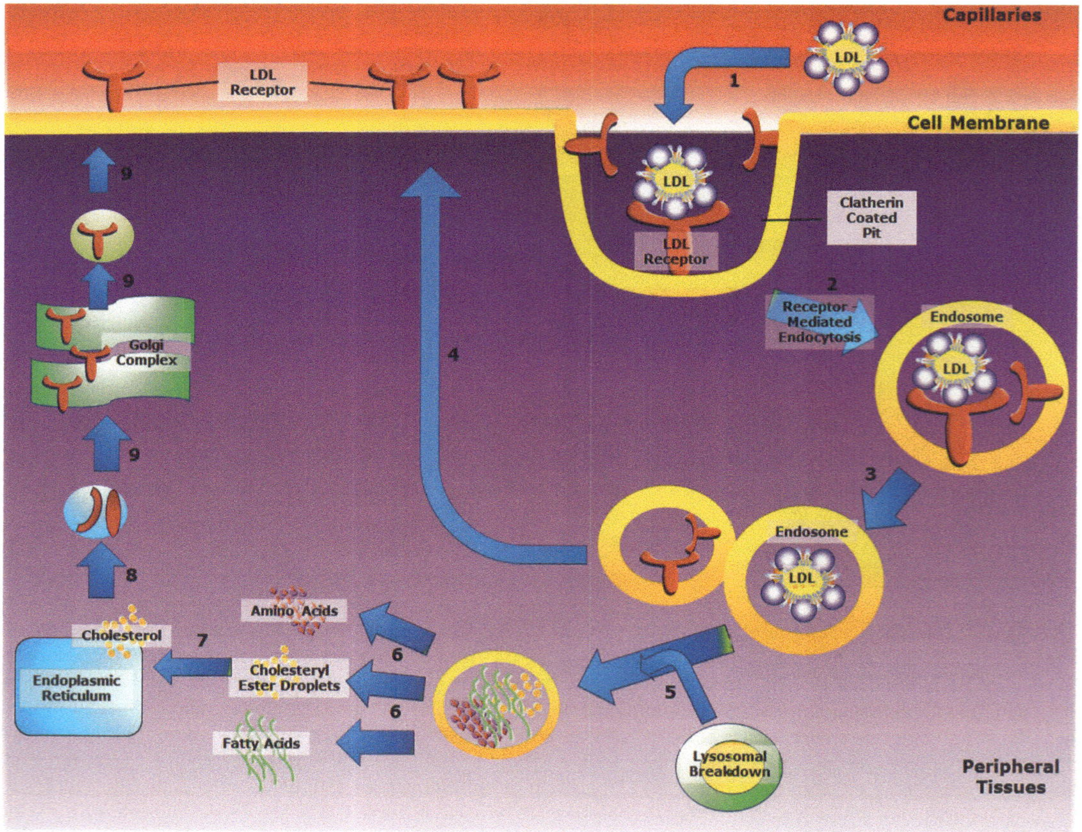

FIG. 10.10 LDL metabolism in peripheral tissues (step 10 expanded from Fig. 10.9). (1) As LDL circulates in the blood, its apoB-100 protein which is embedded in the phospholipid outer layer of LDL particles is recognized by the LDL receptor in the clathrin-coated pit. LDL attaches to the LDL receptor with the help of clathrin. (2) After the LDL-LDL receptor binding, the clathrin-coated pit pinches off from the outer layer and move into the cell in a process called receptor-mediated endocytosis. (3) The LDL and LDL-receptor complex enter the cell as an endosome (intracellular organelle) where the LDL and LDL receptors are separated from each other. (4) LDL receptor is returned to the cell membrane to be reused. (5) Endosome fuses with a lysosome in a process termed "lysosome fusion." (6) Endosome undergoes hydrolysis (breakdown) to release amino acids, cholesteryl esters droplets, and fatty acids. (7) Some cholesterol from LDL is taken up into the endoplasmic reticulum for its processing. (8) Endoplasmic reticulum releases new LDL receptors into the Golgi complex. (9) Golgi complex packages these receptor proteins inserts them into the cell membrane.

hormone-sensitive lipase, located in the adipose tissue. Fatty acids are further oxidized (broken down) by **β-oxidation** (Fig. 10.12) within the mitochondria to form acetic acid molecules, which are subsequently converted into acetyl coenzyme A (acetyl CoA). Acetyl CoA enters the Krebs cycle and produces ATP in the same manner as acetyl CoA derived from pyruvate. Glycerol molecules directly enter the glycolysis pathway once they release their fatty acids.

Since 1 triglyceride molecule yields 3 fatty acid molecules, and each fatty acid is made up of as many as 16 or more carbon atoms, lipid molecules yield much more energy than

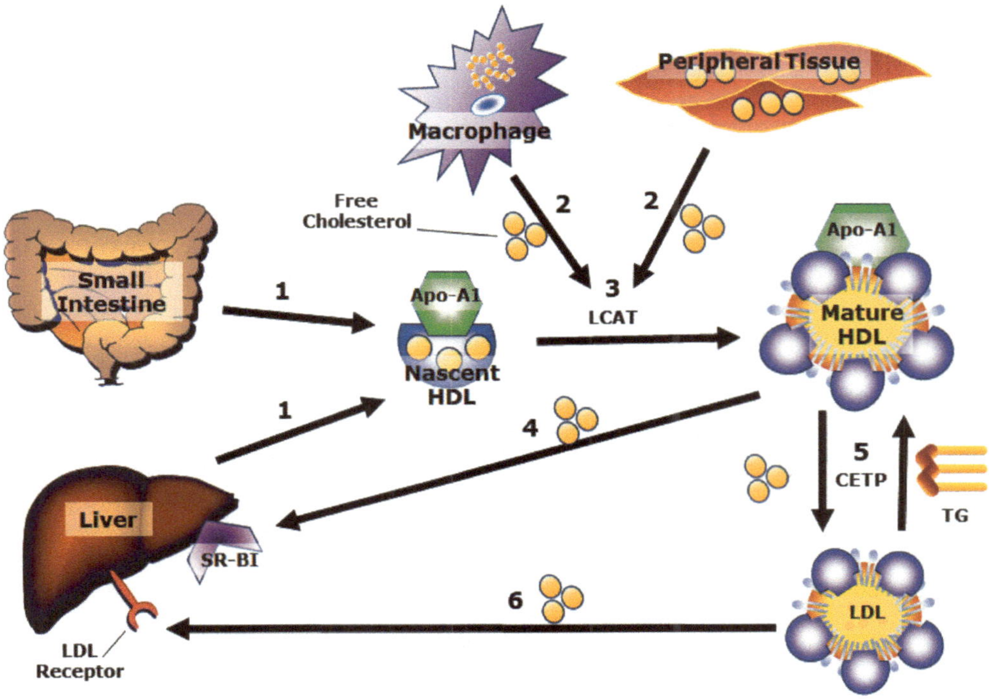

FIG. 10.11 High-density lipoprotein (HDL) metabolism. (1) The liver and the small intestine produce nascent HDLs by secreting Apo-A1 lipoprotein, among others. (2) Free cholesterol is acquired from macrophages (foam cells), other peripheral tissues as well as plasma. (3) Cholesterol is esterified by lecithin cholesterol acyltransferase (LCAT). (4) Nascent HDLs are at this stage referred to as mature HDLs which will transfer their cholesterol to the liver directly via scavenger receptor class B type I (SR-B1). (5) Mature HDL will continue to circulate in the blood, collecting excess cholesterol and offloading it in the liver. In obese condition, mature HDL transfers its cholesterol to the liver indirectly by way of cholesteryl ester transfer protein (CETP)-mediated mechanism to LDL. CETP also causes LDL to exchange triglyceride (TG) with mature HDL. (6) LDL binds to its receptor on the hepatocyte and is internalized.

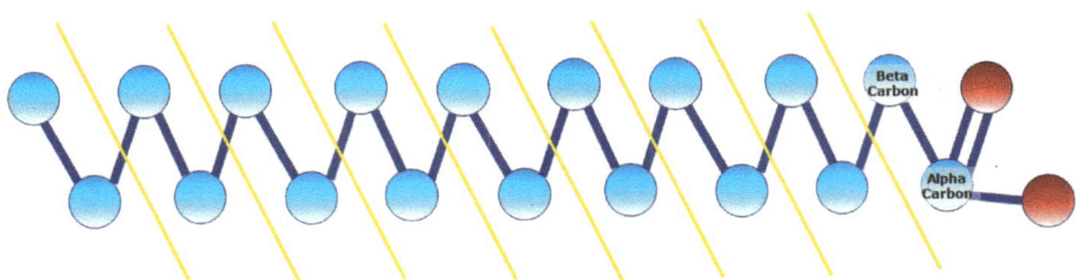

FIG. 10.12 Beta-oxidation of a fatty acid. The long fatty acid is split into several two carbon structures in this process.

carbohydrates and are an essential source of energy. Therefore, when glucose levels are low, triglycerides can be converted into acetyl CoA molecules and used to generate ATP through aerobic respiration pathways.

10.9 Ketogenesis

Ketones are water-soluble molecules consisting of three to four carbon compounds and include **acetoacetate**, **β-hydroxybutyrate**, and **acetone**. Acetoacetate is the most common precursor of the two circulating ketones, acetone and β-hydroxybutyrate (Fig. 10.13). Among these, β-hydroxybutyrate is the most abundant circulating ketone. Ketones are formed when fatty acids are first metabolized to acetyl CoA via beta-oxidation. If the Krebs cycle is overloaded or the ATP levels are met within the cell, any excess acetyl CoA molecules could produce **ketones** as an intermediate step in the synthesis of fatty acids. In this process, acetyl CoA is converted to an intermediate molecule, hydroxymethyl glutaryl CoA, from which acetoacetate is released. They are formed in the liver during fasting, starvation, prolonged exercise, and in cases such as untreated type 1 diabetes mellitus since lipids are targeted in greater amounts during these situations. When needed, ketones could serve as fuel molecules to generate energy by being converted to acetyl CoA.

10.10 Oxidation of ketones

Organs that depend solely on blood glucose, such as the brain, can use ketones as an alternative energy source, especially when glucose levels are less than normal. β-Hydroxybutyrate, as an energy source within the cell, is converted into acetoacetate, acetoacetyl CoA, and subsequently into acetyl CoA (Fig. 10.13). Acetyl CoA turns on the Krebs cycle. Besides its role as an alternate energy source, β-hydroxybutyrate may reduce lipolysis, sympathetic tone, and lower oxidative stress by upregulating antioxidant genes [7].

When ketones are produced faster than being used, the excess levels of ketones than usual are referred to as **ketosis**. One of the products of the breakdown of ketone bodies, CO_2, can

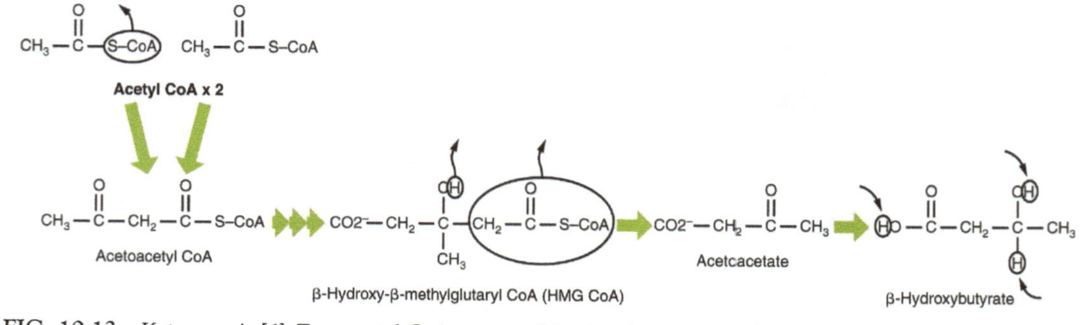

FIG. 10.13 Ketogenesis [6]. Two acetyl CoAs are combined to form various ketones.

acidify the blood, leading to **ketoacidosis**, a dangerous condition often seen in people with type 1 diabetes. In addition, acetone released during the breakdown of ketones may escape through exhalation. One of the symptoms of ketogenesis, therefore, is the fruity smell of the breath.

10.11 Lipogenesis

When glucose levels are high, but energy needs are low, any excess acetyl CoA obtained by glycolysis could be converted into fatty acids, triglycerides, cholesterol, steroids, or bile salts. This process is called **lipogenesis**. Thus the availability of acetyl CoA is significant because it initiates lipogenesis. Once the process of lipogenesis has begun, it advances by adding two carbon atoms from another acetyl CoA, and this process is repeated until fatty acids are of the appropriate length. Creating a triglyceride in this manner is an efficient way of storing the energy available in carbohydrates.

10.12 Management of low blood cholesterol levels

High consumption of cholesterol and saturated fats such as egg yolk, animal fat, and butter raises plasma cholesterol levels. Conversely, diets low in cholesterol and high in polyunsaturated fats such as omega-3 fatty acids, abundant in fish oils, are beneficial in lowering plasma cholesterol levels. Trans unsaturated fats produced by artificial hydrogenation of polyunsaturated oils (used in baked goods and margarine) adversely affect cholesterol profiles.

Exercise is shown to increase the production of HDL, which, in turn, appears to increase the efflux of cholesterol from HDL in the liver and promotes its exchange of cholesterol with LDL (by removing cholesterol from LDL in exchange for triglyceride) [8]. Exercise also reduces the action of LPL which minimizes the breakdown of HDL, thus allowing more HDL to remain in circulation [9]. In addition, hormones released during exercise, such as E, stimulate hormone-sensitive lipase within the adipocyte, which triggers the release of free fatty acids to be used in the energy-producing metabolic pathways.

As high cholesterol is linked to heart disease, patients with high cholesterol are recommended exercises and medication to reduce blood cholesterol levels. **Statins** are cholesterol-lowering drugs. Their actions are in the inhibition of endogenous production of cholesterol within the liver. Other drugs may use mechanisms to inhibit the absorption of cholesterol from the small intestine to lower blood cholesterol levels.

References

[1] 2405 Digestive Process.jpg. Wikimedia Commons, the free media repository. Updated September 8, 2020. Accessed March 20, 2021. https://commons.wikimedia.org/w/index.php?title=File:2405_Digestive_Process.jpg&oldid=451917496.

[2] 2431 Lipid Absorption.jpg. Wikimedia Commons, the free media repository. Updated September 30, 2020. Accessed March 20, 2021. https://commons.wikimedia.org/w/index.php?title=File:2431_Lipid_Absorption.jpg&oldid=476086670.

[3] 2512 Chylomicrons Contain Triglycerides Cholesterol Molecules and Other Lipids.jpg. Wikimedia Commons, the free media repository. Updated October 21, 2020. Accessed March 22, 2021. https://commons.wikimedia.org/w/index.php?title=File:2512_Chylomicrons_Contain_Triglycerides_Cholesterol_Molecules_and_Other_Lipids.jpg&oldid=496159646.

[4] K.R. Feingold, B. Anawalt, Introduction to lipids and lipoproteins, in: K.R. Feingold, A. Boyce, et al. (Eds.), Endotext [Internet], MDText.com, Inc., 2021.

[5] J. Borén, M.-R. Taskinen, M. Adiels, Kinetic studies to investigate lipoprotein metabolism, J. Intern. Med. 271 (2) (2012) 166–173, https://doi.org/10.1111/j.1365-2796.2011.02497.x.

[6] 2514 Ketogenesis.jpg. Wikimedia Commons, the free media repository. Updated September 5, 2020. Accessed March 20, 2021. https://commons.wikimedia.org/w/index.php?title=File:2514_Ketogenesis.jpg&oldid=448495195.

[7] E.N. Dedkova, L.A. Blatter, Role of β-hydroxybutyrate, its polymer poly-β-hydroxybutyrate and inorganic polyphosphate in mammalian health and disease, Front. Physiol. 5 (2014) 260.

[8] J.J. Ruiz-Ramie, J.L. Barber, M.A. Sarzynski, Effects of exercise on HDL functionality, Curr. Opin. Lipidol. 30 (1) (2019) 16–23, https://doi.org/10.1097/mol.0000000000000568.

[9] W. Annema, U.J.F. Tietge, Role of hepatic lipase and endothelial lipase in high-density lipoprotein-mediated reverse cholesterol transport, Curr. Atheroscler. Rep. 13 (3) (2011) 257–265, https://doi.org/10.1007/s11883-011-0175-2.

CHAPTER

11

Obesity and liver disease

OUTLINE

OBJECTIVES

- Describe the anatomical and histological features of the liver.
- Name the two major blood supplies to the liver and the significance of two separate blood supplies.
- List the substances bile is made of.
- Describe the regulation of the release of bile.
- Trace the pathway by which bile is released into the duodenum.
- Describe the major functions of the liver concerning metabolism, detoxification, and excretion of substances.
- Define steatosis, steatohepatitis, and cirrhosis. Compare and contrast the two, including the subtypes of steatohepatitis.

Biology of Cardiovascular and Metabolic Diseases
https://doi.org/10.1016/B978-0-12-823421-1.00009-3

- Explain the mechanism by which alcohol affects liver function.
- Identify the histologic features and lab values in the differential diagnosis of steatohepatitis.
- Define portal hypertension and the cause for the same.
- Explain the conditions that ultimately lead to cirrhosis.

11.1 Introduction

The **liver** is the largest gland in the body, weighing approximately 3 pounds in an adult. It is essential for life since it deals with various metabolic processes, including getting rid of substances that would otherwise be harmful if allowed to accumulate. It is also the first port of entry for most nutrients that are absorbed by the digestive tract. Additionally, the liver is essential in synthesizing many substances, including plasma proteins, clotting agents, and bile, that optimize the absorption of fats. Obesity increases the risk of elevated liver enzymes and causes **steatosis** (**fatty liver**), **cirrhosis** (**scarring**), and **hepatocellular carcinoma**.

11.2 Anatomy of the liver

The liver is the largest gland in the body. It lies inferior to the diaphragm in the right upper quadrant of the abdominopelvic cavity and receives protection from the surrounding ribs (Fig. 11.1). It is stabilized in its position by a peritoneal fold called the falciform **ligament**. The liver is functionally divided into two primary **lobes**: a large right lobe and a much smaller left lobe. The right lobe houses an inferior **quadrate** and a posterior **caudate** lobe, defined by internal features only but not through the external anatomical support. The lobes are further subdivided into smaller compartments called **lobules**. Each lobule has a hexagonal shape filled mainly with hepatocytes arranged in a plate-like manner that is one cell layer thick (Fig. 11.2).

11.3 Blood supply to the liver

The liver receives two separate blood supplies, one from the **hepatic artery** and the second one from the **hepatic portal vein**. The hepatic portal vein is a unique arrangement formed by the fusion of veins from the lower portion of the esophagus, the stomach, the pancreas, the spleen, and the intestines. While the hepatic artery delivers oxygenated blood to the liver, the hepatic portal vein brings partially deoxygenated blood along with nutrients, drugs, and toxins that were collected from the visceral organs with which it is associated. Both of these vessels branch off within the lobule, and these smaller vessels are called **sinusoids**. A hepatic sinusoid is an open, porous blood capillary. Besides processing the blood-borne nutrients and toxins, the liver releases nutrients back into the blood, which continues as the **central vein**. Central veins from the lobules collectively continue as a **hepatic vein**, which in turn joins the inferior vena cava (Fig. 11.3).

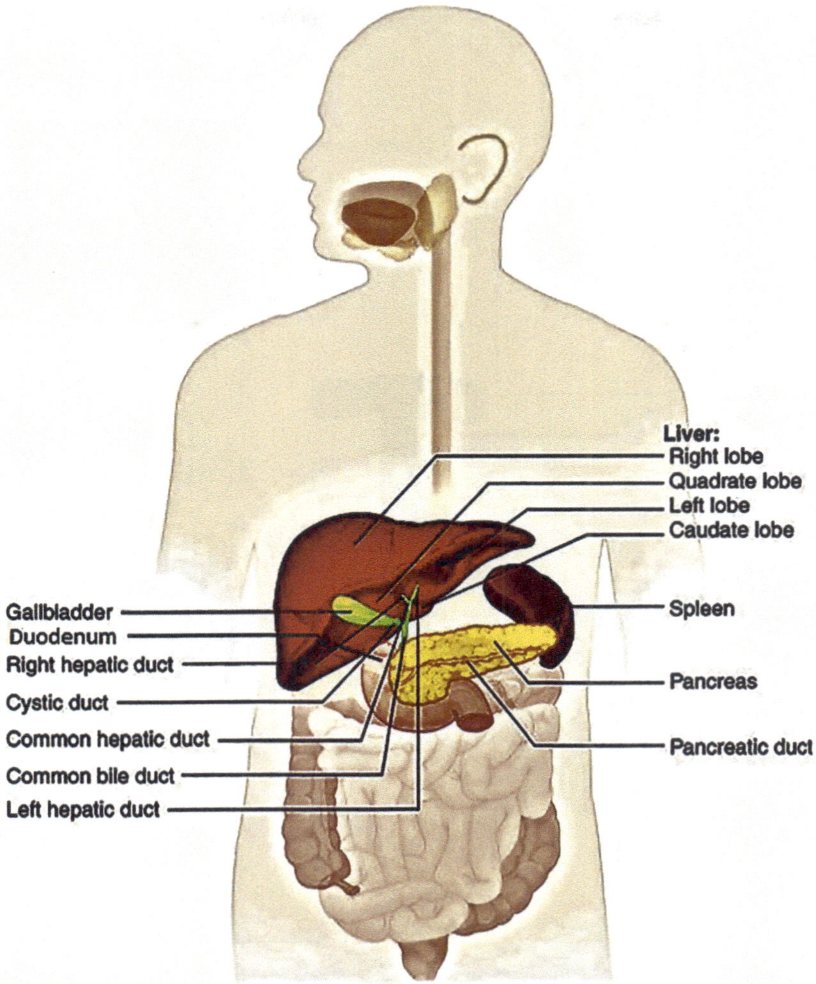

FIG. 11.1 Ventral view of the liver, gallbladder, pancreas, stomach, and duodenum [1]. The pancreatic duct drains pancreatic juice into the duodenum. The cystic duct transports bile from the gallbladder. As the cystic duct joins the common hepatic duct, it is called the common bile duct, which drains bile into the duodenum.

11.4 Hepatocytes and Kupffer cells

Hepatocytes are the most abundant cells in the liver (Fig. 11.2). They account for approximately 80% of the liver's volume. Hepatocytes play a vital role in secretory and metabolic functions. They are tightly packed around the fenestrated endothelium, giving them easy access to the blood. From their central position, hepatocytes process nutrients, toxins, and waste materials carried by the blood. Materials such as **bilirubin**, a bile pigment, are processed and

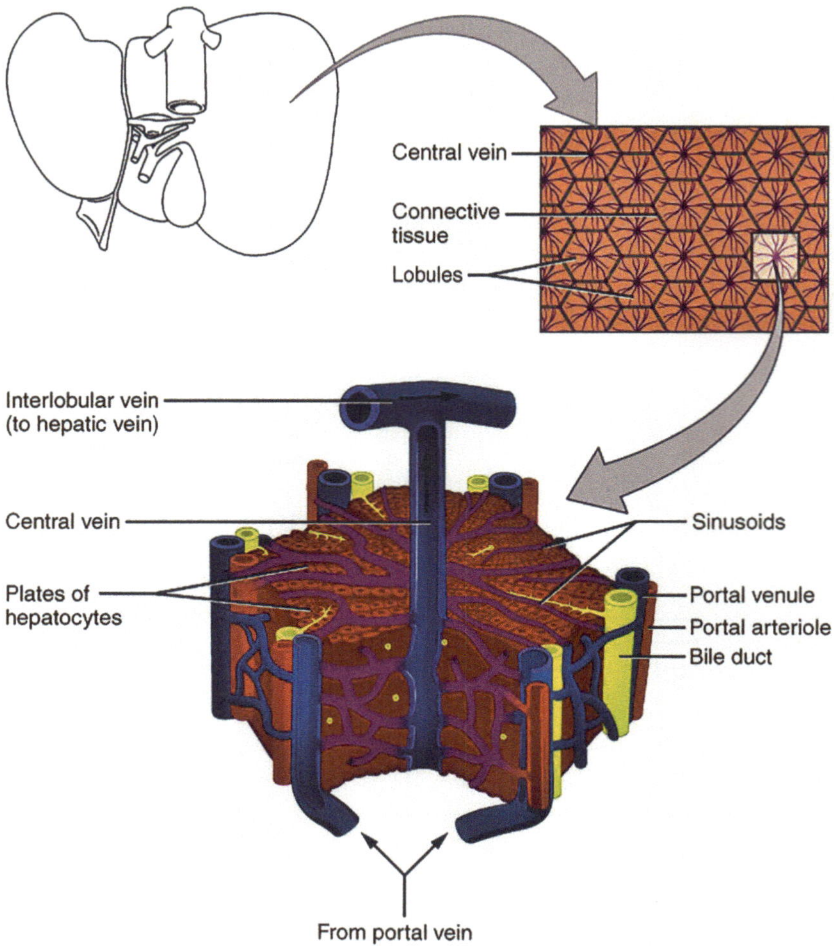

FIG. 11.2 A schematic illustration of the liver showing microscopic details [2]. The lobules are hexagonal structures within each lobe of the liver. The details of a lobule show branches of both hepatic artery (portal arteriole) and hepatic portal vein (portal venule) carrying blood to the liver lobule, which branch off as sinusoids and fuse to become a central vein. Hepatocytes are arranged in a plate-like manner. Bile, synthesized by the hepatocytes, is carried by small bile canaliculi, which drain bile into the bile duct.

excreted into **bile canaliculi**, the smallest bile ducts. Other materials, including proteins, lipids, and carbohydrates, are processed and secreted into the sinusoids or stored in the cells until called upon. The hepatic sinusoids also contain macrophages and are called **Kupffer cells**, which make up approximately 10% of the liver. Kupffer cells are phagocytic cells that remove dead red and white blood cells, bacteria, and other foreign material that enter the sinusoids.

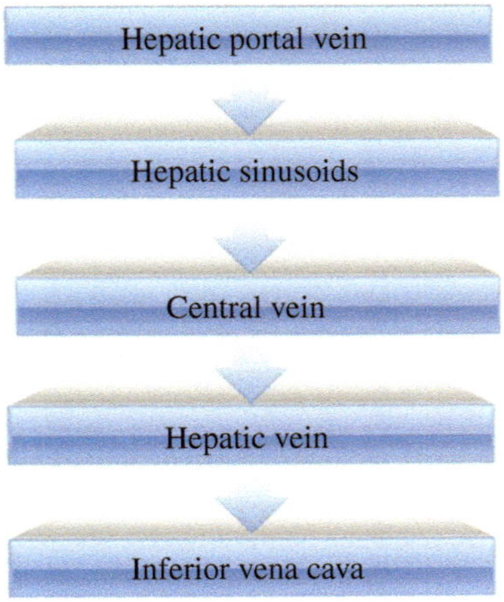

FIG. 11.3 Hepatic portal system. Hepatic artery and hepatic portal vein branch off to become hepatic sinusoids within the lobule. Sinusoids eventually merge to become the central vein. The central vein from each lobule joins to become the hepatic vein which drains into the inferior vena cava.

11.5 Production and circulation of bile

Between adjacent hepatocytes, tiny grooves provide room for each **bile canaliculus** (plural = canaliculi). These microscopic ducts transport bile secreted by hepatocytes first into bile ductules and then into progressively larger bile ducts. The bile ducts fuse to form the larger right and left **hepatic ducts**, merging and exiting the liver as the **common hepatic duct**. The common hepatic duct then joins the **cystic duct** from the gallbladder to become the **common bile duct** which empties bile into the duodenum, the first part of the small intestine (Fig. 11.4). The **portal triad** is a characteristic feature around each corner of hepatic lobules, consisting of three structures: a bile duct, a hepatic artery branch, and a hepatic portal vein branch.

Hepatocytes secrete about 1 L of bile each day. Bile is a yellowish-brown or yellowish-green alkaline solution (pH7.6–8.6) consisting of a mixture of **bile salts**, **bile pigments**, **bile acids**, phospholipids, cholesterol, triglycerides, water, and electrolytes secreted by the liver to accomplish emulsification of lipids in the small intestine. The components most critical to emulsification are bile salts and phospholipids, which have a nonpolar (hydrophobic) and a polar (hydrophilic) region. The hydrophobic portion interacts with the large lipid molecules, whereas the hydrophilic region interacts with the watery chyme in the small intestine. Such interaction results in large lipid globules being pulled apart into many tiny fragments of about 1 μm in diameter, dramatically increasing the surface area available for lipid-digesting enzyme activity just as dish soap works on fats mixed with water.

While most components of bile are eliminated in feces, approximately 95% of bile salts are reclaimed by a special blood vessel arrangement between the large intestine and the liver called the **enterohepatic circulation**. Once bile salts reach the ileum, the last part of the small intestine, they are absorbed and returned to the liver via the hepatic portal vein. The hepatocytes then excrete bile salts into newly formed bile. Thus this precious resource is continuously recycled.

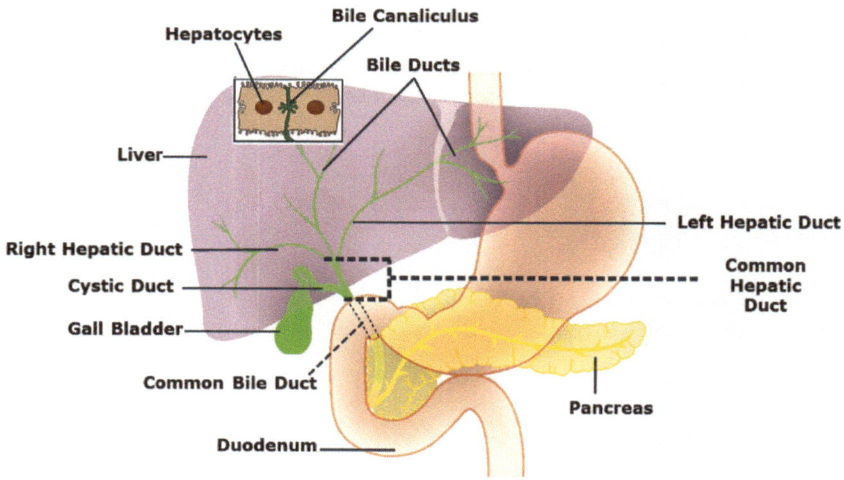

FIG. 11.4 Circulation of bile [3]. Bile from the hepatocytes is transported by bile canaliculi which fuse to become bile ducts. They leave the lobes as right and left hepatic ducts. Eventually, the right and left hepatic ducts join to become the common hepatic duct. The branch of the common hepatic duct that carries bile to the gallbladder is called the cystic duct. The cystic duct and common hepatic duct together become the common bile duct. The common bile duct drains bile into the duodenum.

Bilirubin, the primary bile pigment, is a waste product produced when the liver and spleen remove old or worn-out red blood cells from the circulation [4]. These breakdown products, including proteins, iron, and toxic bilirubin, are transported to the liver from the spleen via the hepatic portal system. In the liver, proteins and iron are recycled, whereas bilirubin is excreted in the bile, accounting for its green color. Intestinal bacteria eventually transform bilirubin into **stercobilin**, a brown pigment that gives stool its characteristic color. In some disease states, bile may not enter the small intestine, resulting in white ("acholic") stools with high-fat content since virtually no fats would be broken down or absorbed. Between meals, bile continues to be produced but is conserved in its concentrated form within the gallbladder.

11.6 Functions of the liver

Hepatocytes have a high rate of metabolism, including synthesis, storage, and degradation of substances. One of the primary functions of the liver is the processing of nutrients and is summarized in Table 11.1.

11.6.1 Carbohydrate metabolism

The liver plays a crucial role in maintaining normal blood glucose levels. It can store excess glucose in the form of glycogen (**glycogenesis**) and break down glycogen into glucose (**glycogenolysis**) when the blood glucose concentrations decrease below normal. In addition,

TABLE 11.1 Metabolic processes of nutrients in the liver.

Protein	Carbohydrate	Lipid
Production of plasma proteins Production of clotting factors Deamination Transamination Urea cycle	Glycogenesis Glycogenolysis Gluconeogenesis Conversion of fructose and galactose to glucose	Synthesis of triglycerides and cholesterol Lipolysis

new glucose is generated from noncarbohydrate sources such as amino acids and glycerol (**gluconeogenesis**) as needed to maintain blood glucose levels. Furthermore, the conversion of galactose and fructose to glucose occurs in the liver.

11.6.2 Protein metabolism

Proteins in the diet are digested into their smallest components, amino acids, absorbed by the blood capillaries and brought to the liver via the hepatic portal vein. In the liver, these amino acids are used to produce various new proteins, such as plasma proteins and **clotting factors**. Other proteins produced by the liver include complement proteins, lipoproteins, and transport proteins. **Deamination** and **transamination** reactions occur when amino acids are utilized for gluconeogenesis, and nonessential amino acids are synthesized. Another vital function of the liver is to remove ammonia, a toxic substance in the form of urea, a nontoxic waste molecule using the urea cycle.

11.6.3 Lipid metabolism

Dietary lipids are released into the blood from the digestive tract, mainly in the form of chylomicrons. Fatty acids are released from these chylomicrons with the help of lipoprotein lipase. New triglycerides and cholesterol are synthesized within the liver using excess glucose and fatty acids. About 80% of the cholesterol produced in the liver is converted into bile salts, which are secreted into the bile. The remaining cholesterol is released into the blood for tissues to use for various purposes. Lipoproteins are required to transport these lipids and are made by the liver. Oxidation of fatty acids to supply energy and synthesize fat from proteins and carbohydrates are the additional metabolic events in the liver.

11.6.4 Other metabolic functions of the liver

The liver is a major detoxification center that excretes drugs and hormones into bile and blood to be removed by the kidneys. Since blood from the visceral organs, including gut bacteria, is drained into the liver via the hepatic portal vein, protection from the enteric bacteria is provided by Kupffer cells lining the sinusoids. Kupffer cells play a significant role in the removal of these bacteria. Breakdown of toxins releases free radicals such as reactive oxygen species. Accumulation of free radicals results in cell injury. The liver, however, is capable of producing **antioxidants. Glutathione** is one of the significant antioxidants produced by the liver. The presence of antioxidants offers protection against cell injury by free radicals.

In addition, the liver is also a storage site for vitamins, and in particular, lipid-soluble vitamins such as **vitamin A** and **vitamin D. Vitamin B12**, a water-soluble vitamin , is stored in the liver. Iron is also stored in the liver as **ferritin**. Nevertheless, another function of the liver is to excrete calcium as part of the bile, which passes into the gut and is eventually lost in the feces.

11.6.5 Cori cycle

Much of the lactic acid produced during anaerobic respiration is converted to carbon dioxide and water when oxygen becomes available, and aerobic respiration is initiated. However, some of the lactic acid produced by the exercising skeletal muscle reaches the liver, where it is converted back into pyruvic acid. Unlike most other organs, the liver contains enzymes that are needed to convert pyruvic acid molecules to glucose 6-phosphate and eventually glucose, which is essentially the reverse of glycolysis (Fig. 11.5).

11.7 Hepatic steatosis (fatty liver disease)

Fatty liver is a liver disease of metabolic origin that results in an abnormal buildup of small vesicles of fat, predominantly triglycerides within hepatocytes. It develops when the body synthesizes too much fat or when the body cannot metabolize fat at a normal rate. Fatty liver is a common condition that affects 10%–20% of the American population and is mainly seen in the age group of 40–60 years. The most common cause of fatty liver disease is excessive consumption of alcohol. Other common causes of fatty liver disease include obesity,

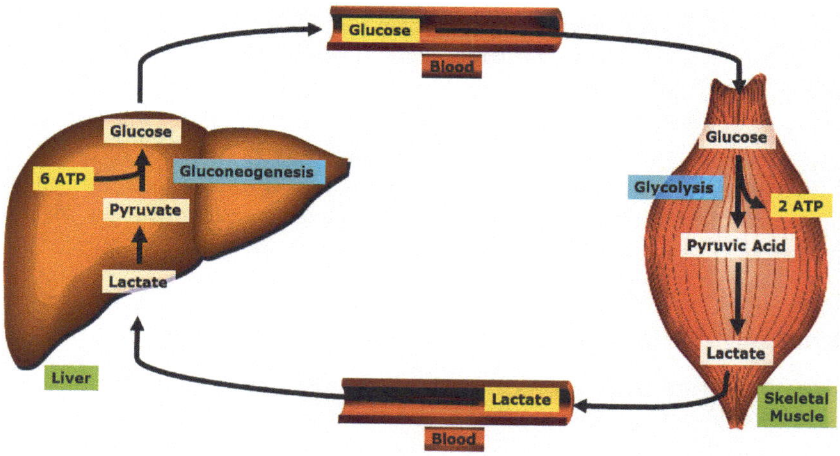

FIG. 11.5 Cori cycle converting lactate to glucose. Lactic acid produced during anaerobic respiration by the muscle cells is brought to the liver, where it is converted back to pyruvic acid and ultimately into glucose [5,6].

hyperlipidemia, insulin resistance, diabetes, and overuse of acetaminophen (Tylenol). Although there is little or no fibrosis in the early stages, fibrous tissue could develop within the liver with continued alcohol intake.

11.7.1 Why does excessive acetaminophen cause liver injury?

Metabolism of acetaminophen by the CYP450 pathway within the liver includes releasing of a toxic intermediate product called ***N*-acetyl-*p*-benzoquinone imine** (**NAPQI**). Usually, glutathione, an antioxidant released as an intermediate product through the normal phase II drug metabolism pathways, neutralizes NAPQI to nontoxic metabolite. However, with excessive doses of acetaminophen, the normal phase II drug metabolism pathways become depleted along with glutathione, and the CYP450 pathway metabolizes a higher portion of the acetaminophen taken, resulting in greater levels of NAPQI formation. Without glutathione, NAPQI concentrations buildup, which can damage proteins, lipids, and nucleic acids. This phenomenon can lead to hepatic injury and hepatocellular death [7].

11.8 Alcoholic fatty liver disease

Alcoholic fatty liver disease is the earliest stage of liver disease. Accumulation of triglycerides in the hepatocytes results from both increased inflow of free fatty acids (FFA) and lipogenesis within the liver. Fat accumulation in the liver is dependent mainly on recirculating FFAs from the adipose tissue pool. The visceral adipose tissue has a greater lipolytic potential than the subcutaneous adipose tissue. The release of FFA from visceral fat depots reaches the liver directly through the portal circulation. Increased oxidation of alcohol within the liver and resulting free radicals may cause injury to hepatocytes, which could compromise many liver functions. Abstaining from alcohol will likely cause the fatty liver to subside. Long-term consequences of this condition, on the other hand, progress to **fibrosis** and **cirrhosis**, a chronic condition of inflammation of the liver or hepatocellular carcinoma.

11.9 Nonalcoholic fatty liver disease (NAFLD)

Nonalcoholic fatty liver disease (**NAFLD**), also called **Nonalcoholic Steatohepatitis** (**NASH**), is the development of steatosis without excessive alcohol intake and resembles alcohol-induced fatty liver disease. It encompasses a spectrum of disease states from steatosis to **steatohepatitis**. There is a strong association between NASH and obesity, dyslipidemia, insulin resistance, type 2 diabetes, and hypertension. NASH develops when the liver breaks down less fat than usual, thus leading to the buildup of fat in the liver tissue (Fig. 11.6). It is diagnosed when more than 10% of the liver is fat.

Steatohepatitis refers to an inflammation within the liver tissue that is often triggered by the increased oxidation of fatty acids. Nonalcoholic steatohepatitis is common and often a "silent" liver disease. Men and women are equally affected. The prominent feature in NASH is an increased amount of fat in the liver, along with injury and inflammation. Most people

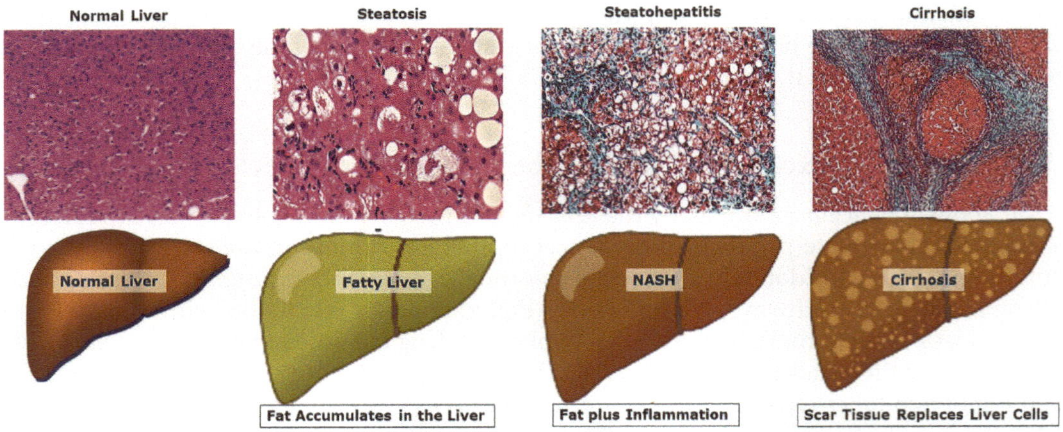

FIG. 11.6 Microscopic view of a healthy liver, steatosis, NASH, and cirrhosis [8–11]. Fatty liver disease (steatosis) is when the liver is larger and yellowish compared to the normal liver. Note the large white lipid droplets found in steatosis. Nonalcoholic steatohepatitis (NASH) contains both excess lipid (white droplets) and inflammatory filtrates. Scar tissue replaces healthy liver in cirrhosis.

with NASH may not be aware that they have a liver problem. NASH affects 2–5% of Americans where the symptoms could include loss of appetite, nausea, vomiting, abdominal pain, and yellowing of the skin (**jaundice**).

11.9.1 Pathogenesis of NASH

NASH is usually first suspected in a person who is found to have elevations in liver enzymes in routine blood test panels, such as **alanine aminotransferase** (**ALT**) or **aspartate aminotransferase** (**AST**) and further examination, using X-ray or imaging studies of the liver, show greater amount of fat than normal. NASH is confirmed when further evaluation shows no apparent reason for liver diseases such as medications, viral hepatitis, or excessive use of alcohol. The diagnosis of NASH and separating it from hepatic steatosis is via liver biopsy where histology reveals fat along with inflammation and damage to hepatocytes.

Although the occurrence of NASH is common, its underlying cause is still not clear. It most often occurs in persons who are middle-aged and overweight or obese. Although many patients with NASH have elevated blood lipids, diabetes, or prediabetes, not every obese person has NASH. Furthermore, some patients with NASH are not obese and have normal blood cholesterol levels. Thus, NASH can occur without any apparent risk factor and can even affect children. While the underlying reason for the liver injury that causes NASH is unknown, several factors are possible candidates. These include insulin resistance, the release of toxic inflammatory agents by adipocytes (cytokines), and **oxidative stress** within hepatocytes.

No specific therapies for NASH exist currently. The most important recommendations given to individuals with this disease are reducing their weight (if obese or overweight), following a balanced and healthy diet, increasing physical activity, and avoiding alcohol and

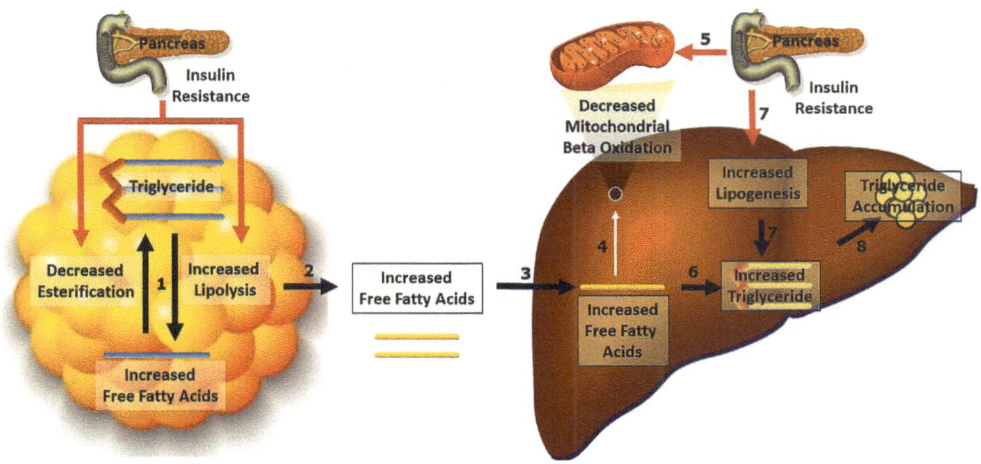

FIG. 11.7 Pathogenesis of NASH [12]. (1) Insulin resistance increases triglyceride (TG) lipolysis (breakdown) and decreases esterification (synthesis and storage) of free fatty acids (FFA) within adipose tissue, increasing FFA release into the blood (step 2). (3) Some of the excess FFAs are taken up by the liver. (4) The liver attempts to break down and remove FFA to prevent their accumulation. . However, there is a decrease in the breakdown and removal of the FFAs due to the liver's insulin-resistant state that prevents the mitochondria from oxidizing (breaking down) excess FFA (step 5). (6) With the the build up of FFAs, the liver increases the synthesis of TG. (7) The insulin-resistant state of the liver will also cause the liver to upregulate proteins that increase lipogenesis (synthesis) of TG. (8) Together, the liver will increase TG accumulation leading to NASH.

unnecessary medications. If NASH is due to coexisting conditions such as diabetes, insulin resistance, hypertension, and elevated cholesterol, these conditions should be treated with appropriate medication and well controlled.

A step-by-step detail of the pathogenesis of NASH is shown in Fig. 11.7.

11.9.2 Progression of NASH to fibrosis

The progression of NASH can take years and even decades. The process can stop and, in some cases, reverse on its own without specific therapy. On the other hand, NASH can slowly worsen, causing scarring or fibrosis to appear and remain in the liver (Fig. 11.8). As fibrosis worsens, the liver becomes scarred, hardened, and unable to function normally (Fig. 11.9). This condition is referred to as cirrhosis. Not every person with NASH develops cirrhosis, but once severe scarring or cirrhosis is present, fewer treatments can halt its progression.

11.10 Cirrhosis (fibrosis)

Cirrhosis is an irreversible fibrotic disease that disrupts liver structure (Fig. 11.9) and function. The most common causes are alcohol abuse and **viral hepatitis**. Cirrhosis accounts for the majority of all medical deaths among alcoholics. The majority of ingested alcohol is metabolized in the liver. Products of alcohol metabolism, most notably **acetaldehyde**, impair

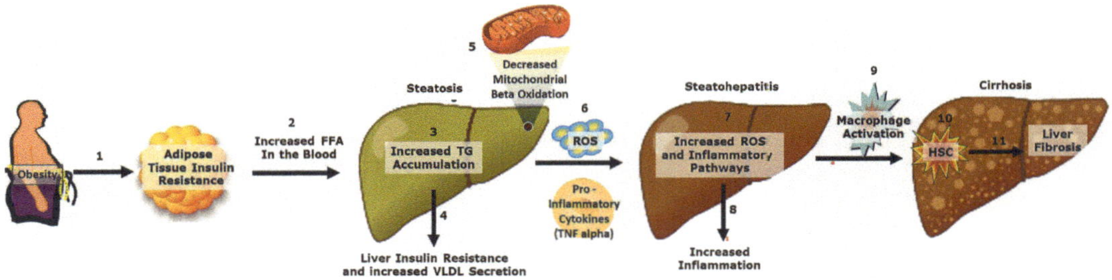

FIG. 11.8 Progression from NASH to fibrosis (aka cirrhosis) [13]. (1) Obesity leads to insulin resistance in the adipose tissue, which increases the release of free fatty acid (FFA) into blood (step 2). (3) The FFAs accumulate in the liver leading to steatosis which will increase the release of very low-density lipoprotein (VLDL; step 4). (5) Decreased beta-oxidation and the use of acetyl coA to derive ATP would lead to a new steady state of the FFAs in the liver resulting in benign steatosis (fatty liver). (6) Processing of lipids in greater amounts results in the build up of toxic metabolites (reactive oxygen species, ROS, and pro-inflammatory cytokines such as tumor necrosis factor-alpha (TNF alpha)). (7) ROS and TNF-alpha will activate inflammatory pathways leading to inflammation of the liver (steatohepatitis (step 8)). (9) Inflammatory pathways result in the activation of macrophages. (10) Macrophages release growth factors which activate hepatic stellate cells (HSC, liver cells which are involved in fibrosis), resulting in the fibrogenic response and the potential progression to cirrhosis (step 11).

FIG. 11.9 The normal liver tissue versus the liver tissue with cirrhosis [14].

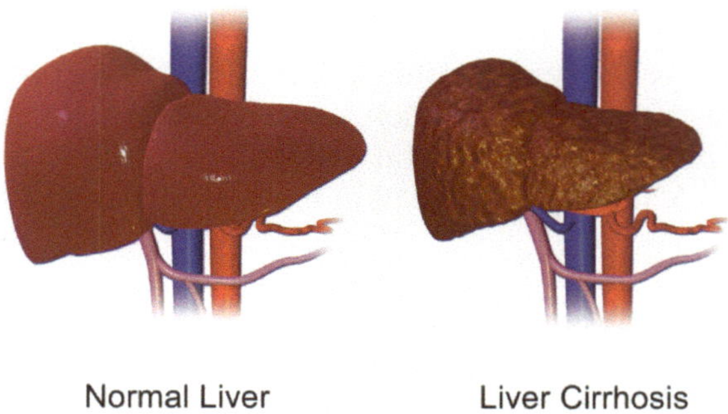

several aspects of hepatocyte metabolic function and produce oxidative stress that leads to cell death. In its early stages, the alcoholic liver disease involves the accumulation of fat in the liver. Obesity and insulin resistance can also contribute to the progression from NASH to cirrhosis (Fig. 11.9). Hepatic **stellate cells** are activated to produce collagen, which occurs chronically if excessive ingestion of alcohol continues. Activation of the stellate cells results from the release of cytokines produced by Kupffer cells in response to toxins reaching from the gut as well as oxidative stress and immunologic activation. Hepatitis will progress to cirrhosis, characterized by fibrous bands stretching from the portal triads to central veins and small, regenerative nodules. Fibrosis interferes with hepatic function. Biliary channels become obstructed, resulting in **portal hypertension**, shunting blood away from the liver, resulting in hypoxic necrosis. A person with cirrhosis may present with a spectrum of symptoms of chronic liver failure, including jaundice, nausea, and malaise and may experience fluid retention (ascites), muscle wasting, and bleeding from the intestines.

Primary treatment for cirrhosis is to secure abstinence from alcohol, although some changes in the liver may be irreversible even after drinking has stopped. Liver transplantation is a treatment for advanced cirrhosis with liver failure. NASH ranks as one of the significant causes of cirrhosis in the United States, behind hepatitis C and alcoholic liver disease.

11.11 Portal hypertension

Portal hypertension is a condition where the resistance to blood flow across the liver is increased. Generally, the liver has a very low resistance vasculature, and pressures increase little as flow increases. However, in several liver disease states, inflammatory responses trigger hepatic stellate cells to increase collagen production, reducing permeability across the sinusoidal endothelium. Fibrosis worsens the situation. The hardening of the liver hinders flow through the sinusoids. Some sinusoids and liver parenchyma may also be destroyed and replaced by fibrous tissue, further impairing liver function.

References

[1] 2422 Accessory Organs.jpg. Wikimedia Commons, the free media repository. Updated August 29, 2020. Accessed March 24, 2021. https://en.wikipedia.org/wiki/File:2422_Accessory_Organs.jpg.

[2] 2423 Microscopic Anatomy of Liver.jpg. Wikimedia Commons, the free media repository. Updated May 22, 2020. Accessed March 24, 2021. https://commons.wikimedia.org/w/index.php?title=File:2423_Microscopic_Anatomy_of_Liver.jpg&oldid=420989806.

[3] Diagram Showing the Position of the Perihilar Bile Ducts CRUK 357.svg. Wikimedia Commons, the free media repository. Updated October 26, 2020. Accessed March 25, 2021. https://commons.wikimedia.org/w/index.php?title=File:Diagram_showing_the_position_of_the_perihilar_bile_ducts_CRUK_357.svg&oldid=502997494.

[4] I. Theurl, I. Hilgendorf, M. Nairz, et al., On-demand erythrocyte disposal and iron recycling requires transient macrophages in the liver, Nat. Med. 22 (8) (2016) 945–951, https://doi.org/10.1038/nm.4146.

[5] C.F. Cori, G.T. Cori, The fate of sugar in the animal body: VIII. The influence of insulin on the utilization of glucose, fructose, and dihydroxyacetone, J. Biol. Chem. 76 (3) (1928) 755–795.

[6] K. Cori, G. Radnitz, Über den Gehalt des menschlichen Blutserums an Komplement und Normalambozeptor für Hammelblutkörperchen, Z. Immun. Exp. Ther. 29 (1920) 445.

[7] V. Gerriets, J. Anderson, T.M. Nappe, Acetaminophen, StatPearls Publishing, 2020.

[8] Stain—Liver Tissue.jpg. Wikimedia Commons, the free media repository. Updated September 28, 2020. Accessed March 24, 2021. https://commons.wikimedia.org/w/index.php?title=File:Stain_-_liver_tissue.jpg&oldid=474270685.

[9] Liver Tissue with NAFLD (48904411988).jpg. Wikimedia Commons, the free media repository. Updated September 19, 2020. Accessed March 24, 2021. https://commons.wikimedia.org/w/index.php?title=File:Liver_Tissue_with_NAFLD_(48904411988).jpg&oldid=465051251.

[10] Steatohepatitis High Mag.jpg. Wikimedia Commons, the free media repository. Updated October 17, 2020. Accessed March 24, 2021. https://commons.wikimedia.org/w/index.php?title=File:Steatohepatitis_high_mag.jpg&oldid=492255903.

[11] Cirrhosis High Mag.jpg. Wikimedia Commons, the free media repository. Updated April 29, 2020. Accessed March 25, 2021. https://commons.wikimedia.org/w/index.php?title=File:Cirrhosis_high_mag.jpg&oldid=415656826.

[12] L.A. Adams, P. Angulo, K.D. Lindor, Nonalcoholic fatty liver disease, CMAJ 172 (7) (2005) 899–905, https://doi.org/10.1503/cmaj.045232.

[13] K. Cusi, Nonalcoholic fatty liver disease in type 2 diabetes mellitus, Curr. Opin. Endocrinol. Diabetes Obes. 16 (2) (2009) 141–149, https://doi.org/10.1097/MED.0b013e3283293015.

[14] Liver Cirrhosis.png. Wikimedia Commons, the free media repository. Updated May 27, 2021. Accessed June 28, 2021. https://commons.wikimedia.org/w/index.php?title=File:Liver_Cirrhosis.png&oldid=565104294.

CHAPTER

12

Endocrine pancreas

OUTLINE

OBJECTIVES

- Locate the prancreas in the abdominal cavity and summarize its gross anatomy.
- Differentiate between the pancreatic islets (islets of Langerhans) and pancretic acini.
- List the cell types found in the islets of Langerhans. Identify the cells that synthesize insulin and those that secrete glucagon.
- Describe the steps involved in the synthesis of insulin. Name the peptide that is released along with insulin and its significance.
- Explain the mechanism by which insulin is released, including the stimuli for its release.
- Describe the structure of insulin receptor and how insulin brings about its action at its target cell.
- Identify the situation that triggers the release of glucagon.
- Compare and contrast the functions of insulin and glucagon in maintaining normal blood glucose concentration.
- Define absorptive and postabsorptive (fasting) states and explain the roles of glucagon and insulin in these states.

Biology of Cardiovascular and Metabolic Diseases
https://doi.org/10.1016/B978-0-12-823421-1.00008-1

12.1 Introduction

The pancreas is a soft and a slender organ located in the upper abdomen directly behind the stomach.(Fig. 12.1). It is between 6 and 10 in. long and 2in. wide. The pancreas is divided into three parts: the head, the body, and the tail (Fig. 12.2). It is primarily an exocrine gland secreting a variety of digestive enzymes and pancreatic juice. It is also a very important endocrine gland secreting hormones, insulin, and glucagon, which are essential in regulating blood glucose concentrations. The pancreatic exocrine function is carried out by **acinar cells**. Pancreatic juice, rich in digestive enzymes, is transported into the small intestine by the **pancreatic duct**.

12.2 Cells and secretions of the pancreatic islets

Approximately 98% of the pancreas is made of **acini** (referred to as **pancreatic acini**, Fig. 12.2). They secrete pancreatic juice that is rich in pancreatic enzymes. Pancreatic juice is released into the duodenum of the small intestine to aid in digestion. The remaining 2% of the tissue consists of **pancreatic islets**, also called **islets of Langerhans** (Fig. 12.2).

The pancreatic islets contain four types of cells:

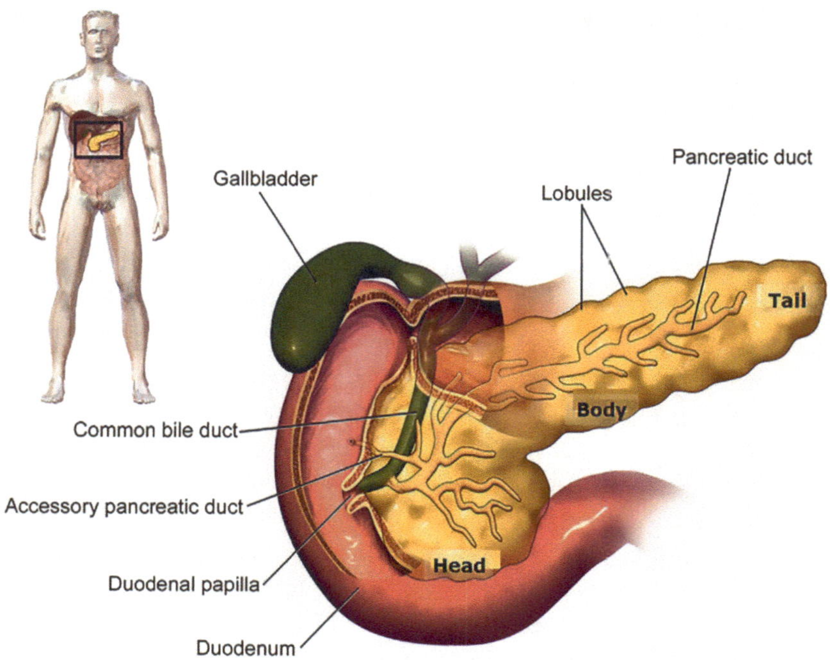

FIG. 12.1 The location of the pancreas in the abdomen [1].

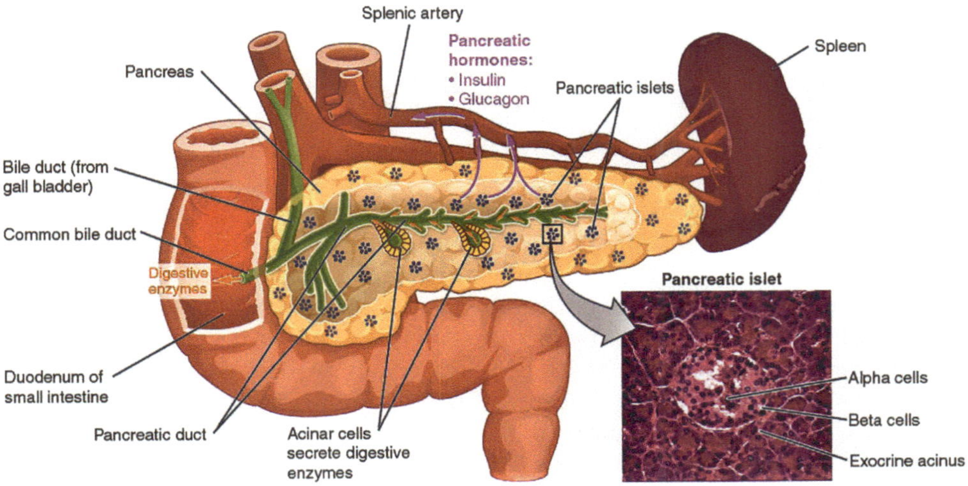

FIG. 12.2 The pancreatic exocrine function involves acinar cells secreting digestive enzymes that are transported into the small intestine by the pancreatic duct [2]. Its endocrine function involves the secretion of insulin (produced by beta cells) and glucagon (produced by alpha cells) within the pancreatic islets. The micrograph reveals pancreatic islets. LM × 760. *Micrograph provided by the Regents of University of Michigan Medical School © 2012.*

The alpha cell accounts for approximately 20% of each islet and produces the hormone **glucagon,** an important hormone regulating blood glucose concentrations. Low blood glucose levels stimulate the release of glucagon, and normal glucose levels in the blood inhibit its secretion.

The **beta cell** makes up approximately 75% of each islet and produces the hormone insulin. Greater than normal blood glucose levels stimulate the release of insulin, whereas the reduction in blood glucose levels decreases its secretion.

The **delta cell** accounts for 4% of the islet population and secretes a peptide hormone called **somatostatin,** which inhibits the secretion of insulin and glucagon.

The **PP cell** makes up about 1% of islet cells and secretes a hormone called the **pancreatic polypeptide**. It is important in the regulation of pancreatic exocrine and endocrine secretions and the storage of glycogen by the liver and gastrointestinal function.

The pancreatic hormones are summarized in Table 12.1.

TABLE 12.1 Hormones of the pancreas.

Associated hormones	Chemical class	Effect
Glucagon (alpha cells)	Protein	Increases blood glucose levels
Insulin (beta cells)	Protein	Reduces blood glucose levels
Somatostatin (delta cells)	Protein	Inhibits insulin and glucagon release
Pancreatic polypeptide (PP cells)	Protein	Regulation of pancreatic exocrine and endocrine secretions and the storage of glycogen by the liver

12.3 Glucose transporters

Insulin serves as a "key" that turns on the glucose transport "ignition" (insulin receptor). When the "key" (insulin) activates the "ignition" (the insulin receptor), the engines (GLUcose Transporters, or GLUTs) move glucose into the cells to keep blood glucose from building up to dangerously high levels. GLUTs transport glucose from an area of higher concentration of glucose to an area of its lower concentration. The characteristics of the five major glucose transporters are summarized in Table 12.2. They are found in all body tissues and differ in their affinity for glucose as glucose levels change. The difference in affinity ensures that tissues critical for life (e.g., nervous tissue) will receive it even at very low blood glucose levels before "less" vital tissue. Although each transporter plays an important role in the uptake of glucose into tissues, GLUT2 and GLUT4 are the most relevant for normal metabolism. GLUT2 is the primary transporter in the liver and pancreatic beta cells and ensures that insulin is secreted only when blood glucose levels are high. It also prevents hepatic uptake of glucose when levels are low such as during fasting. GLUT4 is found in two major insulin-targeted tissues—muscle and adipose tissue. It is not present to a great extent on the cell surface until an insulin signal translocates these transporters to the cell membrane. Thus, GLUT4 functions primarily after a high carbohydrate meal when insulin is secreted, and the need to get glucose from the blood into the skeletal muscle is high.

12.4 Regulation of blood glucose levels by glucagon and insulin

Glucose is an essential source of energy for all body cells. Glucose that is not used immediately by cells as fuel molecules is stored as glycogen in the liver and muscle or converted to triglycerides and stored in the adipose tissue. The alpha and beta cells of the islets of Langerhans sense blood glucose levels and subsequently release glucagon or insulin to maintain normal levels.

TABLE 12.2 The five major glucose transporters and their properties.

Transporter	Major sites of expression (action)	Attraction (affinity) to glucose
GLUT1	Brain vasculature (blood brain barrier), red blood cells, fetal tissues	Even at low glucose levels
GLUT2	Liver, pancreatic beta cells, serosal surfaces of gut, and renal tubular cells of kidney	Only at high glucose levels
GLUT3	Brain neurons, placenta	Even at low glucose levels
GLUT4	(**Insulin dependent**): Adipose tissue and striated muscle	Moderate glucose levels such as after a meal
GLUT5	Jejunum, liver, spermatozoa	Moderate glucose levels

12.4.1 Glucagon

Glucagon is a peptide hormone made by the **alpha cells** of the islets of Langerhans. Receptors on the alpha cell membrane detect any decline in blood glucose levels, such as during prolonged exercise, and release glucagon, which stimulates the liver to convert its stores of glycogen to glucose (glycogenolysis). It increases the uptake of amino acids from the blood to convert them into glucose (gluconeogenesis). Glucagon also stimulates lipolysis, the breakdown of stored triglycerides into free fatty acids (FFA) and glycerol. Some of the free glycerol released into the bloodstream reaches the liver, where it gets converted into glucose (gluconeogenesis). All of these actions increase blood glucose levels. The activity of glucagon is regulated through a negative feedback mechanism; rising blood glucose levels inhibit further glucagon production and secretion and vice versa.

12.4.2 Insulin

Insulin was discovered by a Canadian surgeon, Frederick G. Banting, and his assistant, Charles Best, in 1921. The knowledge of insulin and its function in regulating glucose since its discovery has led to a dramatic turn in the treatment of diabetes, saving millions of lives year after year.

Insulin is a protein hormone. It is made of two polypeptide chains, α and β, joined by two disulfide bonds. The synthesis of insulin begins as **preproinsulin** (Fig. 12.3). Preproinsulin is a long-chain polypeptide produced by mRNA-directed translation in the rough endoplasmic reticulum. Preproinsulin is cleaved immediately after its synthesis to become **proinsulin**. Proinsulin is transported to the Golgi, where it gets packaged into secretory granules. Proteases remove a **connecting peptide** (**C peptide**) to yield insulin within the storage granules. The C peptide is, therefore, secreted along with insulin when the need for insulin arises. Although it has no specific function, the measurement of C peptide is important in monitoring beta cell function in diabetic patients who are receiving exogenous insulin.

The vital function of insulin is to facilitate the uptake of glucose into body cells. Insulin mainly exerts its effects by acting on the liver, adipose tissue, and skeletal muscle. The insulin-independent tissues include the brain, red blood cells, liver, kidneys, and the small intestine because they do not have insulin receptors on their cell membranes and do not require insulin for glucose uptake.

12.5 Mechanism of insulin secretion

The primary function of the beta cells is to release insulin in response to changes in blood glucose concentration and is achieved by GLUT2 and **glucokinase**. GLUT2 activation occurs when blood glucose concentration is about 100 mg/dL. Glucose transport into the beta cell is rapid. The next component of the glucose sensor is glucokinase, the enzyme that initiates glycolysis. Glucokinase activity increases and decreases parallel to changes in blood glucose levels. Similar to glucose, glucokinase is most sensitive to changes in blood glucose concentration above 100 mg/dL.

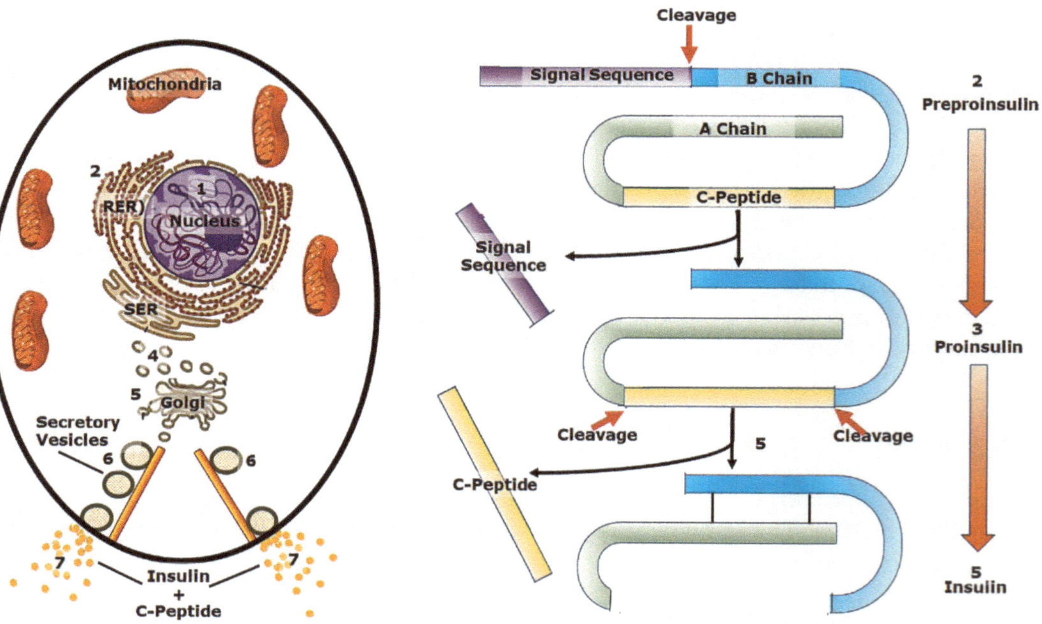

FIG. 12.3 Synthesis of insulin in a beta cell (left panel) and processing of preproinsulin (right panel) [3]. (1) Synthesis of insulin begins in the nucleus. (2) An mRNA is released into the cytoplasm where ribosomes on the rough endoplasmic reticulum complete the process of synthesis of protein which is referred to aspreproinsulin. (3) Further processing of this protein involves splicing of preproinsulin to proinsulin by removing a signal molecule. (4) Transfer vesicles move proinsulin to Golgi Apparatus. (5) C-peptide is cleaved from proinsulin to produce insulin and arepackaged together in the Golgi apparatus as secretory vesicles. (6) The secretory vesicles filled with both insulin and C-peptide are stored near the cell membrane for later use. (7) Insulin is released along with C-peptide from the secretory vesicles as they reach the cell membrane by a process called exocytosis.

The onset of glycolysis within the beta cell involves phosphorylation of glucose and results in an increase in ATP which, in turn, triggers the closing of **ATP-sensitive K^+ channels**. The closing of ATP-sensitive K^+ channels limits the leakage of K^+ from the cell, and, therefore, the membrane potential becomes less negative. The depolarized state of the beta cell stimulates the opening of **voltage-sensitive Ca^{2+} channels** and the subsequent entry of Ca^{2+} into the cell, which, in turn, promotes the **exocytotic** release of insulin from their storage vesicles (Fig. 12.4).

Insulin secretion from cells occurs after glucose ingestion and is reported to occur with precise and biphasic kinetics. The first phase is an **acute phase** (Phase 1) that lasts only a few minutes, followed by a **sustained** second **phase** (Figs. 12.5 and 12.6). Phase 1 release is due to the plasma membrane fusion of a small, readily releasable pool of granules; these granules release insulin in response to an initial rise in glucose from a meal. In this phase, insulin secretion is dependent on ATP generation and a rise in intracellular Ca^{2+}. The phase 1 insulin bolus accounts for $\sim 5\%$ of all insulin released after a meal. The second phase insulin secretion occurs due to a sustained increase in glucose levels (Fig. 12.4). In the second phase, insulin secretion is dependent on a rise in intracellular Ca^{2+}, which helps move the reserve pool

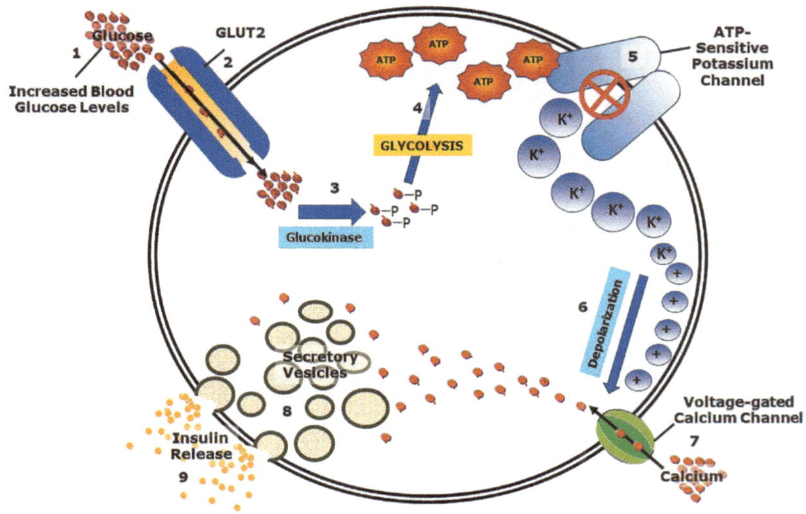

FIG. 12.4 Mechanism of insulin release from beta cell [4]. (1) Rising blood glucose levels following a meal. (2) Glucose is transported into the beta cell by GLUT2. (3) Upon its entry, glucokinase phosphorylates glucose. (4) The phosphorylated glucose will undergo further changes, releasing ATP. (5) The rise in ATP levels within the cytoplasm inactivates (closes) the ATP-sensitive K^+ channel. Positively charged K^+ are now prevented from leaving the beta cell. (6) The rise in positive charges inside the beta cell causes depolarization of the beta cell membrane. (7) Depolarization triggers the opening of voltage-gated Ca^{2+} channels to open, allowing Ca^{2+} to flow inward. (8) Ca^{2+} helps move the secretory vesicles filled with insulin to the beta cell membrane. (9) The secretory vesicles will dock on the beta cell membrane, where insulin is released into capillaries via exocytosis (moves out of the cell). Insulin circulates in the body where it helps move glucose from the blood into the skeletal muscle and liver, which lowers blood glucose levels.

of insulin to the cell membrane for docking and release. Phase 2 accounts for ~95% of all insulin released after a meal. Together, phase 1 and phase 2 ensure that insulin is released quickly and in sufficient amounts to reduce blood glucose levels in a relatively short amount of time (~1–2h) due to glucose uptake into tissues. However, as we will learn later, Phase 1 does not typically occur in Type 2 diabetes, resulting in prolonged elevation of blood glucose levels compared to normal healthy individuals (Fig. 12.5).

The pattern of the glucose-insulin dose-response curve is determined primarily by glucokinase activity, which governs the rate-limiting step for glucose metabolism in the beta cell. Glucose levels below 5mmol/L (90mg/dL) do not affect insulin release; half-maximal stimulation occurs at about 8mmol/L (144mg/dL). Insulinase found in the liver and kidneys breaks down insulin circulating in the plasma leading to insulin degradation, and as a result, insulin has a half-life of only about 6min in the blood. This short duration of action allows rapid changes in the circulating levels of insulin.

Insulin release is stimulated by glucose from the beta cells of the pancreas. It occurs in a biphasic pattern, a fast phase that lasts a few minutes followed by a slower phase 2 (Fig. 12.6). The first phase is caused by the release of a small, readily available pool of secretory vesicles containing insulin, while the second phase is due to secretory vesicles that must move from the inner part of the beta cell to the cell membrane, contributing to the slow increase in insulin

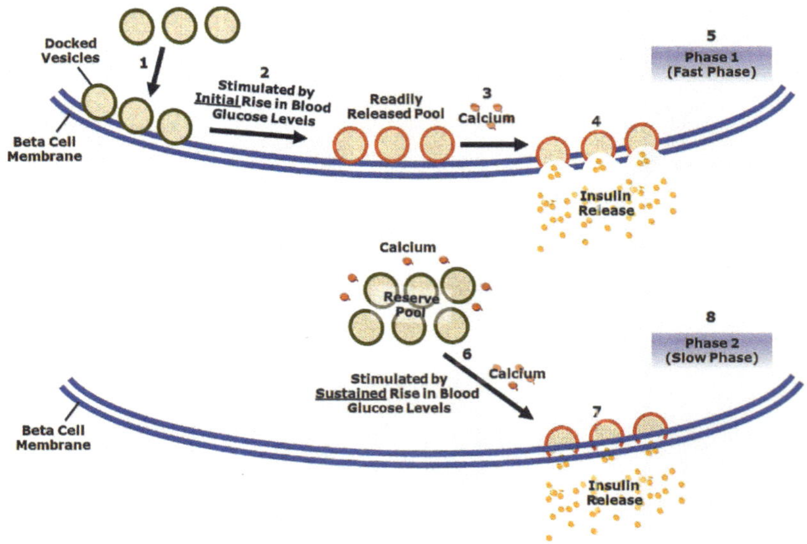

FIG. 12.5 Biphasic release of insulin after a meal [5,6]. (1) In between meals and with the help of Ca^{2+}, a small number of storage vesicles containing insulin dock with the beta cell membrane. (2) After a meal, the blood glucose levels will rise. (3) This causes the docked vesicles to become "activated" with the help of Ca^{2+}. (4) The storage vesicle containing insulin will release insulin into the blood via exocytosis. (5) This phase 1 or fast phase of insulin secretion acts very quickly to release insulin after the initial rise in blood glucose but accounts for only about 5% of the total released. It acts as a bolus of insulin being released into the blood, lasting for only a few minutes after the initial rise in blood glucose. (6) As blood glucose remains high after a meal, the reserve pool located in the center of the beta cell will move to the beta cell membrane with the help of Ca^{2+}. (7) Once docked, insulin is released into the blood via exocytosis. This slow second phase is the major contributor to insulin release after a meal. Maximal quantities of insulin are released after about 60 min. Insulin will continue to be released as long as the blood glucose levels remain above premeal levels.

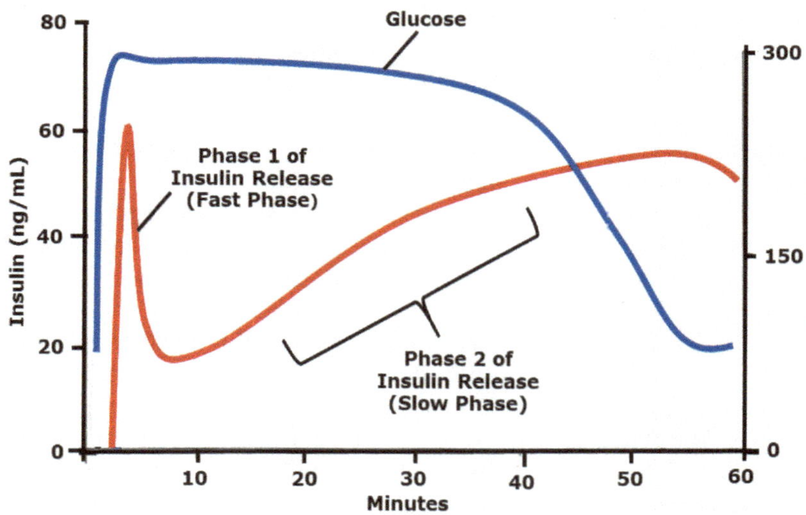

FIG. 12.6 The biphasic glucose-stimulated release of insulin from pancreatic islets [5,6]. Phase 1 is the initial short and rapid release of insulin and phase 2 is the slow and prolonged release of insulin.

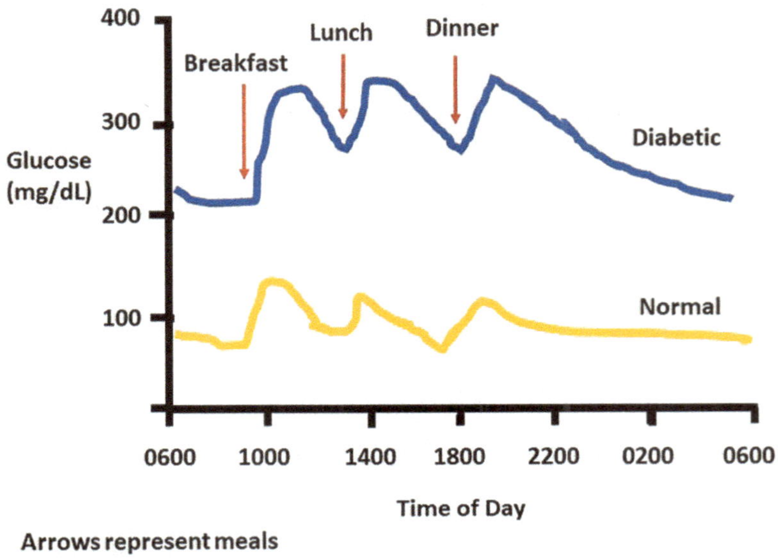

FIG. 12.7 Twenty-four-hour plasma glucose curve in normal and Type 2 diabetic individuals [7]. The fluctuation of blood glucose *(yellow)* and diabetic *(blue)* in humans during a day with three meals. Notice the difference in the glucose curves. After each meal, the normal blood glucose curve will increase quickly, but it returns to premeal levels within 1–2 h. However, in the diabetic patient, the blood glucose level peak is more rounded, and blood glucose concentration does not return to premeal levels until 4–6 h. The difference is that the phase 1 insulin release does not occur in the diabetic, which allows for the blood glucose levels to stay elevated longer than the standard curve. The elevated blood glucose will result in increased complications associated with diabetes.

secretion. The first phase is responsible for priming the system, limiting a rise in blood glucose while also giving phase 2 to start and reach maximum secretion at about 60 min after a meal. As phase 2 continues and insulin concentration increases, blood glucose levels will fall with glucose entering the tissues. With the drop in blood glucose levels, insulin secretion in phase 2 will begin to fall. The fluctuations of blood glucose vary throughout the day in response to meals (Fig. 12.7). The blood glucose levels remain elevated in the diabetic compared to a normal glycemic individual due, in part, to the loss of the phase 1 insulin release in diabetes.

12.6 Normal glucose homeostasis

Under basal conditions, approximately 50% of glucose utilization occurs in the brain and is not insulin dependent. Another 25% of glucose uptake occurs in the liver and gastrointestinal tissues and is also insulin independent. The remaining 25% of glucose metabolism in the postabsorptive state occurs in insulin-dependent tissues, primarily muscle tissue.

The rate of endogenous glucose production precisely matches basal glucose utilization. Approximately 85% of endogenous glucose production is from the liver, and the kidney produces the rest. About half of the basal hepatic glucose production is derived from glycogenolysis and half from gluconeogenesis.

Glucose homeostasis is the body's ability to maintain blood glucose levels within the normal range of 70–110 mg/dL. It is a delicate balance between insulin (lowering blood glucose) and glucagon (raising blood glucose) (Fig. 12.8). Beginning with the rising blood glucose levels after a carbohydrate meal, blood glucose levels rise, resulting in the release of insulin from the beta cells into the blood. The resultant **hyperinsulinemia** and **hyperglycemia** stimulate glucose uptake by splanchnic (liver and gut) and peripheral (primarily muscle) tissues and suppress endogenous glucose production. The cells use glucose for immediate energy needs or it is stored as glycogen for later use. The removal of glucose from the blood into the cells causes the blood glucose levels to decline back to their normal levels. As the blood glucose levels decrease back to normal, the stimulus for insulin release by the pancreas is reduced. If a meal is skipped, the blood glucose levels decline below the normal blood glucose levels stimulating the alpha cells of the pancreas to release glucagon into the blood. Glucagon will act upon the liver to cause glycogen to be broken down and released into the blood as glucose, causing blood glucose values to rise to normal. As the blood glucose levels rise back to normal, the stimulus for glucagon release by the pancreas is reduced.

During **postabsorptive** or fasting periods, blood glucose concentration is tightly maintained between 70 and 110 mg/dL. If blood glucose concentration rises above this range, insulin is released by the beta cells. The arrival of food in the small intestine triggers the release of gastrointestinal hormones such as **glucose-dependent insulinotropic peptide** (previously known as a gastric inhibitory peptide, GIP), which acts as the initial trigger for insulin secretion by the beta cells. Nutrient absorption results in a surge in blood glucose levels which further stimulates insulin secretion. Insulin stimulates the liver to convert excess glucose into glycogen for storage, and it inhibits enzymes involved in glycogenolysis and gluconeogenesis. Insulin also promotes triglyceride and protein synthesis and inhibits the breakdown of proteins and lipids via enzymes involved in lipolysis, glycogenolysis, and gluconeogenesis. As blood glucose levels decrease, further insulin release is inhibited through a negative feedback mechanism. If blood glucose concentration drops below this range, glucagon is released. Glucagon stimulates the liver to release glucose into the blood via glycogenolysis and gluconeogenesis (Fig. 12.8).

12.7 Normal insulin signaling in the skeletal muscle and liver

An insulin receptor is a glycoprotein molecule consisting of two alpha and two beta subunits linked by disulfide bonds (Fig. 12.9). The **alpha subunit** of the insulin receptor is extracellular and contains the insulin-binding domain. On the other hand, the beta subunit has two parts—an extracellular domain and an intracellular domain. The intracellular domain expresses insulin-stimulated kinase activity toward its tyrosine residues.

Upon a rise in blood glucose levels following a meal, insulin is released from the beta cells into the blood. Insulin circulates via the blood throughout the body, with the liver and skeletal muscle being the most significant. In order to initiate signaling, insulin binds to the alpha subunit of the insulin receptor. The binding of insulin to the alpha subunit causes the beta

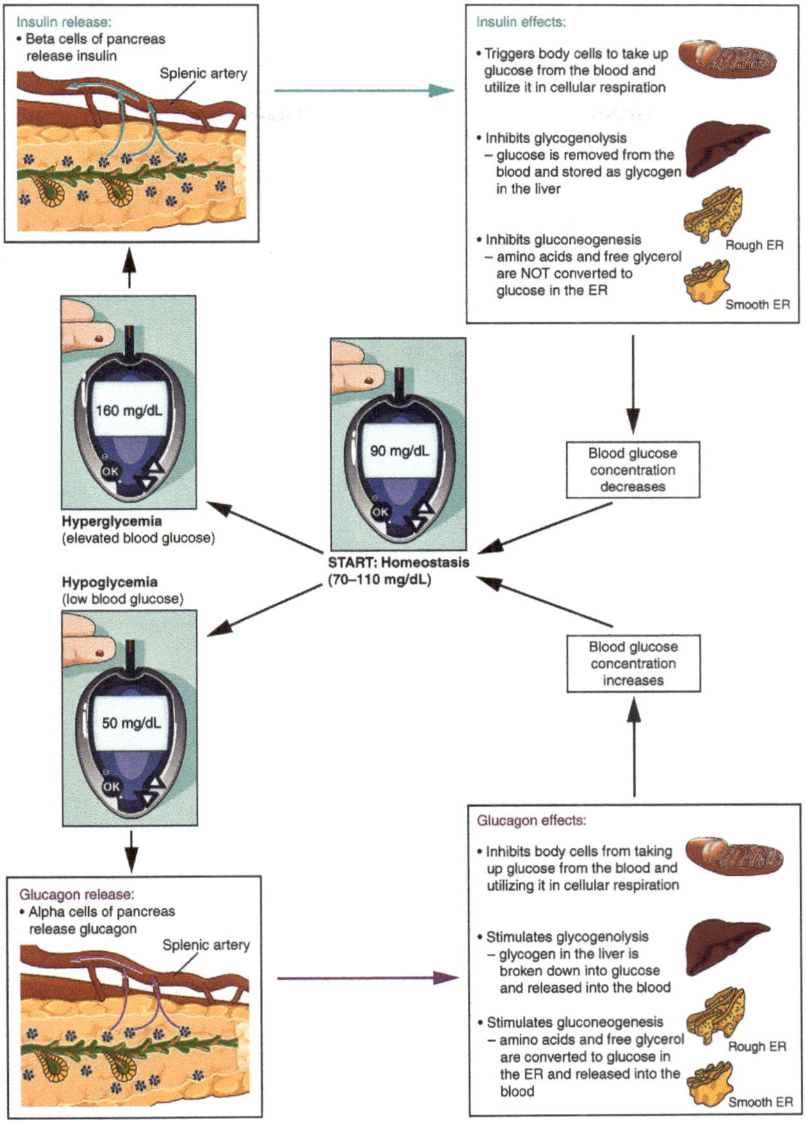

FIG. 12.8 Homeostatic regulation of blood glucose levels [8]. Blood glucose concentration is tightly maintained between 70 and 110 mg/dL. If blood glucose concentration rises above this range such as after a meal, insulin is released, which stimulates body cells to remove glucose from the blood. If blood glucose concentration drops below this range such as skipping a meal, glucagon is released, which stimulates body cells to release glucose into the blood.

subunit to phosphorylate (autophosphorylation—adding a phosphate from ATP to the beta subunit), thus activating the receptor (Fig. 12.10).

Insulin receptor phosphorylation of the beta subunit, with subsequent activation of insulin receptor **tyrosine kinase**, is the first step in the action of insulin on glucose metabolism at its

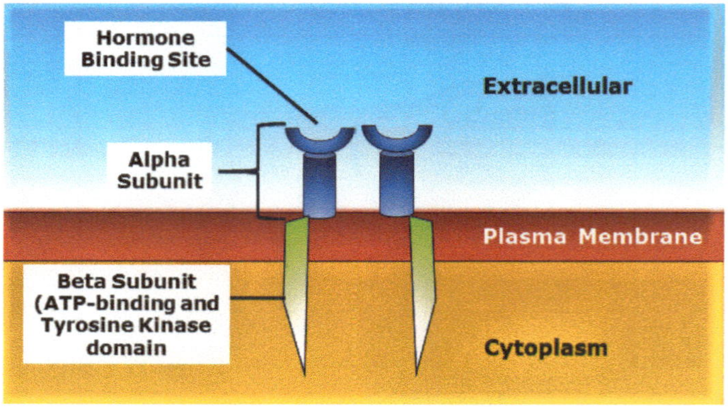

FIG. 12.9 The insulin receptor [9]. The insulin receptor is a transmembrane receptor that is embedded in the plasma membrane. The alpha subunit is extracellular while the beta subunit is extended to the cytoplasm of the cell.

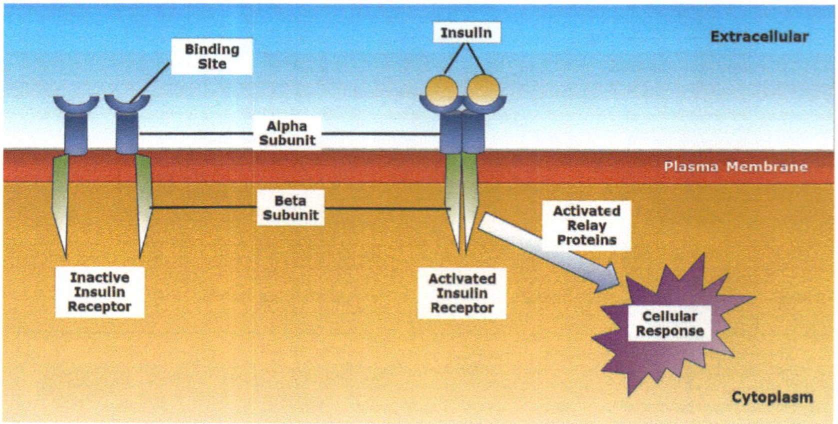

FIG. 12.10 The inactive insulin receptor and the activated insulin receptor after the binding of insulin to the alpha subunit [9].

target cell. Following activation, tyrosine kinase phosphorylates specific intracellular proteins, leading to the movement of intracellular vesicles containing glucose transporters (GLUT4 in skeletal muscle and GLUT 2 in the liver) to the cell membrane (Fig. 12.11). Thus, insulin triggers rapid movement of a pool of glucose transporter vesicles to the cell membrane, where they fuse and expose glucose transporters to the extracellular fluid. The transporters then move glucose into the cell. Such influx of glucose by the action of insulin occurs primarily in the muscle tissue. The glucose, which has entered the cell, is subsequently metabolized by a series of enzymatic steps under insulin regulation. Under anabolic conditions, insulin stimulates **glycogen synthesis**.

The majority (~80%–85%) of glucose taken up by peripheral tissues is by muscle under the influence of insulin, with only a small amount (~4%–5%) being metabolized by adipocytes. Although fat tissue is responsible for only a small amount of total body glucose uptake, it plays a vital role in maintaining total body glucose homeostasis. Insulin is a potent inhibitor of lipolysis, and even small increments in the plasma insulin concentration exert a potent

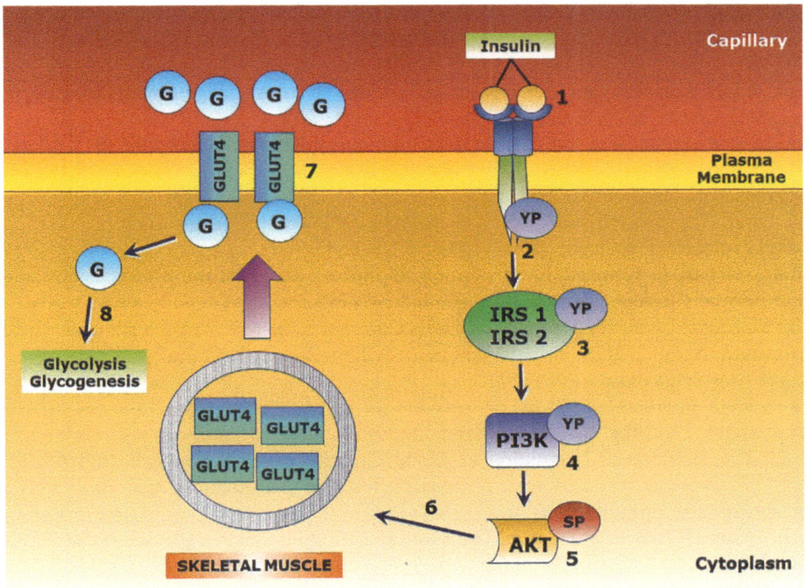

FIG. 12.11 Insulin effects on glucose transport in skeletal muscle (left) and liver (right) [10,11]. (1) Insulin binds to the alpha subunit of the insulin receptor. (2) Insulin binding causes tyrosine (Y) autophosphorylation (P) of the beta subunit of insulin receptor (adenosine triphosphate, ATP, in the cell adds P to the tyrosine amino acid on the beta subunit). (3) This causes insulin receptor substrate 1 (IRS1 and IRS2) to be tyrosine phosphorylated. (4) This leads to Phosphatidylinositol-3′-kinase (PI3K) being tyrosine phosphorylated. (5) This causes AKT (Protein Kinase B) to be serine (S) phosphorylated (phosphate added to serine amino acid). (6) A signal is then sent to the storage vesicles containing GLUT4 of the skeletal muscle to translocate (move) to the cell membrane. (7) Glucose attaches to GLUT4 protein of the skeletal muscle . (8) Glucose travels through the GLUT4 of the skeletal muscle at the cell membrane and into the cell to be either broken down (glycolysis) or stored as glycogen (glycogenesis). The influx of glucose will lower blood glucose levels which will cause GLUT4 and GLUT2 transporters to be taken back into the cell by endocytosis to be used the next time.

antilipolytic effect, leading to a marked reduction in the plasma FFA level. The decline in plasma FFA concentration results in increased glucose uptake by the muscle tissue. Thus, changes in the plasma FFA concentration in response to increased plasma levels of insulin and glucose play an essential role in maintaining normal glucose homeostasis.

In the absence of insulin, GLUT4 and GLUT2 usually are recycled slowly between the cell membrane and the cell interior. Mutagenesis experiments have shown that insulin receptors lacking tyrosine kinase activity are entirely ineffective in mediating insulin stimulation of cellular metabolism. Similarly, mutagenesis of any of the three major phosphorylation sites impairs insulin receptor kinase activity, resulting in a decrease in the acute metabolic and growth-promoting effects of insulin.

References

[1] Blausen 0699 PancreasAnatomy2.png. Wikimedia Commons, the free media repository. Updated October 25, 2020. Accessed March 29, 2021. https://commons.wikimedia.org/w/index.php?title=File:Blausen_0699_PancreasAnatomy2.png&oldid=501280308.

[2] 1820 The Pancreas.jpg. Wikimedia Commons, the free media repository. Updated September 17, 2020. Accessed March 25, 2021. https://commons.wikimedia.org/w/index.php?title=File:1820_The_Pancreas.jpg&oldid=462487953.

[3] Y. Yang, Q.-X. Hua, J. Liu, et al., Solution structure of proinsulin: connecting domain flexibility and prohormone processing, J. Biol. Chem. 285 (11) (2010) 7847–7851.

[4] Glucose Insulin Release Pancreas.svg. Wikimedia Commons, the free media repository. Updated September 26, 2020. Accessed March 30, 2021. https://commons.wikimedia.org/w/index.php?title=File:Glucose_Insulin_Release_Pancreas.svg&oldid=472497886.

[5] A. Caumo, L. Luzi, First-phase insulin secretion: does it exist in real life? Considerations on shape and function, Am. J. Physiol. Endocrinol. Metabol. 287 (3) (2004) E371–E385.

[6] D.C. Robbins, J. Jaspan, B. Vasquez, E. Van Cauter, Biphasic patterns of peripheral insulin and glucose levels after lunch in normal subjects, Diabetes Care 10 (3) (1987) 293–299.

[7] Suckale08 fig 3 Glucose Insulin Day.png. Wikimedia Commons, the free media repository. Updated September 30, 2020. Accessed April 2, 2021. https://commons.wikimedia.org/w/index.php?title=File:Suckale08_fig3_glucose_insulin_day.png&oldid=476789626.

[8] 1822 The Homostatic Regulation of Blood Glucose Levels.jpg. Wikimedia Commons, the free media repository. Updated September 17, 2020. Accessed March 29, 2021. https://commons.wikimedia.org/w/index.php?title=File:1822_The_Homostatic_Regulation_of_Blood_Glucose_Levels.jpg&oldid=462468447.

[9] T.M. Batista, C.T. Cederquist, C.R. Kahn, The insulin receptor goes nuclear, Cell Res. 29 (7) (2019) 509–511, https://doi.org/10.1038/s41422-019-0185-0.

[10] L. Sylow, V.L. Tokarz, E.A. Richter, A. Klip, The many actions of insulin in skeletal muscle, the paramount tissue determining glycemia, Cell Metab. 33 (4) (2021) 758–780, https://doi.org/10.1016/j.cmet.2021.03.020.

[11] P. Iozzo, F. Geisler, V. Oikonen, et al., Insulin stimulates liver glucose uptake in humans: an 18F-FDG PET Study, J. Nucl. Med. 44 (5) (2003) 682–689.

CHAPTER

13

Diabetes mellitus

OUTLINE

OBJECTIVES

- Name the two main types of diabetes mellitus.
- Summarize the pathogenesis of type 1 diabetes mellitus.
- Describe the causes of type 2 diabetes mellitus.
- Link the role of genetics in the development of type 1 and type 2 diabetes mellitus.

Biology of Cardiovascular and Metabolic Diseases
https://doi.org/10.1016/B978-0-12-823421-1.00015-9

- Explain the development of insulin resistance in the adipose tissue, liver, and muscle.
- Describe how obesity is tied to the pathogenesis of type 2 diabetes mellitus.
- Define advanced glycation end products and their effects.
- Summarize the signs and symptoms of diabetes mellitus.
- Define diabetic ketoacidosis and explain its manifestation in a patient.
- List acute signs and symptoms of diabetes mellitus.
- List the major chronic complications of diabetes mellitus.
- Explain how oral glucose tolerance test works.
- Relate the role of exercise in the management of diabetes.

13.1 Introduction

Diabetes mellitus, commonly referred to as diabetes, is a metabolic disorder characterized by high blood glucose levels over a prolonged period. While 70–110 mg/dL **fasting blood glucose** is considered normal, blood glucose levels between 100 and 125 mg/dL is considered **prediabetes**, and 126 mg/dL or higher is defined as diabetes. There are two major forms of diabetes mellitus, **type 1** and **type 2**.

Diabetes is often associated with abnormal production or secretion of insulin and the lack of responsiveness to insulin (insulin resistance) at the target cell. The latest numbers estimated as per Diabetes Statistics Report for 2020 is 34.2 million (10.5%) in the United States. It is estimated that approximately 7.3 million people have the condition but have not been diagnosed [1]. Moreover, approximately 79 million people in the United States are estimated to have prediabetes, a condition in which blood glucose levels are high but not high enough to be classified as having diabetes.

13.2 Pathogenesis of type 1 diabetes mellitus

Type 1 diabetes mellitus, also referred to as **juvenile** or **insulin-dependent diabetes**, is caused by decreased insulin production. Some of the causes (Fig. 13.1) for type 1 diabetes may include:

(1) an **autoimmune** disease destroying beta cells of the pancreatic islets leading to insulin deficiency. Specific genes that encode class II major histocompatibility (MHC) complex molecules are recognized to increase susceptibility for this condition. It may also be triggered by viral infection where similarities of viral proteins to pancreatic proteins may confuse the immune cells to destroy beta cells. A virus is processed by the antigen-presenting cell, which activates the helper T cell. Helper T cell, in turn, activates cytotoxic T cells through the release of specific cytokines. The activated cytotoxic T cells destroy beta cells by releasing harmful chemical agents (Fig. 13.2). Additionally, antibodies may

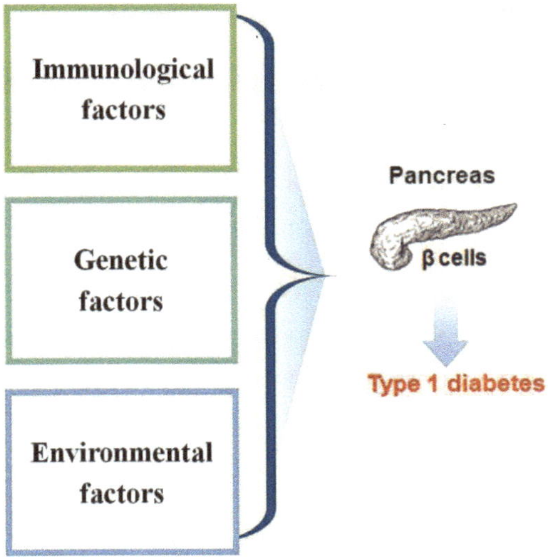

FIG. 13.1 Pathogenesis of type 1 diabetes mellitus.

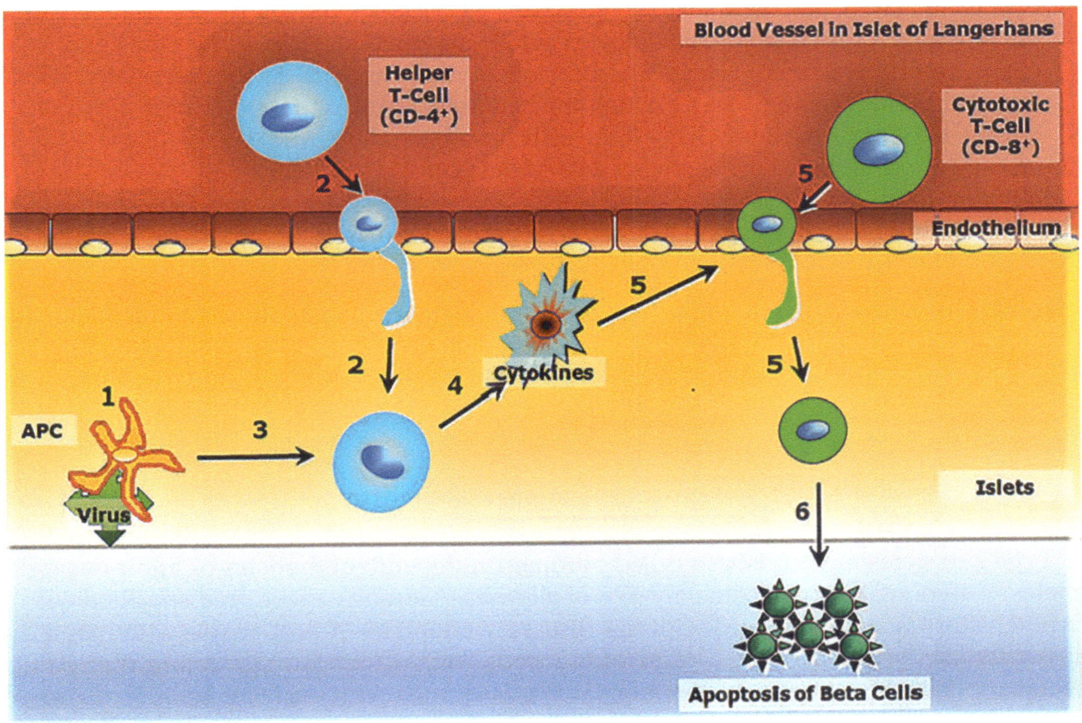

FIG. 13.2 Autoimmune attack of the beta cells in type 1 diabetes [2]. (1) A virus within the Islets of Langerhans (Islets) is processed by the antigen-presenting cells (APCs). (2) Helper T cells (CD4) interact with the APCs. (3) The APC activates helper T cell ($CD4^+$). (4) $CD4^+$ release cytokines (e.g., TNF alpha and IFN gamma). (5) These cytokines activate cytotoxic T cells (CD8). (6) The release of harmful chemicals by the $CD8^+$ cells causes apoptosis (destruction and death) of the beta cells.

TABLE 13.1 Comparison of type 1 and type 2 diabetes mellitus.

	Type 1	Type 2
Body weight	Not necessarily obese	Obese (>80%)
		Obese (>80%)
Racial differences	Higher in person of Northern European descent	Higher in person of African descent and in native Americans
Genetic component	Yes	Yes
Treatment with insulin	Always necessary	Often not necessary
Insulin secretion	Severe deficiency	Variable- hyperinsulinemia sometimes present
Insulin resistance	Rare	Usual
Circulating autoantibodies	Yes	No
Ketosis	Common	Rare

mistakenly target beta cells, insulin, glutamic acid decarboxylase or islet cell antigens and destroy which ultimately leads to lower insulin production. *Helicobacter pylori* and exposure to cow's milk proteins are some other environmental factors associated with type 1 diabetes (Table 13.1). Although the beta cells are affected, symptoms may not appear until approximately 70% of the cells have been destroyed;

(2) chronic pancreatitis where there is a loss of beta cells;

(3) a nonimmune condition where there is excess production of glucagon. Glucagon is typically released when blood glucose levels are below normal, but abnormally high glucagon concentrations could lead to hyperglycemia.

Type 1 diabetes occurs predominantly in younger age group including children. Synthetic insulin must be administered by injection or infusion. Type 1 diabetes accounts for less than 5% of all diabetes cases.

13.3 Type 2 diabetes mellitus

Type 2 diabetes melitus is also called **noninsulin-dependent diabetes** or **adult-onset diabetes**. It accounts for approximately 95% of all cases. A strong genetic predisposition with a family history is also associated with this disease. About 80%–90% of people with type 2 diabetes are overweight or obese. The increase in body weight, high fat, high-calorie diets, and a sedentary lifestyle are the most common factors associated with the increased prevalence of type 2 diabetes. Although older adults are most likely to develop type 2 diabetes, the age at onset has been falling in recent years. Type 2 diabetes is now common among teenagers and young adults.

13.3.1 Genetic factors

Genetic factors play an essential role in the development of diabetes (Table 13.1). Genetic studies have identified over 40 genetic variants that increase the risk for type 2 diabetes, but these variants account for only about 10% of the heritability of the disorder. Individually, persons with these variants have an increased risk of diabetes by 10%–15% compared with persons without the variants. Since there are a significant number of genes that appear to be associated with type 2 diabetes, it is difficult to determine the risk precisely or to develop selective preventive or therapeutic strategies based on the genetic profile.

13.4 Pathogenesis of type 2 diabetes mellitus

In a normal situation, following a meal (absorptive state), insulin is secreted into the hepatic portal vein and is brought to the liver, where it suppresses hepatic glucose output and stimulates the uptake of glucose by body cells. If the liver fails to receive a signal to turn off hepatic glucose output, either due to insulin resistance or decreased insulin production, it produces glucose. Thus there will be two sources of glucose in the body, one from the liver and the other from the digestion of food in the gastrointestinal tract. Thus marked hyperglycemia ensues.

Postbinding defects in insulin action are mainly responsible for insulin resistance. These include diminished insulin receptor tyrosine kinase activity, insulin signal transduction abnormalities, decreased glucose transport, reduced glucose phosphorylation, and impaired **glycogen synthase** activity. Enhanced lipid oxidation appears to be involved in the cause of such defects. Impaired glycogen synthesis represents a major pathway responsible for insulin resistance in type 2 diabetes, and a defect in the glycogen synthetic pathway represents the earliest detectable abnormality in type 2 diabetes.

Insulin resistance is commonly seen in the pathogenesis of type 2 diabetes. In response to insulin resistance, the pancreas increases its insulin output, but the beta cells become exhausted and begin to die over time. Pathogenesis of type 2 diabetes is complicated since patients often exhibit a combination of varying degrees of insulin resistance and relative insulin deficiency. Furthermore, each clinical feature can arise through genetic or environmental factors, making it difficult to determine the exact cause in an individual patient. Hyperglycemia itself can impair beta cell function and aggravate insulin resistance, leading to a vicious cycle of hyperglycemia, causing a deteriorating metabolic state. Hyperinsulinemia occurs in response to insulin resistance and may contribute to the pathogenesis of diabetes. Thus, in type 2 diabetes with mild to moderate fasting hyperglycemia (140–200 mg/dL), basal hepatic glucose production (HGP) is increased and is the primary abnormality responsible for the elevated fasting plasma glucose.

In the postabsorptive (fasting) state, the fasting plasma insulin concentration is two- to fourfold greater in type 2 diabetic patients than in nondiabetics. Since hyperinsulinemia is a potent inhibitor of HGP, hepatic resistance to the action of insulin must be present in the postabsorptive state to explain the excessive output of glucose by the liver. The mechanisms responsible for the increase in hepatic glucose output not only include decreased sensitivity to insulin but also hyperglucagonemia (increased production of glucagon); increased

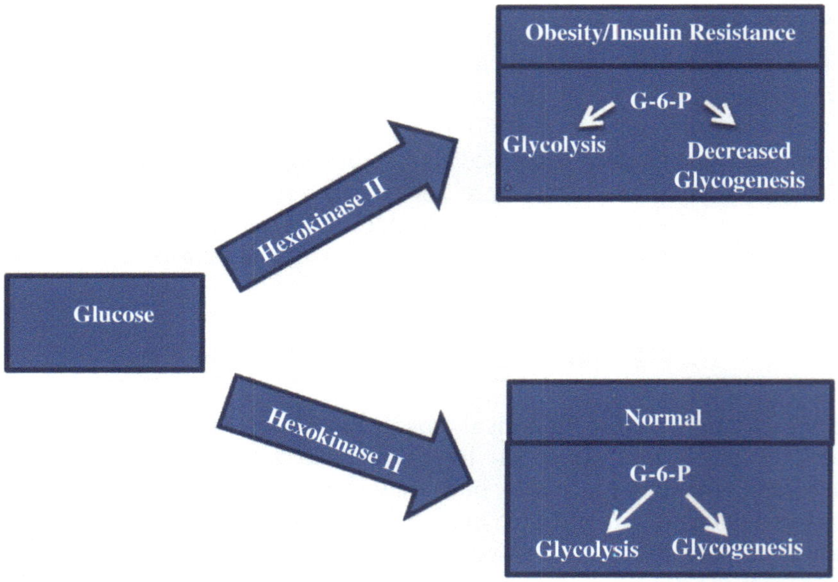

FIG. 13.3 Glycogen synthesis in normal and obese/insulin-resistant individuals. The conversion of glucose to glucose-6-phosphate by hexokinase II is regulated by insulin.

circulating levels of gluconeogenic precursors such as lactate, alanine, glycerol; increased FFA oxidation; and enhanced sensitivity to glucagon.

Glucose phosphorylation is the conversion of glucose to glucose-6-phosphate (G-6-P). Isoenzymes of hexokinase (HKI-HKIV) catalyze the first step of glucose metabolism. Insulin regulates the action of hexokinase II (HKII) (Fig. 13.3). HKII activity increased by 50%–200% in response to hyperinsulinemia. However, in type 2 diabetics, there is reduced expression of HKII, which, in turn, limits glycogenesis.

Once glucose is phosphorylated to G-6-P, it can either be converted to glycogen or enter the glycolytic pathway. Reduced insulin-stimulated glycogen synthesis is a characteristic finding in all insulin-resistant states, including obesity, diabetes, and the combination of obesity and diabetes. The inability of insulin to promote glycogen synthesis is a characteristic and early defect in the development of insulin resistance in both obesity and type 2 diabetes. Additionally, glycogen synthase is an essential insulin-regulated enzyme that controls the rate of muscle glycogen synthesis. Under fasting conditions, total glycogen synthase activity in type 2 diabetics is reduced, and the ability of insulin to activate glycogen synthase is severely impaired.

13.4.1 Dyslipidemia and insulin resistance

Dyslipidemia in patients with insulin resistance is characterized by an elevated level of total triglyceride, a reduced level of HDL cholesterol, and the presence of heterogeneity within LDL particles, with an increased proportion of LDL found as small dense, cholesterol-poor particles (Fig. 13.4). Low-density lipoprotein (LDL) consists of multiple distinct subclasses differing in size and lipid content. The small dense LDL is more atherogenic than the large dense LDL because small dense LDL can penetrate the endothelial layer of the

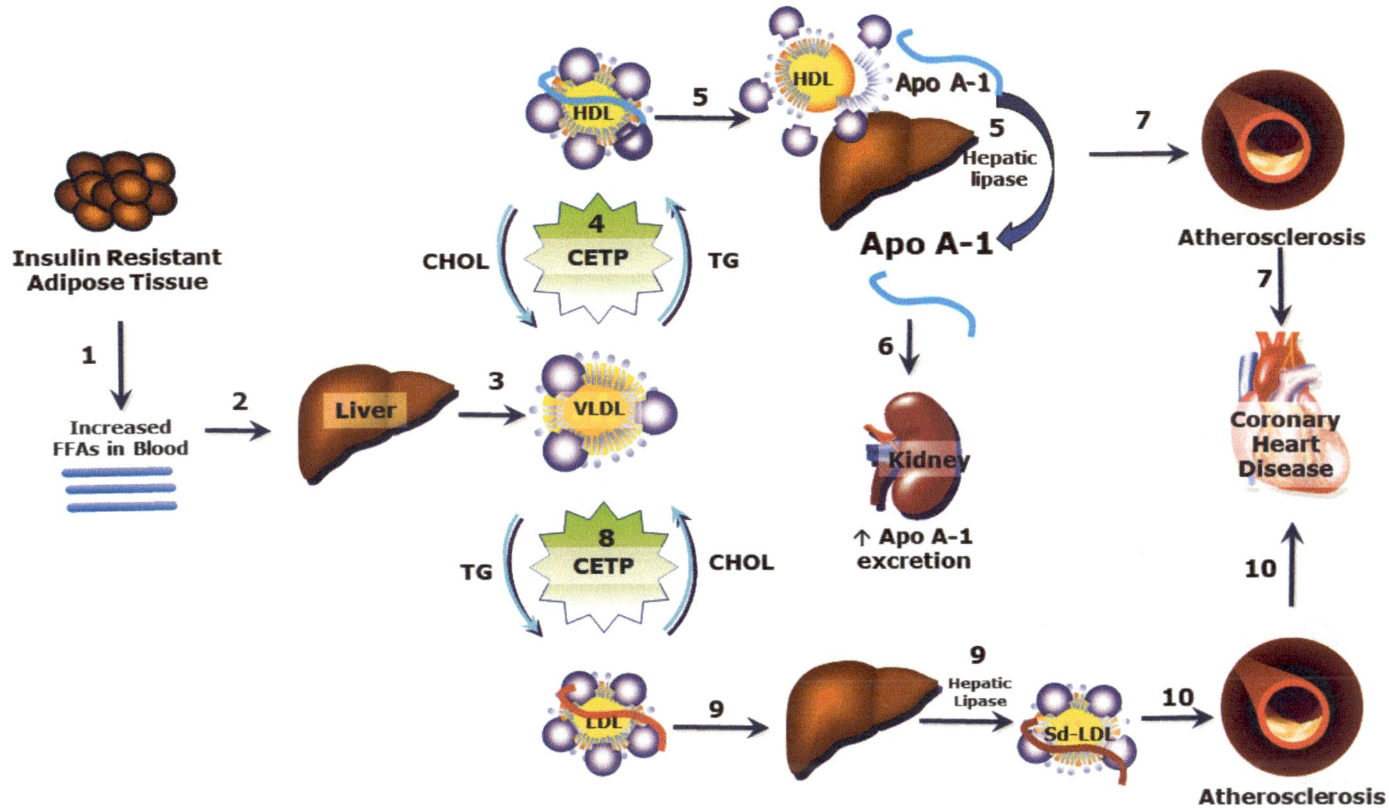

FIG. 13.4 Insulin resistance and dyslipidemia (abnormally elevated cholesterol or fats (lipids) in the blood) [3]. (1) Increased lipolysis. (2) Elevated free fatty acid (FFA) levels in the plasma leading to increased fatty acid uptake by the liver. (3) Synthesis and release of VLDL into blood. (4) Cholesteryl ester transfer protein (CETP) allows exchange of VLDL triglycerides for one of the cholesterol ester molecules (CHOL) from the HDL. (5) Triglyceride-rich and cholesterol-depleted HDL travels to the liver and undergoes hydrolysis (breakdown) of its triglyceride by hepatic lipase which results in the dissociation of the structurally essential protein apo A-1 from the HDL particle. (6) The free apo A-1 travels to the kidney and is cleared. (7) Decreased number of HDLs in the blood leads to atherosclerosis and coronary heart disease. (8) Transfer of triglyceride to LDL in exchange for CHOL. (9) LDL undergoes hydrolysis by hepatic lipase resulting in small, dense, cholesterol-depleted LDL particle (Sd-LDL). (10) Increased levels of Sd-LDL could lead to chronic conditions such as atherosclerosis and CAD.

arterial wall more easily and are more susceptible to oxidation to become modified LDL in coronary endothelium. Also, the apo B molecule in small, dense LDL undergoes a conformational change that leads to a decreased affinity for the LDL receptor, allowing LDL particles to remain in circulation longer and be more liable to oxidative modification and uptake into the vessel.

Interestingly, a diet that is high in carbohydrates will tend to promote small dense LDL formation. This occurs due to increased fatty acid and apolipoprotein B production in the liver, which leads to increased VLDL production (Fig. 13.4). Carbohydrates prolong the time that triglyceride-rich VLDL particles persist in the blood. Greater VLDL availability allows triglyceride-rich VLDL to interact with other particles, namely HDL and LDL, enriching them in triglycerides (via the action of cholesteryl-ester transfer protein, CETP). VLDL is converted to LDL. Triglyceride-rich LDL is "remodeled" by enzymes like hepatic lipase, which create small LDL.

13.4.2 Role of adipocytes in the pathogenesis of type 2 diabetes mellitus

Obese individuals and in the majority (>80%) of type 2 diabetic subjects, there is an expanded fat cell mass. Adipocytes are resistant to the antilipolytic effects of insulin. Most obese and diabetic individuals are characterized by visceral adiposity and visceral fat cells have a high lipolytic rate. Not surprisingly, both type 2 diabetes and obesity are characterized by an elevation in the plasma free fatty acid (FFA) concentration. Elevated plasma FFA levels as well as increased triglyceride/acetyl CoA content in the muscle, liver, and beta cells lead to the development of muscle/hepatic insulin resistance and impaired insulin secretion.

The majority (>80%) of type 2 diabetics in the United States are overweight. Both lean and especially obese type 2 diabetics are characterized by elevated FFA concentration in the plasma. FFAs are stored as triglycerides in adipocytes and serve as an essential energy source during conditions of fasting. In healthy individuals, insulin has two key roles: increased uptake and storage of FFA in the form of triglyceride (lipogenesis) and the prevention of triglyceride breakdown (lipolysis) into FFAs by inhibiting the enzyme hormone-sensitive lipase (Fig. 13.6). As a result, insulin prevents FFA and triglycerides from accumulating in tissues outside the adipocyte (ectopic fat). Ectopic fat can lead to diabetic complications.

Chronically elevated plasma FFA concentrations are linkedto insulin resistance in the muscle and liver and impair insulin secretion. Thus, high plasma FFA levels can cause three major pathogenic disturbances responsible for impaired glucose homeostasis in type 2 diabetics, the muscle, liver, and beta cell, to be joined by the fourth, the adipocyte. In addition to FFA that circulates in plasma in increased amounts, type 2 diabetics and obese non-diabetics have increased stores of triglycerides in the muscle and liver tissues. Triglycerides in the liver and muscle are in a state of constant turnover, and the metabolites, including inflammatory cytokines of intracellular FFAs and acetyl CoAs, have been shown to impair insulin action in both the liver and muscle tissues. This sequence of events has been referred to as **lipotoxicity**. There is evidence to implicate lipotoxicity as an essential cause of beta cell dysfunction (Fig. 13.5).

In summary, healthy individuals with insulin-sensitive adipocytes use FFAs for triglyceride synthesis, preventing FFA accumulation in the blood. In type 2 diabetics, the ability of

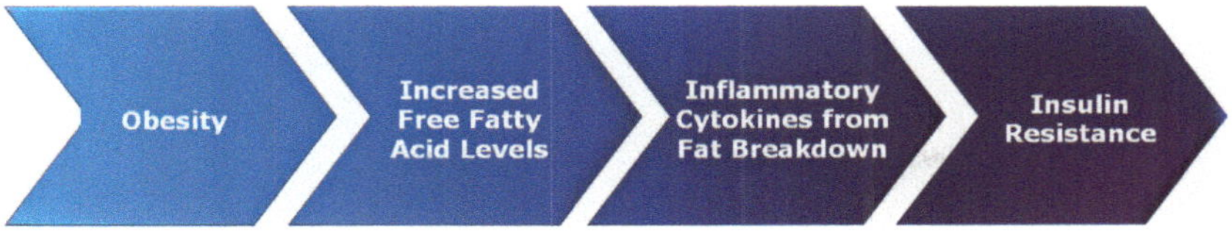

FIG. 13.5 Development of insulin resistance is depicted in a step-by-step manner.

insulin to inhibit lipolysis and increase storage is markedly reduced. Obesity is associated with an upregulation of inflammatory pathways within the adipose tissue which is linked to decreases triglyceride storage (lipogenesis) and increases lipolysis, releasing FFA into blood (Fig. 13.6).

13.4.3 Free fatty acids and hepatic glucose metabolism

The liver plays a pivotal role in the regulation of glucose metabolism. Following carbohydrate ingestion, the liver suppresses its basal rate of glucose production and takes up approximately one-third of the glucose in the ingested meal. Collectively, suppression of HGP and amplification of hepatic glucose uptake accounts for the maintenance of about half of the rise in plasma glucose concentration following the ingestion of a carbohydrate meal.

Several factors regulate HGP—insulin inhibits HGP, whereas glucagon and FFA stimulate the same (Fig. 13.7). It has been documented that a significant portion of the suppressive effect of insulin on HGP is mediated via inhibition of lipolysis and a reduction in circulating plasma FFA concentrations. FFAs enhance the activity of glucose-6-phosphatase, the enzyme that ultimately controls the release of glucose by the liver. In normal subjects, an increase in plasma FFA levels stimulates gluconeogenesis.

In healthy individuals, HGP is promptly suppressed during the absorptive phase in response to insulin within 20 min. However, after glucose ingestion, hepatic glucose uptake increases. On the other hand, in the type 2 DM patients, there is diminished hepatic glucose uptake suggesting impairment in glucose tolerance in type 2 diabetes. On the other hand, there is a net release of glucose from the liver (negative balance) in both healthy and type 2 diabetes subjects reflecting glucose production by the liver in the postabsorptive state. HGP continues to occur during fasting hyperinsulinemia in people with type 2 diabetes suggesting insulin resistance.

Insulin resistance involving both muscle and liver is characteristic of glucose intolerance in type 2 diabetes. The liver represents an important site of insulin resistance, as is reflected by the overproduction of glucose despite fasting hyperinsulinemia and hyperglycemia. This accelerated rate of hepatic glucose output is the primary determinant of the elevated fasting plasma glucose concentration in type 2 diabetics. After glucose ingestion (i.e., in the insulin-stimulated state), both decreased muscle glucose uptake and impaired suppression of HGP contribute to insulin resistance. Following ingestion, the defects in insulin-mediated glucose uptake by muscle and the suppression of HGP by insulin contribute to the

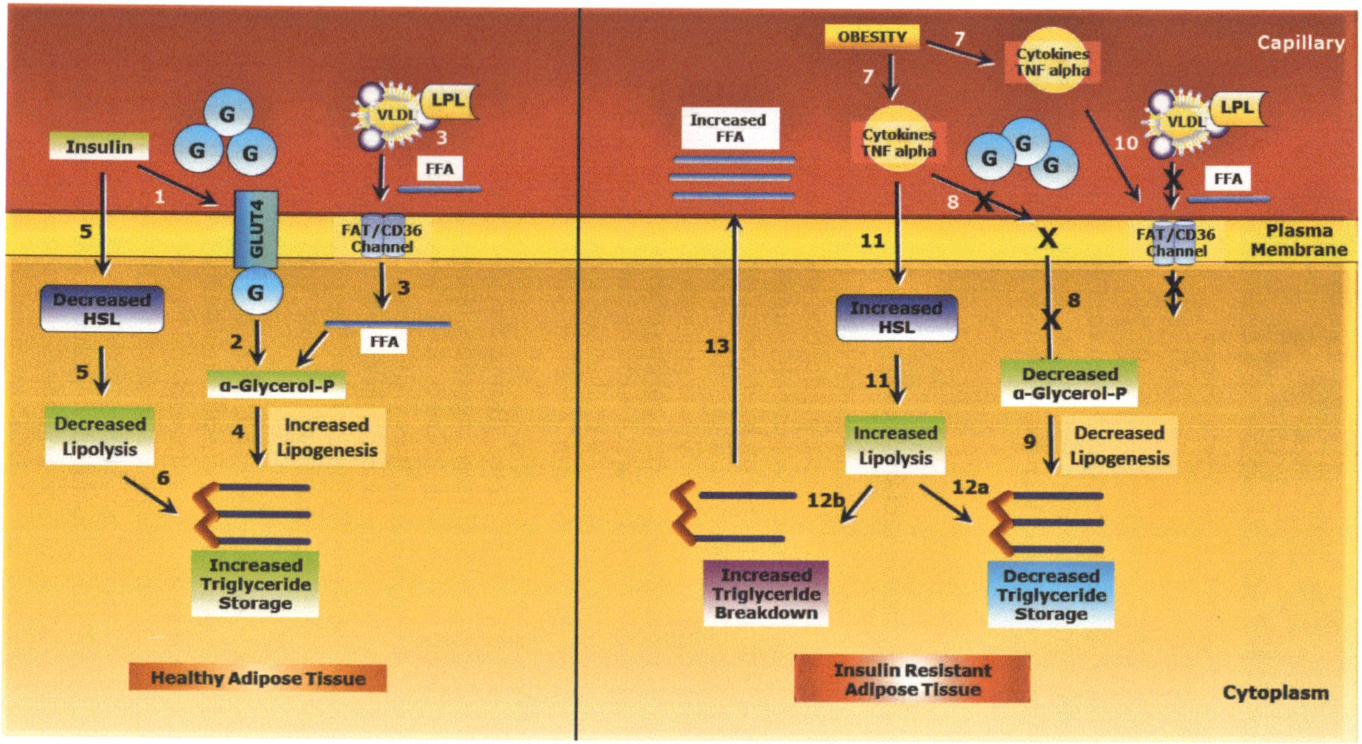

FIG. 13.6 Healthy (left) and insulin-resistant (right) adipose tissue metabolism [4]. (1) In healthy adipose tissue (left), insulin promotes glucose entry into the adipocyte through GLUT4. (2) Glucose is converted to alpha-glycerol phosphate (α-Glycerol-P), the primary source of the glycerol backbone of triglycerides. (3) Insulin promotes the uptake of FFA into the adipocyte through the fatty acid transport channel (FAT/CD36). (4) FFA combines with alpha-glycerol phosphate to form triglycerides (increased lipogenesis). (5) Insulin signaling decreases triglyceride lipolysis because of the decreased activity of hormone-sensitive lipase (HSL). (6) Decreased lipolysis will further promote triglyceride synthesis and storage. (7) Inflammation in overweight and obese individuals leads to the buildup of cytokines such as TNF-alpha. (8) TNF-alpha blocks insulin signaling and prevent glucose from entering the adipocytes. (9) Decreased glucose entry into the adipocytes decreases the synthesis of α-Glycerol-P (used to form glycerol backbones of triglyceride) and ↓ lipogenesis (triglyceride synthesis). (10) Cytokines block FFAs from entering the adipocyte through the FAT/CD36 causing increased FFAs in the blood. (11) TNF-alpha increases the activity of HSL and therefore greater lipolysis. (12a) Decreased triglyceride storage and (12b) increased lipolysis. (13) Higher than normal triglyceride breakdown increases FFAs secreted into the blood, further worsening inflammation and adipose tissue insulin resistance.

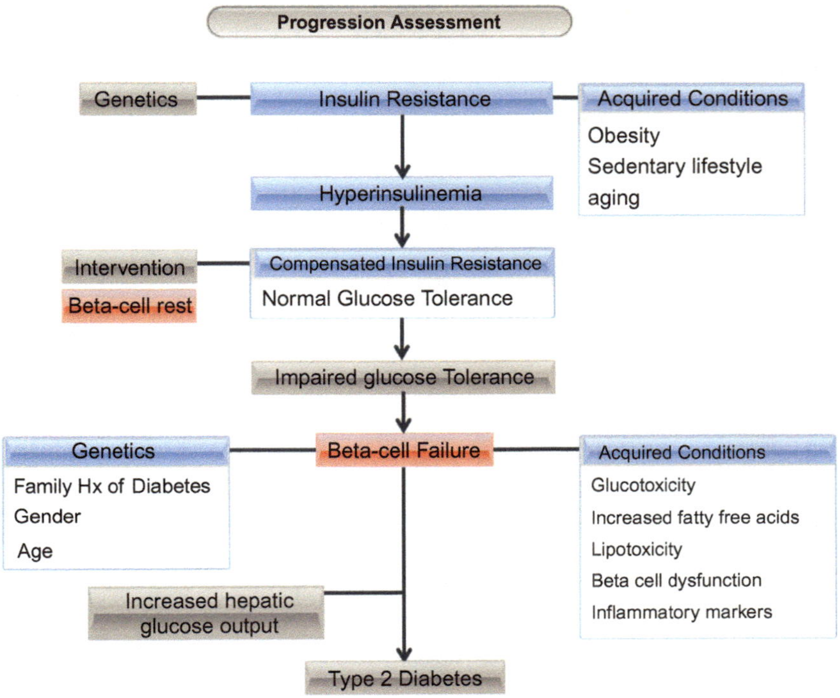

FIG. 13.7 Progression to type 2 diabetes mellitus.

disturbance in whole-body glucose homeostasis in type 2 diabetes. People with type 2 diabetes manifest abnormalities in tissue (muscle, fat, and liver) sensitivity to insulin and insulin secretion. Insulin resistance, therefore, is an early and characteristic feature of the natural history of type 2 diabetes in high-risk populations.

In type 2 diabetics, there is a strong correlation between elevated fasting plasma FFA concentration and lipid oxidation rate and fasting plasma glucose concentration and basal rate of HGP. The elevated plasma FFA levels

(1) increase FFA uptake by hepatocytes leading to accelerated lipid oxidation and accumulation of acetyl CoA. The increased concentration of acetyl CoA, in turn, stimulates pyruvate carboxylase, the rate-limiting enzyme in gluconeogenesis, as well as glucose-6-phosphatase, the rate-controlling enzyme for glucose release from the hepatocyte;
(2) increase the rate of FFA oxidation releasing ATP and fewer nucleotides (NADH) to drive gluconeogenesis;
(3) induce hepatic insulin resistance by inhibiting the insulin signal transduction system.

In type 2 diabetics, these deleterious effects of elevated plasma FFA concentrations correlate with increased plasma glucagon levels, increased hepatic sensitivity to glucagon, and increased hepatic uptake of circulating gluconeogenic precursors (Fig. 13.8).

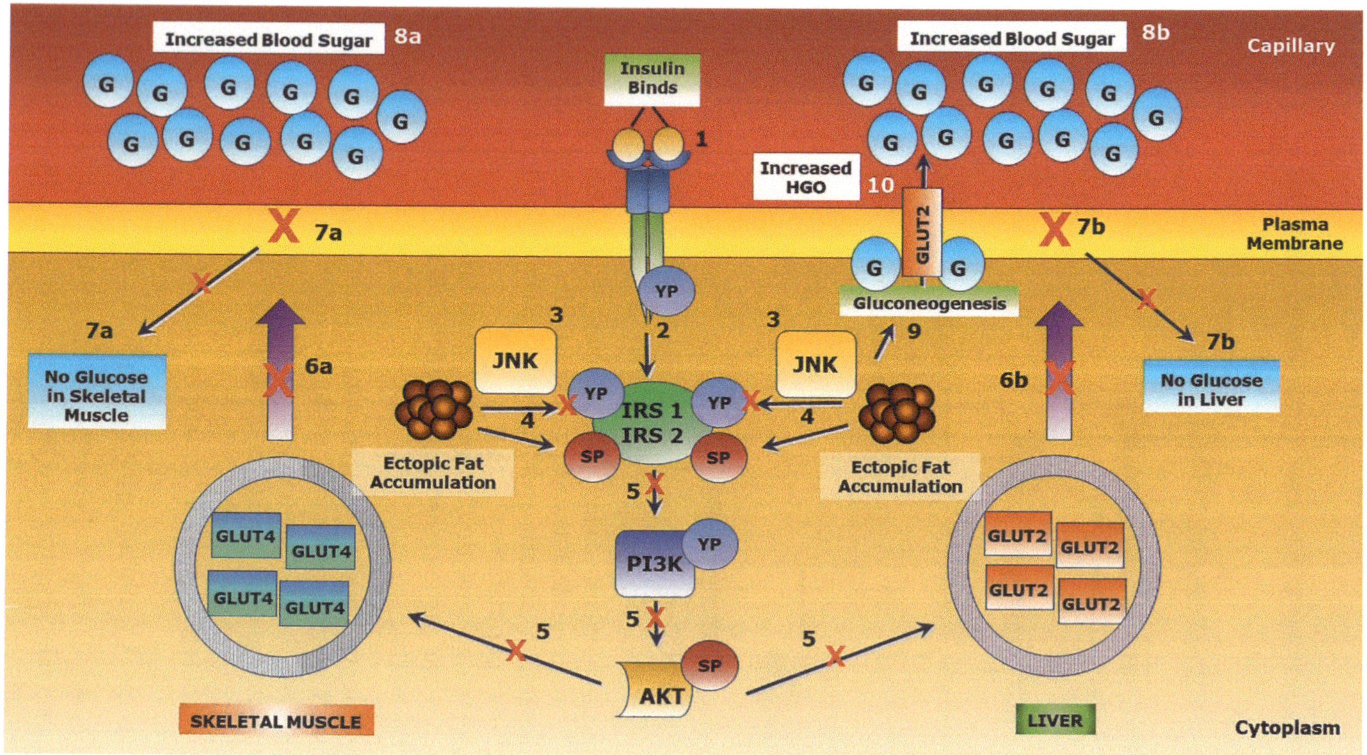

FIG. 13.8 Insulin resistance in skeletal muscle (left) and liver (right) [5,6]. (1) Insulin binds to its receptor. (2) Tyrosine (Y) autophosphorylation (P). (3) Ectopic fat accumulation in skeletal muscle and liver leading to the activation of inflammatory protein c-Jun N-terminal kinase (JNK). (4) JNK causes insulin receptor substrate 1 (IRS1/2) to be serine phosphorylated (SP) instead of tyrosine-phosphorylated. (5) Decreased insulin signaling cascade. (6) Decreased GLUT4 (skeletal muscle) and GLUT2 (liver) protein translocation to the plasma membrane of skeletal muscle (a) and liver (b). (7) Decreased glucose uptake by the skeletal muscle (a) and liver (b). (8) Hyperglycemia. (9) Increased gluconeogenesis. 10) Glucose moves out of the liver via the GLUT2 receptor [increased hepatic glucose output (HGO)].

13.4.4 High levels of free fatty acids and muscle glucose metabolism (Randle cycle)

Elevated plasma FFA concentrations can induce insulin resistance in muscle via mechanisms involving alterations in a variety of intracellular lipid signaling molecules, which, in turn, exert their inhibitory effects on the insulin signal transduction system, glucose transport, glucose phosphorylation, glycogen synthase, pyruvate dehydrogenase, Krebs cycle involved in glucose metabolism. Excessive FFA oxidation leads to increased intracellular accumulation of acetyl CoA. Additionally, it increases the NADH/NAD ratio, causing a slowing of the Krebs cycle and accumulation of citrate, a potent inhibitor of phosphofructokinase. Increased levels of citrate lead to the buildup of G-6-P, which in turn inhibits HKII. The block in glucose phosphorylation causes the accumulation of intracellular free glucose, which restrains glucose transport into the cell (insulin resistance). Decreased glucose uptake will result in less glycogen synthesis.

Another mechanism involved in decreased glycogen synthesis is via a direct inhibitory effect of acetyl CoAs on glycogen synthase. Such accelerated plasma FFA oxidation resulting in muscle glucose transport, glucose oxidation, and glycogen synthesis is called the **Randle cycle**.

13.4.5 Insulin resistance in the skeletal muscle

Under normal conditions, approximately 80% of total body glucose uptake occurs in the skeletal muscle. In response to an increase in plasma insulin concentration, muscle glucose uptake increases linearly. In type 2 diabetics, the onset of insulin action is delayed for approximately 40 min, and the ability of insulin to stimulate glucose uptake is markedly blunted (Fig. 13.8). The elevated FFA in the circulation can lead to an increase in fat accumulation within the skeletal muscle. If enough FFA has accumulated in the skeletal muscle, it can impair normal insulin signaling pathway, leading to increased blood glucose levels. Skeletal muscle insulin resistance is a hallmark of type 2 diabetes (Fig. 13.9).

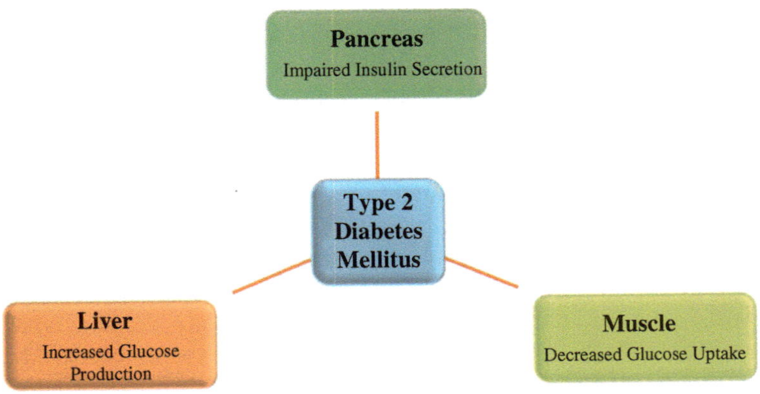

FIG. 13.9 The role of the pancreas, liver, and muscle tissue in the onset of type 2 diabetes.

In the lean state, adipocytes efficiently store fatty acids as triglycerides (TG), causing the FFA levels within muscle cells to remain low and allowing their rapid oxidation in mitochondria during periods of caloric need. Insulin-stimulated glucose uptake under these conditions is expected. In hyperphagia (overeating)-induced obesity, as caloric intake increases, adipocytes experience increased triglyceride deposition [7]. In the early stages of high caloric intake, adipocytes continue to actively store additional triglycerides and maintain near normal lipolytic rates during fasting. Due to adipocytes' ability to store higher amounts of triglyceride, the skeletal muscle can maintain rapid oxidation of triglyceride. Fatty acid levels can rise slightly, but skeletal muscle maintains high insulin sensitivity in this early stage of high caloric intake [8]. On further overloading with triglyceride, hypertrophy of adipocytes occurs. Enlarged adipocytes have been found to secrete large amounts of monocyte chemoattractant protein-1 (MCP-1) into circulation [4]. MCP-1 functions as a chemoattractant that enhances macrophage infiltration into adipose tissue. Consistent with this role of MCP-1, adipose tissue of lean subjects usually consists of approximately 5%–10% macrophages, whereas in obese patients, macrophage content can be as high as 50% of the total number of cells. Increased macrophage entry into adipocytes results in the increased release of chemoattractants such as MCP-1 and TNF-alpha [8]. Increased TNF-alpha results in increased lipolysis. These actions result in increased levels of circulating FFAs and triglycerides to be taken up directly by skeletal muscle through the action of lipoprotein lipase [4]. Excess circulating FFAs cause the accumulation of triglycerides and activated lipids in the form of long-chain fatty acyl-CoA esters in the skeletal muscle. The excess triglyceride, called ectopic fat, reduces mitochondrial oxidative phosphorylation of fatty acids, thus reducing beta-oxidation and increasing triglyceride accumulation in the skeletal muscle. Excess triglyceride in skeletal muscle disrupts insulin-stimulated glucose transport, thus triggering insulin resistance in skeletal muscle (Fig. 13.8).

13.4.6 Glycation and reactive oxygen species

Glycation is the first step in the complex series of prolonged reactions in the production of damaging molecules known as Amadori reactions, Schiff base reactions, and Maillard reactions. These reactions occur when simple sugar molecules such as fructose or glucose become attached to proteins or lipids without the moderation of an enzyme. It is a chaotic process that damages the function of biomolecules such as proteins that can lead to various diseases. All the reactions lead to **advanced glycation end products (AGEs)**. In healthy individuals, glucose attaches to a protein or fat for only a short time. When glucose levels rise, such as in type 2 diabetes, glucose and protein or glucose and fat will not detach, causing AGE formation. AGEs can come from exogenous (made outside the body) or endogenous (made by the body) sources. For example, food producers add AGEs to everyday foods to improve appearance and taste. Foods with significant browning, caramelization, or added AGEs can be exceptionally high in these pro-inflammatory and disease initiating compounds. Some examples are donuts, BBQ meats, cake, and dark-colored sodas. Endogenous glycation, which forms AGEs, occurs in the body (mainly the bloodstream) and is associated with increased oxidative damage that is linked to many age-related diseases. Elevated levels of AGEs contribute to many diabetes-related complications, including neuropathy, retinal disease, and kidney failure. Elevated levels of AGEs also deplete nitric oxide levels, which promotes vascular damage and an increased risk for heart disease. Another drawback related to AGEs is that as glucose molecules bind to high-density lipoprotein molecules, the "good" cholesterol is blocked from binding to receptor sites in the liver. The result is that the liver is duped into believing that there is a shortage of cholesterol and keeps manufacturing more to deposit into the bloodstream, thus increasing the risk of cardiovascular disease.

Reactive oxygen species (ROS) are highly reactive chemical compounds generated when O_2 loses an electron. Although ROS may serve as signaling molecules in lower concentrations, they could cause cell death leading to disease states such as diabetes and cardiovascular diseases in greater amounts. Antioxidants act as "free radical scavengers" and prevent and repair the damage done by free radicals by giving one of their electrons to the free radical to stabilize. Antioxidant substances include vitamins A, C, and E, among others. They are found in fruits and vegetables, nuts, grains, certain meats, poultry, and fish.

Fig. 13.10 highlights the consequences of excessive ROS production in response to hyperglycemia in the vasculature, where ROS-driven changes in cell phenotype are mediated by a range of signaling pathways and transcription factors. Cells of the kidneys, eyes, and nervous system also undergo cell- and organ-specific phenotypic changes due to hyperglycemia-mediated ROS production.

13.5 Acute consequences of diabetes mellitus

Diabetes often progresses undiagnosed for years. Early detection and subsequent treatment will help minimize the chance of developing more severe complications of diabetes. Kidneys filter glucose and help regulate blood glucose concentration in a normal situation. However, when glucose levels are high, some of it spills into the urine, drawing water. Loss of fluid by the kidneys increases thirst. Decreased glucose uptake signals the body by increasing appetite. Thus, the three early manifestations of diabetes are excessive urination (**polyuria**), thirst (**polydipsia**), and increased appetite (**polyphagia**). Excessive blood glucose levels also increase blood osmolality, which draws water into the intravascular space. The excess blood volume could lead to **diabetic coma**. Excessive glucose levels in the blood would limit the kidneys from conserving all of it. Therefore, glucose spills into the urine and is referred to as **glycosuria**.

During insulin deficiency, cells are deprived of glucose, forcing them to rely on fat stores for fuel. In this situation, the liver is forced to use an alternative lipid metabolic pathway that results in increased ketone bodies (or ketones), which are acidic. The buildup of ketones in the blood is referred to as ketosis. Uncontrolled diabetes can lead to ketoacidosis, a dangerous form of metabolic acidosis. Acetoacetate is one example of ketone, which can be spontaneously decarboxylated to acetone, a highly fat-soluble compound excreted slowly by the lungs and is responsible for the fruity odor of the breath of individuals with diabetic ketoacidosis.

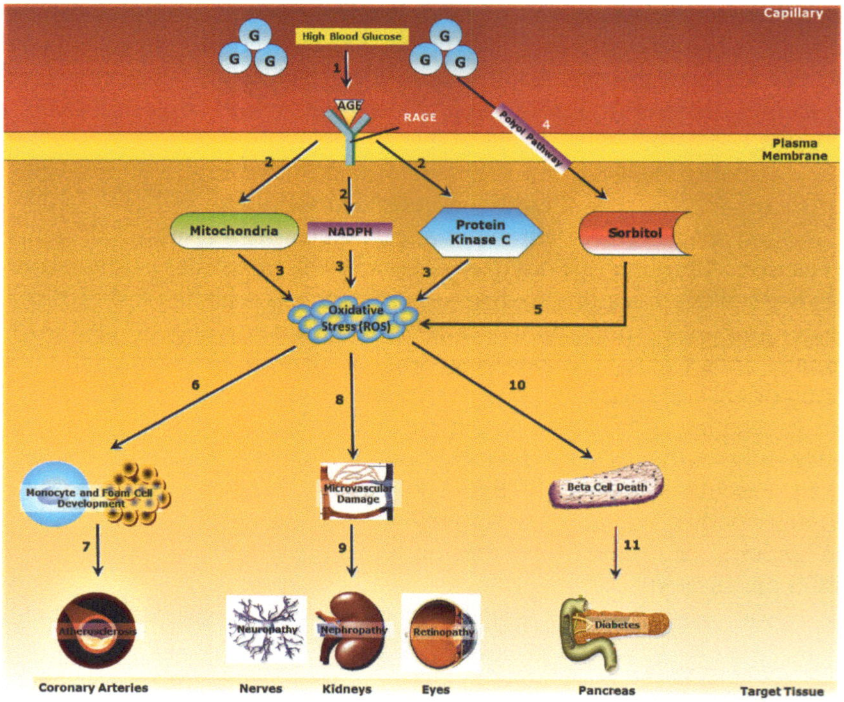

FIG. 13.10 High blood glucose (hyperglycemia) increases advanced glycation end products (AGEs) and reactive oxygen species formation leading to cardiovascular and metabolic disease [8]. (1) Hyperglycemia stimulates the buildup of advanced end products (AGE) that interact with the receptor for AGE (RAGE) on the plasma membrane. (2) Interaction of AGE with the RAGE triggers the production of ROS which in turn causes: (a) NADPH oxidase and (b) PKC activation. (3) Increased oxidative stress. (4) Hyperglycemia and increased intracellular glucose levels would turn on polyol pathway in greater amounts and the buildup of sorbitol. (5) All of these steps contributes to oxidative stress. (6) Oxidative stress serves as the trigger in the development of atherosclerotic lesion via attraction of monocytes to the endothelium of coronary arteries. (7) Plaque formation. (8) Microvascular injury. (9) Neuropathy, nephropathy, retinopathy. (10) Loss of beta cells. (11) Diabetes mellitus.

The effects of a lack of insulin on protein metabolism result in a net shift toward protein catabolism. The breakdown of muscle proteins leads to muscle wasting and muscle weakness, and weight loss. Decreased amino acid uptake coupled with increased protein degradation results in excess amino acids in the blood. These amino acids can be used as substrates for gluconeogenesis, which further aggravates hyperglycemia. Despite increased food intake, however, progressive weight loss is typical from the effects on fat and protein metabolism.

13.6 Chronic complications of diabetes mellitus

Over time, persistently high glucose levels in the blood injure tissues throughout the body, especially those of the blood vessels and nerves. Abnormal G-protein activation such asprotein kinase C pathway in diabetic patients may initiate uncharacteristic cellular events, ultimately leading to the buildup of free radicals, triggering inflammation, and local injury.

Although the exact mechanism that leads to cellular damage and organ dysfunction is unknown, the following may contribute to microvascular complications.

- **Increased polyol pathway:** When glucose levels are high, some of the glucose enters the polyol pathway, resulting in the production of sorbitol as an intermediate product. Increased sorbitol concentration increases cellular osmolality, generates ROS, and likely leads to other types of cellular dysfunction (Fig. 13.10).
- **Increased AGE formation:** AGEs cross-link proteins such as collagen and extracellular matrix proteins affecting their structure and function, accelerate atherosclerosis, promote glomerular dysfunction, reduce nitric oxide synthesis, and induce endothelial dysfunction.
- **Activation of protein kinase C (PKC):** PKC alters the transcription of genes, thus affecting normal functions, especially of the proteins in endothelial cells and neurons.
- **Increased hexosamine pathway:** The hexosamine pathway may alter the functions of proteins such as endothelial nitric oxide synthase, change in gene expression of transforming growth factor or plasminogen activator inhibitor-1.

Inflammation of the lining of arteries triggers the process of atherosclerosis, heart disease, and stroke. Damage to the glomerular capillaries and the thickening of the basement membrane impairs kidney function and can lead to kidney failure. Chronic microvascular complications result in diabetic nephropathy, a leading cause of end-stage renal disease. Albuminuria, which is the presence of albumin (a protein) in urine, is the earliest manifestation consistent with progressive nephropathy. Similarly, damage to blood vessels that serve the eyes and thickening of the basement membrane can cause hemorrhage, ischemia, fibrosis, and blindness. Diabetic retinopathy is the leading cause of blindness. Microaneurysms of the retinal capillaries are the earliest manifestation of nonproliferative retinopathy. Macular edema is also common in retinopathy, caused by neovascularization, and can lead to retinal detachment.

Blood vessel damage also reduces circulation to the limbs, whereas nerve damage leads to a loss of sensation, called neuropathy, particularly in the hands and feet (peripheral neuropathy), which is another complication of long-term diabetes. Autonomic neuropathy may result in constipation/diarrhea, erectile dysfunction, and incontinence. Demyelination of peripheral nerves where the myelin coat around axons is damaged is another complication where symptoms typically begin distally and move proximally. Burning or prickling and pain in extremities progress over time to numbness. Poor blood circulation to distal areas such as the limbs increases the risk of injury, infection, and tissue death (necrosis), contributing to a high rate of the toe, foot, and lower leg amputations in people with diabetes (Fig. 13.11). Diabetes-related alterations occur in neutrophil function and impaired wound healing. Recurrent infections are also more common in diabetic patients as diabetes is an immunocompromised state. Taken together, two reasons why people with diabetes have diminished ability to heal are a reduced ability by white blood cells in blood to remove bacteria and a reduction in the amount of O_2 to wound due to destruction of blood cells by the process of glycation, which destroys cells. Urinary tract infection, fungal, and periodontal disease are a few examples. Together, these complications make diabetes the seventh leading cause of death in the United States.

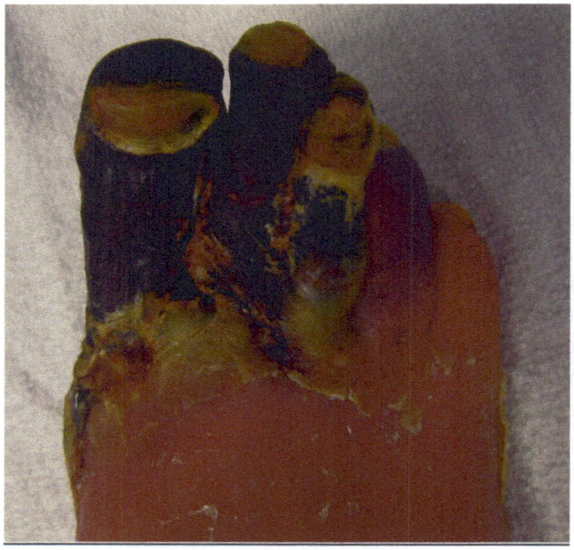

FIG. 13.11 Gangrene, one of the chronic complications of diabetic patients [9].

13.7 Diagnosis of diabetes mellitus

Diabetes is diagnosed when lab tests reveal that blood glucose levels are higher than normal, a condition called hyperglycemia. In everyday life, the gastrointestinal tract represents the normal route of glucose entry into the body. However, the assessment of tissue glucose uptake following glucose ingestion presents a challenge because of the difficulties in quantitating the rate of glucose absorption, suppression of HGP, and glucose uptake. Moreover, because the plasma glucose and insulin concentrations are changing simultaneously, it is difficult to draw conclusions about insulin secretion or insulin sensitivity.

13.7.1 Hemoglobin A1c test

One of the proteins to which glucose commonly binds is hemoglobin. The average amount of glucose in the blood can be found by measuring the hemoglobin A1c (HbA1c) level. HbA1c test or glycated hemoglobin test is an important blood test in determining diabetes and regulating it. In a normal subject, the HbA1c is 4%–5.6%. In uncontrolled diabetes, a higher concentration of blood glucose combines with hemoglobin, and therefore the HbA1c would be greater than 6.5% (Fig. 13.12 and 13.13). People with HbA1c between 5.7% and 6.4% are considered as having a risk for diabetes. The HbA1c test is a reliable indicator of diabetes as it reflects glucose levels over a 3–4-month period as the life span of red blood cells is approximately 120 days. For every 1% unit reduction in A1c, there is a 10% decrease in relative risk. For example, if a patient has an HbA1c of 10.0 that drops to 8.0, though they are not yet at goal, they have managed to decrease their risk of microvascular complications by about 20% [10].

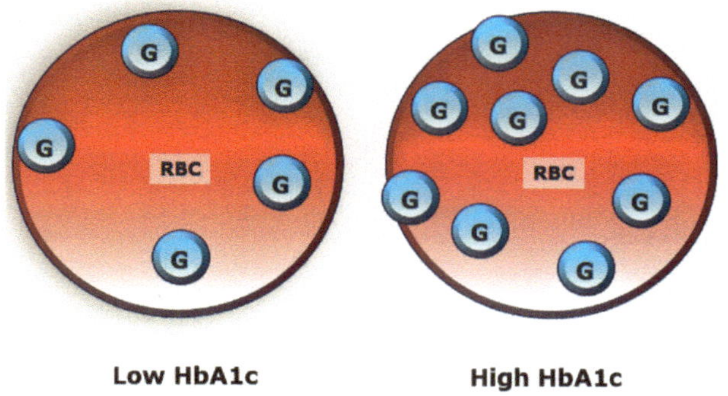

FIG. 13.12 Low and high HbA1c levels [11]. Low HbA1c has a low amount of glucose (G) attached while high HbA1c contains high levels of glucose attached, indicating high levels of glucose in the blood over the previous 90–120 days.

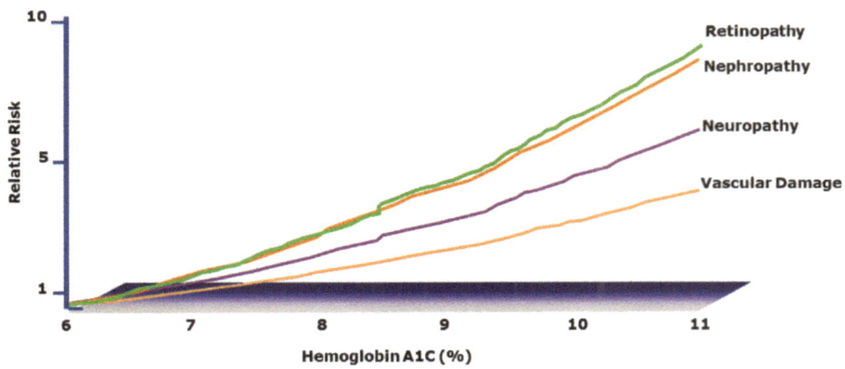

FIG. 13.13 HbA1C and microvascular complications [12]. The relationship between HbA1C levels and the development of retinopathy, nephropathy, neuropathy, and vascular damage. With higher HbA1C levels, there is an increased risk of the development of these complications, and any reduction in HbA1C results in a decrease in the risk for these complications.

13.7.2 Glucose uptake during oral glucose tolerance test (OGTT)

Administration of a known amount of oral glucose to healthy subjects to quantitate blood glucose levels subsequently to determine glucose uptake. During the 4 h after glucose (75 g) ingestion, it is expected that the liver takes up 28% of the oral load; 72% is taken up by peripheral tissues, of which the brain (an insulin-independent tissue) accounts for 22% of the total glucose load. HGP decreases to about 53%. The contribution of skeletal muscle to the uptake of an oral glucose load is approximately 45%.

On the other hand, in type 2 diabetic individuals, an oral glucose load disposition is significantly altered (Fig. 13.14). The change in glucose metabolism is accounted for by two factors: decreased tissue glucose uptake and abnormal suppression of HGP. While inappropriate suppression of HGP accounts for approximately one-third of the defect in total body glucose homeostasis, reduced glucose uptake accounts for the remaining two-thirds. Since hyperglycemia enhances hepatic glucose uptake in proportion to the increase in plasma

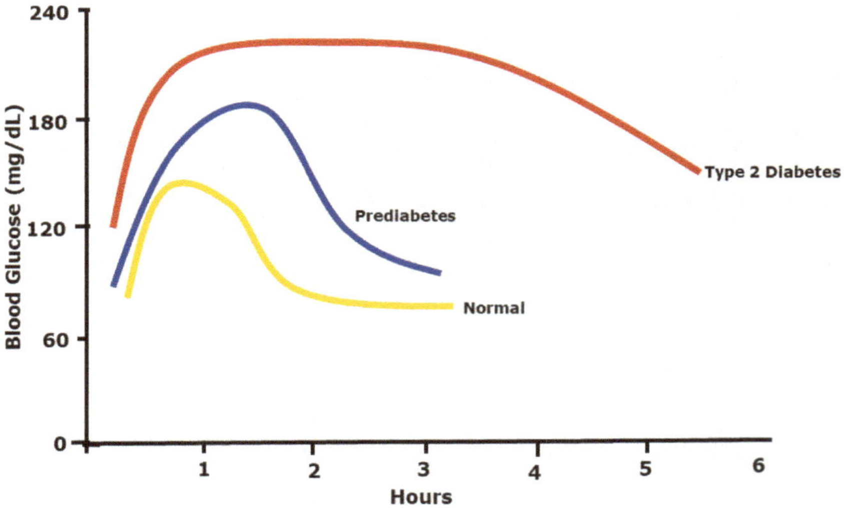

FIG. 13.14 Oral glucose tolerance test in normal, prediabetes, and diabetic conditions [13]. In normal healthy individuals, the 75g glucose load is cleared quickly after an initial increase in blood glucose. In prediabetes, the blood glucose levels rise quickly but take longer than usual to return to resting levels. In type 2 diabetes, fasting glucose levels are at elevated levels, followed by an elevation of blood glucose after glucose bolus. It remains elevated hours after ingestion due to insulin resistance of the skeletal muscle, adipose tissue, and liver.

glucose concentration, the liver glucose uptake is markedly reduced in all type 2 diabetic subjects following glucose ingestion.

Treatment of diabetes depends on the type, severity of the condition, and the ability of the patient to make lifestyle changes. Regular exercise and consumption of a healthy diet can reduce blood glucose levels to normal levels. Some patients with type 2 diabetes may not be able to control the disease with lifestyle changes. Some patients, although able to control diabetes with lifestyle changes, they do require medication over time.

The prevention and treatment of diabetes have been transformed dramatically since the discovery of insulin. Insulin has led to the rapid development of new lifesaving treatments that are widely available. It has initiated advancements that have fundamentally enhanced patients' daily lives with diabetes and dramatically extended their life expectancy. The use of human biosynthetic insulin has virtually eliminated local reactions at injection sites using hypodermic needles and insulin syringes. The diabetic patient on insulin must balance food intake along with the dosage of insulin. In addition to HbA1C levels, home glucose monitoring allows patients to alter the different methods of insulin therapy based on accurate assessments of glucose control. Insulin pumps are now available, which utilize patient-specific computer-driven algorithms that adjust insulin doses in real time in correlation with fluctuations in blood glucose to maintain a normal physiologic range. Although the first-line treatment of type 2 diabetes was insulin, research advances have resulted in alternative options, including medications that enhance pancreatic function.

13.7.3 Diabetes and exercise

Type 2 diabetes may be reversed by moderate weight loss, regular physical activity, and consumption of a healthy diet. However, if blood glucose levels cannot be controlled, the diabetic patient will eventually require insulin. Approximately 150 min a week of moderately intense physical activities is recommended for diabetic patients. Increased physical activity is expected to have reduced abdominal fat, decreased blood pressure, and an improved lipid profile. Endurance exercise such as running, swimming, or cycling reduces platelet adhesiveness, lowering the risk of coronary artery disease. Exercise is effective in reducing the side effects of the disease. However, exercise is not recommended unless diabetes is under control. The most immediate effect of exercise is hypoglycemia since glucose use is much greater during exercise than at rest. Those with insulin or other appropriate medication must check blood glucose 30 min before exercise. The ideal blood glucose level for exercise is between 100 and 250 mg/dL. If blood glucose concentration is lower than 100 mg/dL, fruit or crackers before exercise helps prevent hypoglycemia. Two hundred and fifty mg/dL is the caution zone where a urine test for ketones must be followed; 300 mg/dL is too high for exercise as it could result in dehydration and ketoacidosis. Symptoms such as feeling shaky, weak, sweaty, cold, clammy, and confused suggest hypoglycemia, at which time exercise must be terminated.

References

[1] National Diabetes Statistics Report, Centers for Disease Control and Prevention, US Dept of Health and Human Services, 2020.
[2] P. Pirot, A.K. Cardozo, D.L. Eizirik, Mediators and mechanisms of pancreatic beta-cell death in type 1 diabetes, Arq. Bras. Endocrinol. Metabol. 52 (2008) 156–165.
[3] M.A. Austin, K.L. Edwards, Small, dense low density lipoproteins, the insulin resistance syndrome and noninsulin-dependent diabetes, Curr. Opin. Lipidol. 7 (3) (1996) 167–171, https://doi.org/10.1097/00041433-199606000-00010.
[4] A. Guilherme, J.V. Virbasius, V. Puri, M.P. Czech, Adipocyte dysfunctions linking obesity to insulin resistance and type 2 diabetes, Nat. Rev. Mol. Cell Biol. 9 (5) (2008) 367–377, https://doi.org/10.1038/nrm2391.
[5] A. Chadt, H. Al-Hasani, Glucose transporters in adipose tissue, liver, and skeletal muscle in metabolic health and disease, Pflug. Arch. Eur. J. Phys. 472 (9) (2020) 1273–1298, https://doi.org/10.1007/s00424-020-02417-x.
[6] D.B. Savage, K.F. Petersen, G.I. Shulman, Disordered lipid metabolism and the pathogenesis of insulin resistance, Physiol. Rev. 87 (2) (2007) 507–520.
[7] C. Gill, R. Mestril, A. Samali, Losing heart: the role of apoptosis in heart disease—a novel therapeutic target? FASEB J. 16 (2) (2002) 135–146, https://doi.org/10.1096/fj.01-0629com.
[8] N.A. Calcutt, M.E. Cooper, T.S. Kern, A.M. Schmidt, Therapies for hyperglycaemia-induced diabetic complications: from animal models to clinical trials, Nat. Rev. Drug Discov. 8 (5) (2009) 417–430, https://doi.org/10.1038/nrd2476.
[9] GangreneFoot.JPG. Wikimedia Commons, the free media repository. Updated December 12, 2020. Accessed July 29, 2021. https://commons.wikimedia.org/w/index.php?title=File:GangreneFoot.JPG&oldid=518380142.
[10] M.C. Riddle, H.C. Gerstein, R.R. Holman, et al., A1C targets should be personalized to maximize benefits while limiting risks, Diabetes Care 41 (6) (2018) 1121–1124, https://doi.org/10.2337/dci18-0018.
[11] American Diabetes Association, 2. Classification and diagnosis of diabetes: standards of medical care in diabetes—2018, Diabetes Care 41 (Suppl. 1) (2018) S13–S27.
[12] J.S. Skyler, Diabetic complications: the importance of glucose control, Endocrinol. Metab. Clin. 25 (2) (1996) 243–254.
[13] Q. Wang, J. Jokelainen, J. Auvinen, et al., Insulin resistance and systemic metabolic changes in oral glucose tolerance test in 5340 individuals: an interventional study, BMC Med. 17 (1) (2019) 217, https://doi.org/10.1186/s12916-019-1440-4.

Index

Note: Page numbers followed by *f* indicate figures and *t* indicate tables.

Kolibri 360 GmbH
Ferdinand-Jühlke-Straße 7
99095 Erfurt, Deutschland
produktsicherheit@kolibri360.de